BIOINSTRUMENTATION

BIOINSTRUMENTATION

John G. Webster, Editor
University of Wisconsin

John Wiley & Sons, Inc.

ACQUISITIONS EDITOR   Bill Zobrist

SENIOR PRODUCTION EDITOR   Caroline Sieg

SENIOR MARKETING MANAGER   Katherine Hepburn

SENIOR DESIGNER   Dawn Stanley

This book was set in 10/12 Times New Roman by the author and printed and bound by Malloy Lithographing. The cover was printed by Brady Palmer.

This book is printed on acid free paper. ∞

ISBN: 0-471-26327-3
WIE ISBN: 0-471-45257-2

Printed in the United States of America

10 9 8 7 6 5 4 3 2 1

Preface

As a first course in bioinstrumentation for undergraduate biomedical engineers, this book describes measurement methods used in the growing fields of cell engineering, tissue engineering, and biomaterials in medicine and biology. While many books on medical instrumentation only cover hospital instrumentation, this text also encompasses measurements in the growing fields of molecular biology and biotechnology. It will also serve as a text for medical and biological personnel who wish to learn measurement techniques.

Chapter 1 covers the basics of instrumentation systems, errors, and the statistics required to determine how many subjects are required in a research study. Because many biomedical engineers are not specialists in electronics, Chapter 2 provides the necessary background in circuits, amplifiers, filters, converters, and signal processing.

Chapter 3 describes clinical measurements of molecules, such as oxygen and glucose, to biotechnology measurements such as DNA sequencing. For the fields of biomaterials and tissue engineering, Chapter 4 covers measurements on polymers, using surface analysis, protein adsorption, and molecular size.

Measurements on blood components are the most common measurements on cells, as described in Chapter 5. Cells are counted and identified using changes in impedance and light scattering. Chapter 6 covers cellular measurements in biomaterials and tissue engineering, such as cellular orientation, rolling velocity, pore size, deformation, shear stress, adhesion, migration, uptake, protein secretion, proliferation, differentiation, signaling, and regulation.

Chapter 7 describes measurements on the nervous system—action potentials, EEG, ERG, EOG, EMG and audiometry—and brain imaging using X ray, CT, MRI, nuclear imaging, SPECT, PET, and biomagnetism. Chapter 8 covers heart and circulation, with measurements of cardiac biopotentials, pressures, output, sounds, viability—as well as blood flow and pressure in the periphery.

Measurements of pulmonary volume, flow, diffusion, and airway resistance are described in Chapter 9, which also includes kidney clearance, bone mineral, and skin water loss. Chapter 10 covers measurements of body temperature, fat, and movement.

Each chapter has references for further study as well as homework problems to test comprehension. The web site www.wiley.com/college/webster contains complete laboratory instructions for 12 laboratories, examination questions, quiz questions, and PowerPoint figures.

Suggested prerequisite courses are calculus, biology, chemistry, and physics. I welcome suggestions for improvement of subsequent editions.

John G. Webster
webster@engr.wisc.edu
June 2003

Contents

x   Contents

Chapter 1

Measurement Systems

Kevin Hugo

1.1 Studying Biomedical Engineering

Edna Jones is a 67-year-old retired teacher who is having difficulty with her vision. She can't remember when it started, but her eyesight has increasingly blurred over the course of the last year. It has now become so poor that last week she drove her car off the road. She hasn't been to a doctor for many years, but now she's decided to see her family physician.

The nurse measured and recorded Edna's vital signs. Although her weight, 90 kg (198 lb.), was measured with an electronic scale, her height, 173 cm (5 ft. 8 in.), was measured manually. She marveled at the device that took her blood pressure (118/76) and pulse (63 beats per minute) simultaneously. Another automated instrument took her body temperature (98.6 °F or 37 °C) from her ear in a few seconds, instead of having to place a thermometer under her tongue for a minute.

Edna related her vision problems and accident to the doctor, and she is certain that she did not suffer any injuries. She finds it strange that the doctor asked about how much water she drinks, but she admitted that she has been rather thirsty lately and that she drinks and passes a lot of water. The doctor also noted that Edna has gained a fair amount of weight.

The doctor began the physical examination by observing Edna's eyes with an ophthalmoscope. There were cotton-like blotches on the normally orange–pink retina. The physician then felt her hands and feet; they were cold, and she has lost some of the sensation of touch in her feet. Finally, samples of urine and blood were taken.

It takes several minutes for the laboratory at the clinic to analyze the urine and blood samples; they use an automated instrument to measure the blood sugar (glucose) in each. The doctor is concerned about the results, and orders an electrocardiogram and further tests on the blood and urine samples before returning to Ms. Jones to discuss her condition.

1.1.1 Scales of Biological Organization

The patient in the above story has noninsulin-dependent diabetes mellitus (NIDDM), a disease in which afflicted persons have high levels of blood glucose. Diabetes mellitus is a very common chronic disease with many complications, and is one of the most expensive diseases to treat in the United States. However, with early detection and diligent treatment, the risk of complications can be reduced (Diabetes, 1993).

In diagnosing the patient's disease and managing the future treatment, it is necessary to take measurements using biomedical instruments, which is the subject of this book. One way to classify the various types of bioinstrumentation is to consider the level of biological organization involved in taking the measurement. For example, it may involve molecules, cells, tissues, organ systems, or the entire body. The chapters of this book are presented in the order of these biological levels.

This chapter discusses how and why measurements are taken. For example, every patient has vital signs taken not only because they may be useful in analyzing the problem that the patient is concerned with, but also because they indicate that person's general health. We must also understand that the results of measurements will vary due to errors in taking the measurement as well as individual differences.

An automated thermometer measured Ms. Jones' temperature in seconds. This is accomplished by using a temperature-sensing element as well as electronics to derive the temperature based on a mathematical algorithm. This type of electronic signal processing is explained in Chapter 2.

Measurements of concentrations of glucose and other molecules are performed by spectrophotometry, as described in Chapter 3. Ms. Jones may have coronary artery disease and need an implanted graft vessel. Chapter 4 describes the electron microscopes used to test proposed biomaterials and the adhesion of protein molecules to them.

Methods used to analyze the blood sample that was drawn are described in Chapter 5. Diabetes causes a number of complications. To study the urinary casts (cylindrical protein molds) passed in the urine, Chapter 6 describes the optical microscopes used. Ms. Jones reported a loss of feeling in her legs. Chapter 7 describes the nervous system and the test used to diagnose nerve disease.

Chapter 8 describes the electrocardiogram, which is used to diagnose Ms. Jones' cardiac problems, as well as methods to obtain blood pressure measurements. Her kidney function is diagnosed by measuring creatinine clearance, as described in Chapter 9. NIDDM patients are frequently overweight. Chapter 10 describes measurements on the body, including measurements of body fat.

1.1.2 Fields of Biomedical Engineering

Biomedical engineering is a cross-disciplinary field that incorporates engineering, biology, chemistry, and medicine. This involves both the acquisition of new information and technology, but also of the development of new devices, processes, and algorithms in regards to biological systems.

Table 1.1 shows that instrumentation can be classified by the field of biomedical engineering responsible for its design. Each of these fields is briefly described in the following paragraphs in this section. The Whitaker Foundation describes various fields of biomedical engineering at www.whitaker.org/glance/acareer.html.

Table 1.1 Biomedical engineers work in a variety of fields.

Bioinstrumentation
Biomaterials
Biomechanics
Biosignals
Biosystems
Biotransport
Cellular engineering
Clinical engineering
Tissue engineering
Rehabilitation engineering

Bioinstrumentation applies the fundamentals of measurement science to biomedical instrumentation. It emphasizes common principles and unique problems associated with making measurements in living systems. A physiological variable originates from a molecular, cellular, or systemic process whose nature may be described by a mechanical, electrical, chemical, optical, or other event. A variable must be carefully specified before being measured. The sensor should be designed to minimize the disturbance to the measured variable and its environment, comply with the requirements of the living system, and maximize the signal-to-noise ratio, i.e. the clarity of the signal. The signal, typically after being converted into an electrical form, is then conditioned using linear and nonlinear signal processing, and delivered to an appropriate output device. Bioinstrumentation includes methods for obtaining invasive and noninvasive measurements from the human body, organs, cells, and molecules; electronic instrumentation, principles of analog and digital signal processing, and typical output devices. It includes measurement concepts such as accuracy, reproducibility, noise suppression, calibration methods, and safety requirements.

Biomaterials is the application of engineering materials to the production of medical devices and diagnostic products. In the last decade, advances in biology, especially at the molecular level, have lead to the design and development of new classes of materials derived from natural sources. These include molecularly engineered materials, hybrid materials and devices, biomimetic or synthetic biological materials, and other biologically related materials. Biomaterials covers current and traditional applications for biologically and pharmacologically active materials as well as materials used in such applications as tissue and organ engineering, diagnostic products, and drug delivery.

Biomechanics covers the behavior of biological tissues and fluids to incorporate complexities ranging from nonlinear viscoelasticity to non-Newtonian flows. Biomechanics includes both biofluid mechanics and biosolid mechanics at the molecular, cellular, tissue, and organ-system levels. Molecular and cellular biomechanics is integrated with the macroscopic behavior of tissues and organs. This is put into practice by the modeling of various biological systems. An example of biomechanics is ergonomics, wherein everyday devices, such as chairs and desks, are designed to reduce stress and injury to the body.

Biosignals covers time series analysis in biomedical studies and concentrates on use of data to uncover the nature of the underlying phenomena, the mechanisms of signal production, and the fundamental origins of the variability in the signal. Standard methods of signal analysis have been used for characterizing rather than for elucidating the mechanisms. Biosignals treats signals from a variety of sources; standard methods of signal analysis, including transform and statistical techniques and their relationships; methods of signal generation, fractal analysis methods for signals with different characteristics, methods for analyzing chaotic signals, and approaches to reviewing data to determine or distinguish among possible origins.

Biosystems, which uses modern biology, has given us the tools to identify and characterize molecules and cells, the fundamental building blocks of organ systems. Understanding how molecules and cells function in tissues, organs and organ systems is the domain of biological systems analysis. Whereas molecular biology has focused on reductionism, biomedical engineering is in a unique position to move biology to the next frontier where synthesis will lead to an understanding of the function of complex systems. Biosystems analysis integrates properties of biological systems with current tools of systems analysis.

Biotransport phenomena covers transport processes from the organ to the subcellular level. Quantitative understanding of biological processes is based on the fundamentals of the complementary processes of mass, momentum, and energy transport. Transport of ions, substrates, proteins, viruses, and cells is a central issue for the quantitative understanding of biological systems. It builds an integrated picture of the convection, diffusion, permeation, and reaction kinetics in the circulation, through capillary beds and the tissues of the body. While using the fundamentals of transport phenomena for understanding membrane transport, cellular and tissue energetics, enzymatic regulation and metabolism, it also considers how data from microsensors, tissue samples, or regional concentrations over time can be acquired, quantified, and interpreted. Transport phenomena considers methods suitable for interpreting data from intact systems.

Cellular engineering develops and communicates quantitative biochemical and biophysical design principles that govern cell function. An engineering perspective integrates underlying molecular mechanisms. Relevant topics are fundamental kinetics, mechanics, and transport processes that use calculus, differential equations, and cell biology. Models analyze cell functions, such as metabolism, signaling and regulation, biomolecular uptake and secretion, proliferation, adhesion, migration, and differentiation characterizing molecular and cellular properties. Techniques from modern biochemistry and molecular biology alter these parameters in order to test model predictions and demonstrate how the underlying design principles can be applied to manipulate cell function.

Clinical engineering deals with managing diagnostic and laboratory equipment in the hospital. Clinical engineers work with health care workers to determine equipment needs, search for optimal equipment, specify equipment, perform incoming inspection, train health care workers in proper operation, provide inventory of equipment, decide whether to perform maintenance in house or purchase vendor contracts, and perform safety inspections.

Rehabilitation engineering is a newer area of biomedical engineering. A rehabilitation engineer works directly with patients such as disabled individuals to assist in achieving a better standard of life. The rehabilitation engineer modifies or designs new equipment for such an individual. A rehabilitation engineer could be involved with an individual requiring a prosthetic limb, by designing the limb to suit the individual's personal needs. Such a limb would be required of an individual who has lost a leg, but still desires to run on his or her own power.

1.1.3 Fields of Biological Science

Table 1.2 shows that instrumentation can be organized by the biological discipline that utilizes it. Each of these disciplines can be divided into several subspecialties. For example, medicine can be subdivided into anatomy, anesthesiology, biomolecular chemistry, biostatistics and medical informatics, family medicine, medical microbiology and immunology, medical physics, neurology, neurophysiology, obstetrics-gynecology, oncology, ophthalmology, pathology and laboratory science, pediatrics, physiology, preventive medicine, psychiatry, radiology, rehabilitation medicine, and surgery.

Table 1.2 Biomedical engineers work in a variety of disciplines. One example of instrumentation is listed for each discipline.

Agriculture - Soil monitoring
Botany - Measurements of metabolism
Genetics - Human genome project
Medicine - Anesthesiology
Microbiology - Tissue analysis
Pharmacology - Chemical reaction monitoring
Veterinary science - Neutering of animals
Zoology - Organ modeling

1.1.4 Workplace Environments for Biomedical Engineers

Biomedical engineers may become a part of the workplace in a variety of different areas, as shown in Table 1.3. Industry has a demand for biomedical engineers for development of new technologies and also for testing the theories that others develop. They could also be technology representatives for existing products. Positions in the government often deal with testing of proposed products for safety and reliability, to safeguard the public from harm. In a clinical setting, the biomedical engineer may be the only individual with working knowledge of some of the instruments, and is needed to allow for their efficient use. Research institutions are also the place of developing new ideas, and often involve the dissemination of techniques and practices to others to further the general understanding.

Table 1.3 Biomedical engineers may work in a variety of environments

Industry
Government
Clinical institutions
Academic research

1.2 The Need for Bioinstrumentation

1.2.1 The Scientific Method

Figure 1.1 shows that researchers often need to use instrumentation to obtain data as part of the scientific method. For example, we might hypothesize that exercise reduces high blood pressure yet experimentation and analysis are needed to support or refute the hypothesis. Experiments are normally performed multiple times. Then the results can be analyzed statistically to determine the probability (hopefully less than 0.05) that the results might have been produced by chance. Results are reported in scientific journals with enough detail so that others can replicate the experiment to confirm them.

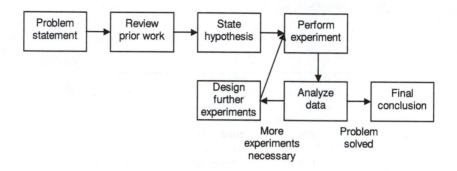

Figure 1.1 In the scientific method, a hypothesis is tested by experiment to determine its validity.

1.2.2 Clinical diagnoses

Figure 1.2 shows that physicians often need instrumentation to obtain data as part of the scientific method. For example, a physician obtaining the history of a patient with a complaint of poor vision would list diabetes as one possibility on a differential diagnosis.

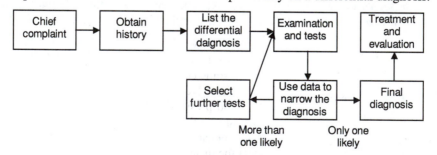

Figure 1.2 The physician obtains the history, examines the patient, performs tests to determine the diagnosis and prescribes treatment.

1.2.3 Feedback in Measurement Systems

Figure 1.3 shows that the measurand is measured by a sensor, which converts the variable to an electrical signal, which can undergo signal processing. Sometimes the measurement system provides feedback through an effector to the subject. As an example, light flashes stimulate the eye when measuring visual evoked responses from electrodes placed over the visual cortex of the brain.

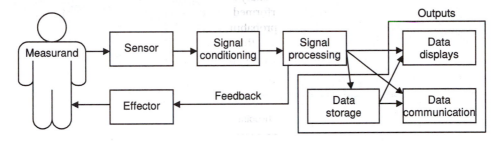

Figure 1.3 A typical measurement system uses sensors to measure the variable, has signal processing and display, and may provide feedback.

Figure 1.4(a) shows that a patient reading an instrument usually lacks sufficient knowledge to achieve the correct diagnosis. Figure 1.4(b) shows that by adding the clinician to form an effective feedback system, the correct diagnosis and treatment result.

In certain circumstances, proper training of the patient and a well-designed instrument can lead to self-monitoring and self-control, one of the goals of bioinstrumentation. This is shown in Figure 1.5. An example of such a situation is the day-to-day monitoring of glucose by people suffering from diabetes. Such an individual will contact a clinician if there is an alert from the monitoring instrument.

1.3 Instrumentation

1.3.1 Measurands

The measurand is the measured quantity. It may be voltage as when measuring the electrocardiogram, pressure as when measuring the blood pressure, force as when measuring weight, or other quantities. Table 1.4 lists some common biomedical measurands.

Measurands can be obtained by invasive or noninvasive methods. For example, if we want to accurately measure the blood pressure in the aorta in vivo (i.e. in a living organism), we must place a pressure sensor on the tip of a catheter (a long narrow tube), cut into a peripheral vessel, and advance the catheter to the aorta. An alternative invasive method is to place the pressure sensor at the distal end of the catheter outside the body.

A noninvasive method to measure blood pressure is to place a pressure cuff around the arm and listen for the Korotkoff sounds due to turbulence of the blood under the cuff. Urine can be analyzed noninvasively by collecting a sample and analyzing it in vitro (maintained in an artificial environment outside the living organism) in the clinical chemistry laboratory. Weight and electrocardiography are also examples of noninvasive measurement methods.

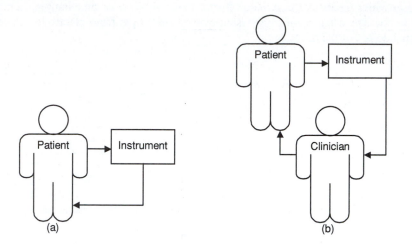

Figure 1.4 (a) Without the clinician, the patient may be operating in an ineffective closed loop system. (b) The clinician provides knowledge to provide an effective closed loop system.

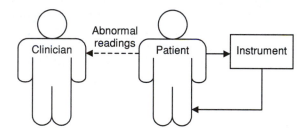

Figure 1.5 In some situations, a patient may monitor vital signs and notify a clinician if abnormalities occur.

1.3.2 Sensors

A variety of sensors have been adapted from industry for use in measuring biomedical variables such as pressure, flow, and temperature. Table 1.5 shows the specifications for a typical blood pressure sensor. Note that several specifications are designed to prevent patient electrocution.

Linearity is the degree to which the sensor stays within its theoretical output for some input. Section 1.3.3 discusses linearity in more detail. Hysteresis is a phenomenon in which increasing the pressure on the sensor gives one output but decreasing the pressure to the previous value gives a different output, as shown in Figure 1.6. Risk current is the amount of current flowing through the sensor at 120 V. Defibrillator withstand is the amount of energy (in joules) the sensor can withstand if a defibrillator is used while the sensor is taking measurements.

Table 1.4 Common medical measurands.

Measurement	Range	Frequency, Hz	Method
Blood flow	1 to 300 mL/s	0 to 20	Electromagnetic or ultra-sonic
Blood pressure	0 to 400 mmHg	0 to 50	Cuff or strain gage
Cardiac output	4 to 25 L/min	0 to 20	Fick, dye dilution
Electrocardiography	0.5 to 5 mV	0.05 to 150	Skin electrodes
Electroencephalography	5 to 300 μV	0.5 to 150	Scalp electrodes
Electromyography	0.1 to 5 mV	0 to 10000	Needle electrodes
Electroretinography	0 to 900 μV	0 to 50	Contact lens electrodes
pH	3 to 13 pH units	0 to 1	pH electrode
pCO$_2$	40 to 100 mmHg	0 to 2	pCO$_2$ electrode
pO$_2$	30 to 100 mmHg	0 to 2	pO$_2$ electrode
Pneumotachography	0 to 600 L/min	0 to 40	Pneumotachometer
Respiratory rate	2 to 50 breaths/min	0.1 to 10	Impedance
Temperature	32 to 40 °C	0 to 0.1	Thermistor

Table 1.5 Sensor specifications for blood pressure sensors are determined by a committee composed of individuals from academia, industry, hospitals, and government.

Specification	Value
Pressure range	−30 to +300 mmHg
Overpressure without damage	−400 to +4000 mmHg
Maximum unbalance	±75 mmHg
Linearity and hysteresis	±2% of reading or ±1 mmHg
Risk current at 120 V	10 μA
Defibrillator withstand	360 J into 50 Ω

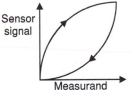

Figure 1.6 A hysteresis loop. The output curve obtained when increasing the measurand is different from the output obtained when decreasing the measurand.

Optimal sensors are designed specifically for the desired measurand, immune to interference from other variables, and have high gain as shown in Figure 1.7(b). A sensor with low gain, as shown in Figure 1.7(a), would not change its signal as much for a small change in the measurand as one with a high gain.

Most sensors are analog and provide a continuous range of amplitude values for output, as shown in Figure 1.8(a). Other sensors yield the digital output shown in Figure 1.8(b). Digital output has poorer resolution, but does not require conversion before being input to digital computers and is more immune to interference. Digital and analog signals will be examined with more detail in Chapter 2.

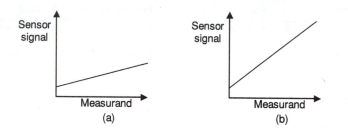

Figure 1.7 (a) A low-sensitivity sensor has low gain. (b) A high sensitivity sensor has high gain.

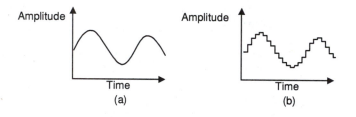

Figure 1.8 (a) Analog signals can have any amplitude value. (b) Digital signals have a limited number of amplitude values.

1.3.3 System Characteristics

Specific Characteristics

Bioinstrumentation should be designed with a specific signal in mind. Table 1.6 shows a few specifications for an electrocardiograph (a device that records the electrical activity generated by the heart). The values of the specifications, which have been agreed upon by a committee, are drawn from research, hospitals, industry, and government.

The electrocardiogram (ECG), the recording of the electrical activity of the heart, can have a voltage value of anywhere between 0.5 mV and 5 mV (Table 1.4). The electrocardiograph should be able to handle this range, therefore its dynamic input signal range is designed to handle signals from –5 mV to 5 mV. If the measurand exceeds the input range of the instrument, the output will saturate and cause an error. Figure 1.9 shows an input signal that exceeds the ±5 mV input range (a) and the resulting amplified signal (b).

Table 1.6 Specification values for an electrocardiograph are agreed upon by a committee.

Specification	Value
Input signal dynamic range	±5 mV
Dc offset voltage	±300 mV
Slew rate	320 mV/s
Frequency response	0.05 to 150 Hz
Input impedance at 10 Hz	2.5 MΩ
Dc lead current	0.1 µA
Return time after lead switch	1 s
Overload voltage without damage	5000 V
Risk current at 120 V	10 µA

Direct current (dc) offset voltage is the amount a signal may be moved from its baseline and still be amplified properly by the system. Figure 1.10 shows an input signal without (a) and with (b) offset.

Slew rate is the maximum rate at which the system can observe a changing voltage per unit time without distortion. In the specifications above, if the signal changes more than 320 mV/s then the resulting output signal will be distorted.

The ECG signal is composed of components that vary in frequency from 0.05 Hz to 150 Hz (Table 1.4). The electrocardiograph is designed to accommodate this as is shown by its frequency response range. The frequency response of a device is the range of frequencies of a measurand that it can handle. Frequency response is usually plotted as gain versus frequency. Figure 1.11 shows the plotted frequency response of this electrocardiograph. Frequency response is discussed in greater detail in Section 2.2.6.

Input impedance is the amount of resistance the signal sees coming into the electrocardiograph. It is desirable to have very high input impedances, in this case 2.5 MΩ, so that the majority of the signal will enter the system to be processed and not be lost in the skin impedance.

Dc lead current is the amount of current that flows through the leads to the patient being observed by the electrocardiograph. It is important that current flowing to the patient not exceed very miniscule amounts (on the order of microamperes). This is because larger currents can polarize electrodes, resulting in a large offset voltage at the amplifier input.

When recording the electrocardiogram, several lead configurations can be used. When switching between the different lead configurations, the electrocardiograph should take no more than 1 s to adjust itself to correctly display the signal again. Lead configurations, as well as the ECG in general, are discussed in greater detail in Chapter 8.

Overload voltage is the maximum voltage the electrocardiograph can withstand without damage from defibrillators. A defibrillator is a device that attempts to "jump start" the heart by using a large amount of voltage (on the order of a few thousand volts). Defibrillators have two paddles that are placed on the patient's chest. When a large voltage difference exists between the two paddles, current flows through the body from one paddle to the other. This current is what causes the heart to begin beating again.

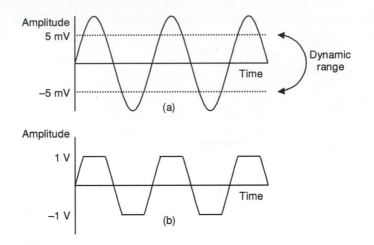

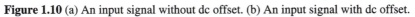

Figure 1.9 (a) An input signal that exceeds the dynamic range. (b) The resulting amplified signal is saturated at ±1 V.

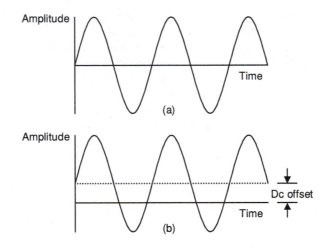

Figure 1.10 (a) An input signal without dc offset. (b) An input signal with dc offset.

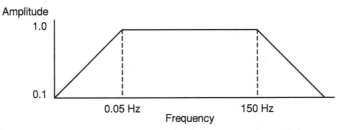

Figure 1.11 Frequency response of the electrocardiograph.

When placing leads on a patient that will be hooked up to instrumentation, it is of paramount importance to make sure the instrumentation does not shock the patient. Risk current is the amount of current that would flow through the patient if they came into contact with 120 V from the power system.

General Characteristics

Another characteristic that all systems have is whether or not they are linear. Linearity is highly desirable for simplifying signal processing. For example, if you measure some input (measurand) you get some output. Now if you double that input you would expect to double the output. This is not the case for nonlinear systems. Doubling the input in a nonlinear system might yield triple the output. Figure 1.12 shows the input/output relation of a linear system (a) and of a nonlinear system (b).

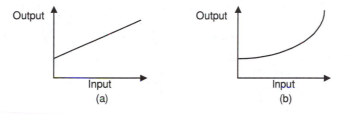

Figure 1.12 (a) A linear system fits the equation $y = mx + b$. Note that all variables are italic. (b) A nonlinear system does not fit a straight line.

Finally, Figure 1.13 shows that all bioinstrumentation observes the measurand either continuously or periodically. Continuous signal measurement is most desirable because no data are lost. However, computer-based systems require periodic measurements, since by their nature, computers can only accept discrete numbers at discrete intervals of time.

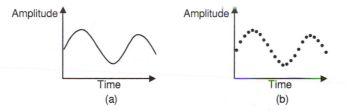

Figure 1.13 (a) Continuous signals have values at every instant of time. (b) Discrete-time signals are sampled periodically and do not provide values between these sampling times.

1.3.4 Outputs

Bioinstrumentation can direct its output three ways: display(s), storage, and communication. A display is simply a monitor or some other device attached to the instrumentation that gives the user a sense of what the input is doing. The term display is a bit misleading since output does not necessarily have to be visible (e.g. a stethoscope uses sound for its output). Storage is increasingly common as computers are integrated into bioinstrumentation. Storage usually takes the form of saving data onto a hard drive or some other type of permanent record. Communication involves transmitting the output to some other location for analysis. Modern imaging systems make extensive use of communication via computer networks and wireless transmission.

1.3.5 Panels and Series

Certain groups of measurements are often ordered together because they are very commonly used or because they are related. This may occur even if the measurements are based on different principles or are taken with different sensors. Table 1.7 is an example of one of these groups of measurements, which are called panels or series.

Table 1.7 Complete blood count for a male subject.

Laboratory test	Typical value
Hemoglobin	13.5 to 18 g/dL
Hematocrit	40 to 54%
Erythrocyte count	4.6 to 6.2×10^6/μL
Leukocyte count	4500 to 11000/μL
Differential count	Neutrophil 35 to 71%
	Band 0 to 6%
	Lymphocyte 24 to 44%
	Monocyte 1 to 10%
	Eosinophil 0 to 4%
	Basophil 0 to 2%

Hemoglobin is the protein which caries oxygen in the bloodstream. Hematocrit is the percent of solid material in a given volume of blood after it has been centrifuged. An erythrocyte is a red blood cell. A leukocyte is a white blood cell. The differential count tells the proportions of each type of white blood cell in the blood. Unusual values for different leukocytes can be indicative of the immune system fighting off foreign bodies.

1.4 Errors in Measurements

1.4.1 Sources of Error

When we measure a variable, for example, the voltage of the electrocardiogram, we seek to determine the true value, as shown in Figure 1.14(a). This true value may be corrupted by a variety of errors. For example, movement of electrodes on the skin may cause an undesired added voltage called an artifact. Electric and magnetic fields from the power lines may couple into the wires and cause an undesired added voltage called interference, as shown in Figure 1.14(b). Thermal voltages in the amplifier semiconductor junctions may cause an undesired added random voltage called noise. Temperature changes in the amplifier electronic components may cause undesired slow changes in voltage called drift. We must evaluate each of these error sources to determine their size and what we can do to minimize them. Usually we try to minimize the largest error, since that is the weakest link in the measurement.

Figure 1.14 (a) Signals without noise are uncorrupted. (b) Interference superimposed on signals causes error. Frequency filters that can be used to reduce noise and interference are discussed in Chapter 2.

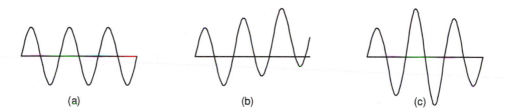

Figure 1.15 (a) Original waveform. (b) An interfering input may shift the baseline. (c) A modifying input may change the gain.

Figure 1.15(b) shows that an interfering input may be added to the original waveform and shift it away from the baseline. Figure 1.15(c) shows that a modifying input may multiply the original waveform to change its gain.

1.4.2 Accuracy and Precision

Resolution is the smallest incremental quantity that can be reliably measured. For example, a voltmeter with a larger number of digits has a higher resolution than one with fewer digits. A speedometer with more tick marks between each numbered speed has a higher resolution than one with no tick marks. However, high resolution does not imply high accuracy.

Figure 1.16 shows that precision is the quality of obtaining the same output from repeated measurements from the same input under the same conditions. High resolution implies high precision.

Repeatability is the quality of obtaining the same output from repeated measurements from the same input over a period of time.

Accuracy is the difference between the true value and the measured value divided by the true value. Figure 1.17(a) shows low accuracy and high precision. Figure 1.17(b) shows high accuracy and high precision.

Obtaining the highest possible precision, repeatability, and accuracy is a major goal in bioinstrumentation design.

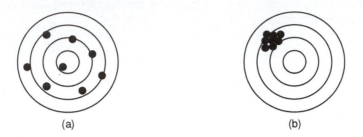

(a)　　　　　　　　　　　　　(b)

Figure 1.16 Data points with (a) low precision and (b) high precision.

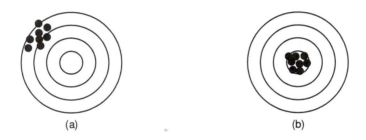

(a)　　　　　　　　　　　　　(b)

Figure 1.17 Data points with (a) low accuracy and (b) high accuracy.

1.4.3 Calibration

Measuring instruments should be calibrated against a standard that has an accuracy 3 to 10 times better than the desired calibration accuracy. The accuracy of the standard should be traceable to the National Institute of Standards and Technology (www.nist.gov/).

If the instrument is linear, its output can be set to zero for zero input. Then a one-point calibration defines the calibration curve that plots output versus input. Figure 1.18 shows that if the linearity is unknown, a two-point calibration should be performed and these two points plus the zero point plotted to ensure linearity. If the resulting curve is nonlinear, many points should be measured and plotted to obtain the calibration curve.

If the output cannot be set to zero for zero input, measurements should be performed at zero and full scale for linear instruments and at more points for nonlinear instruments.

Calibration curves should be obtained at several expected temperatures to determine temperature drift of the zero point and the gain.

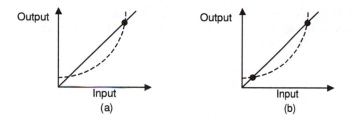

Figure 1.18 (a) The one-point calibration may miss nonlinearity. (b) The two-point calibration may also miss nonlinearity.

1.5 Statistics

1.5.1 Distributions

Mean

If we make n measurements of x, for example of the weights of a population, we may wish to report an estimate of them in a condensed way. The simplest statistic is the estimated sample mean

$$\bar{x} = \frac{\sum x_i}{n} \tag{1.1}$$

where $i = 1, 2, \ldots n.$

Standard Deviation

A measure of the spread of data about the mean is the estimated sample standard deviation.

$$s = \sqrt{\frac{\sum (x_i - \bar{x})^2}{n-1}} \tag{1.2}$$

The standard deviation of the mean s is

$$s_{\bar{x}} = \frac{s}{\sqrt{n-1}} \tag{1.3}$$

Example 1.1 If Person 1 weighs 68 kg, Person 2 weighs 90 kg and Person 3 weighs 95 kg, find the mean, the standard deviation, and the standard deviation of the mean.
 Here $n = 3$. The mean of the data is

$$\bar{x} = \frac{\sum x_i}{n} = \frac{68 + 90 + 95}{3} = 84.3\,\text{kg}$$

The standard deviation of the data is

$$s = \sqrt{\frac{\sum (x_i - \bar{x})^2}{n-1}} = \sqrt{\frac{(68 - 84.3)^2 + (90 - 84.3)^2 + (95 - 84.3)^2}{3-1}} = 14.4\,\text{kg}$$

The standard deviation of the mean is

$$s_{\bar{x}} = \frac{s}{\sqrt{n-1}} = \frac{14.4}{\sqrt{3-1}} = 10.2\,\text{kg}$$

Gaussian Distribution

The spread (distribution) of data may be rectangular, skewed, Gaussian, or other. The Gaussian distribution is given by

$$f(x) = \frac{e^{-(x-\mu)^2/(2\sigma^2)}}{\sqrt{2\pi}\,\sigma} \tag{1.4}$$

where μ is the true mean and σ is the true standard deviation of a very large number of measurements. Figure 1.19 shows the familiar "bell curve" shape of the Gaussian distribution.

Poisson Probability

The Poisson probability density function is another type of distribution. It can describe, among other things, the probability of radioactive decay events, cells flowing through a counter, or the incidence of light photons. The probability that a particular number of events K will occur in a measurement (or during a time) having an average number of events m is

$$p(K;m) = \frac{e^{-m}\,m^K}{K!} \tag{1.5}$$

The standard deviation of the Poisson distribution is $\sqrt{m}$. Figure 1.20 shows a typical Poisson distribution.

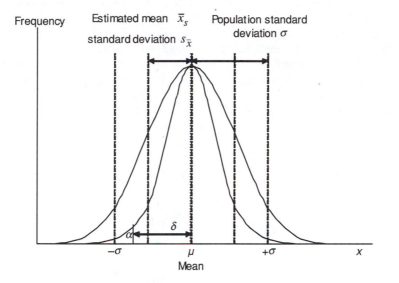

Figure 1.19 For the normal distribution, 68% of the data lie within ±1 standard deviation. By measuring samples and averaging, we obtain the estimated mean $\bar{x}_s$, which has a smaller standard deviation $s_{\bar{x}}$. α is the tail probability that $\bar{x}_s$ does not differ from μ by more than δ.

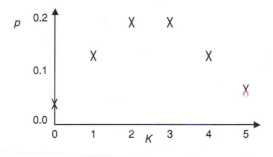

Figure 1.20 A typical Poisson distribution for $m = 3$.

Example 1.2 Assume that the average number of cells passing through a counter in one sampling period is 10. Calculate the probability that 15 cells will be counted in a sampling period.

Since 10 is the average number counted cells, $m = 10$. We want to find the probability of 15 cells being counted so $K = 15$.

$$p(K;m) = \frac{e^{-m}m^K}{K!} = \frac{e^{-10}10^{15}}{15!} = 0.035$$

1.5.2 Statistical Estimations

Suppose we want to know the average weight of adults. It is not feasible to weigh every single adult and then take the average of all the weights. All adults are called the population. Instead, we decide to take a small fraction of the adults, say 1 out of every 1000, and average these weights. This small fraction is called our sample population. Now we have an average weight for our sample population. We want to know if our sample population average weight is a good estimation of the population average weight. There are two ways for us to find out.

If we represent our sample population average weight as a single number, $\bar{x}_s$, then we have made a point estimation of the population average weight. Let's call μ the population average weight and the difference between μ and $\bar{x}_s$ is δ. The standard deviation of the population is σ. The tail probability that $\bar{x}_s$ will not differ from μ by more than δ is α. We can control the probability of how much $\bar{x}_s$ differs from μ by adjusting the sample size, N, using the formula

$$N = Z_\alpha^2 \left(\frac{\sigma}{\delta} \right)^2 \tag{1.6}$$

where Z_α is a multiple of the standard deviation. This number is obtained from a statistical table (Distribution Tables, 2002). The confidence is how sure one wants to be that the sample size, N, calculated will actually meet the specifications for δ

$$\alpha = \frac{1 - \text{percent confidence}}{2} \tag{1.7}$$

If we know the standard deviation, σ, of the population average weight, we can use it. Often, the standard deviation of a population can be found in the literature or from some other source. If it cannot, then the sample standard deviation, s, must be used instead. Also, the Student t distribution, t_α, must be used instead of Z_α.

$$N = t_\alpha^2 \left(\frac{s}{\delta} \right)^2 \tag{1.8}$$

Example 1.3 Suppose we know that the standard deviation of the adult average weight is 3 kg. Using Eq. (1.6), we will set δ to 2, which means our estimate does not deviate from the population average weight by more than ±2 kg. Lastly, we want a 95% confidence that the sample size N we calculate will meet our specifications for δ.

We calculate $\alpha = (1 - 0.95)/2 = 2.5\% = 0.025$. We look up in a statistical table (Distribution Tables, 2002) the value for $Z_{.025}$ and find that it equals 1.96. We can then find N by using

$$N = (1.96)^2 \left(\frac{3}{2}\right)^2 = 8.6 \approx 9$$

which indicates that we would need to take the average of nine weights to get an estimation of the population average weight within ±2 kg with a confidence of 95%. We could then say that the average weight of the population was some value, w.

Another way to represent estimations is with a confidence interval. When dealing with confidence intervals, δ is twice as large because it represents the difference between the upper limit and lower limit instead of between μ and $\overline{x}_s$. The sample size needed for an interval estimation is twice that needed for a point estimation and can be found with the formula

$$N = 2Z_\alpha^2 \left(\frac{\sigma}{\delta}\right)^2 \tag{1.9}$$

Using the conditions stated above, we could solve for N and find it equal to approximately 18. We would need 18 samples to be 95% confident that our weight represented the average weight of the population. We could then report our average weight, w, as

[Lower limit on average weight] $\leq w \leq$ [Upper limit on average weight]

In the event that we do not know the standard deviation of the population, we can use the sample standard deviation, s, and the Student t distribution t_α.

1.5.3 Hypothesis Testing

In hypothesis testing, there are two hypotheses. H_0, the null hypothesis, is a hypothesis that assumes that the variable in the experiment will have no effect on the result and H_a is the alternative hypothesis that states that the variable will affect the results. For any population, one of the two hypotheses must be true. The goal of hypothesis testing is to find out which hypothesis is true by sampling the population.

In reality, H_0 is either true or false and we draw a conclusion from our tests of either true or false. This leads to four possibilities, as shown in Table 1.8.

Usually, the probability of making a Type I error is designated as α and the probability of making a Type II error is designated as β. Common values for α, the significance level, are 0.01, 0.05, and 0.1. The statistical power of a test is the probability that H_0 will be rejected when it is false. Statistical power is given by $1 - \beta$. Common power levels are above 0.8, indicating that when H_0 is false it will be correctly rejected more than 80% of the time.

Table 1.8 The four outcomes of hypothesis testing.

Conclusion	Real situation	
	H_0 true	H_a true
Accept H_0	Correct decision	Type II error, $p = \beta$
Reject H_0	Type I error, $p = \alpha$	Correct decision

Many scientific studies involving a hypothesis test will report the p value of the test. A very small p value means that the null hypothesis is unlikely to be true. When p is below an arbitrary cut-off value, e.g. 0.05, the result is called statistically significant. A study that reports $p = 0.05$ means that since the result obtained would happen only once in 20 times if H_0 were true, then H_0 should be rejected.

Table 1.9 Equivalent table of Table 1.8 for results relating to a condition or disease.

Test result	Has condition?	
	No	Yes
Negative	True negative (TN)	False negative (FN)
Positive	False positive (FP)	True positive (TP)

Test results about a condition or disease use the terminology in Table 1.9. In this case H_0 states that an individual does not have a condition or a disease, whereas H_a states that an individual does have a condition or disease.

The terms in Table 1.9 are useful for defining measures that describe the proportion of, for example, a disease in a population and the success of a test in identifying it.

Incidence is the number of cases of a disease during a stated period, such as x cases per 1000 per year.

Prevalence is the number of cases of a disease at a given time such as y cases per 1000. It is all diseased persons divided by all persons.

$$\text{Prevalence} = \frac{\text{TP} + \text{FN}}{\text{TN} + \text{TP} + \text{FN} + \text{FP}} \qquad (1.10)$$

Sensitivity is the probability of a positive test result when the disease is present. Among all diseased persons, it is the percent who test positive.

$$\text{Sensitivity} = \frac{\text{TP}}{\text{TP} + \text{FN}} 100\% \qquad (1.11)$$

Specificity is the probability of a negative diagnostic result in the absence of the disease. Among all normal persons, it is the percent who test negative.

$$\text{Specificity} = \frac{\text{TN}}{\text{TN} + \text{FP}} 100\% \tag{1.12}$$

Considering only those who test positive, positive predictive value (PPV) is the ratio of patients who have the disease to all who test positive.

$$\text{PPV} = \frac{\text{TP}}{\text{TP} + \text{FP}} 100\% \tag{1.13}$$

Considering only those who test negative, negative predictive value (NPV) is the ratio of nondiseased patients to all who test negative.

$$\text{NPV} = \frac{\text{TN}}{\text{TN} + \text{FN}} 100\% \tag{1.14}$$

In order to estimate if a proposed test is useful or not, we must run trials on a sample of subjects who are known normal and diseased. We can calculate the number of subjects required in our trial by consulting books on biostatistics (Dawson-Saunders and Trapp 1990). Clinical trials are prospective double blind randomized treatments with additional requirements on sample selection and ethical issues.

Tests to diagnose disease are not perfect. Figure 1.21 shows that positive test results may occur for only a portion of the normal population and only a portion of the diseased population. We set the threshold for the test to maximize true results and minimize false results.

The sample size needed to test a hypothesis depends on the level of significance and power desired. We can increase the threshold to make it very difficult to reject H_0 in order to reduce Type I errors (false positives). But then we risk not rejecting H_0 when it is false and making a Type II error (false negative). If we decrease the threshold, the converse is also true.

For a test comparing the means m_1 m_2 of two Gaussian populations with known and equal standard deviations, σ, the size of each sample needed to test H_0: $|m_1 - m_2| < \delta$, H_a: $|m_1 - m_2| > \delta$, is (Selwyn, 1996, Appendix B)

$$N = 2(Z_\alpha + Z_\beta)^2 \left(\frac{\sigma}{\delta}\right)^2 \tag{1.15}$$

where Z_α and Z_β are the values for the standard Gaussian distribution corresponding to, respectively, the tail probabilities of size α and β. For example, if $\alpha = 0.05$, $\beta = 0.1$, and $1 - \beta = 0.9$, $Z_\alpha = 1.96$ and $Z_\beta = 1.65$ and $N = 26.06(\sigma/\delta)^2$. If the result of the test suggests that we accept H_0, then there is a 5% probability that H_0 is false. If the result of the test suggests that we reject H_0, then there is a 10% probability that H_0 is true.

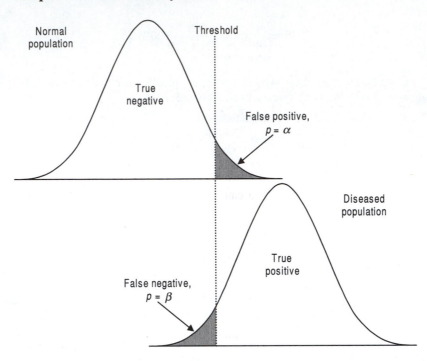

Figure 1.21 The test result threshold is set to minimize false positives and false negatives.

For a test about the mean m of a normal population, with H_0: $|m| < \delta$ and H_a: $|m| > \delta$, the sample size needed is

$$N = \left(Z_\alpha + Z_\beta\right)^2 \left(\frac{\sigma}{\delta}\right)^2 \tag{1.16}$$

where Z_α and Z_β have the same meaning as in Eq. (1.15). Kanji (1993) provides a fine summary of statistical tests.

A final word of caution about N in the above equations. All the sample sizes calculated are estimates based on several assumptions, some of which may be difficult to prove. N is not an exact number. Hence, it may be convenient to modify some of the relevant factors involved in the equations in order to seek their effect on N. In any case, a high value for N indicates the need for pilot studies aimed to reduce the experimental variability.

1.6 Lifelong Learning

Most biomedical engineers have had formal education from a college through a program of instruction designed by others. This education should continue throughout their career through a program of instruction designed by themselves. New technology appears constantly and must be learned to promote a productive career. A good place to start is by reading books, which review a large amount of material, present it in an organized fashion, and provide references for further study. To review books and conference proceedings, go to a technical library electronic card catalog or www.amazon.com and search on "biomedical engineering."

To obtain recent information, go to www.bmenet.org/BMEnet/ or www.bmes.org/, join societies such as the Biomedical Engineering Society or the Institute of Biomedical Engineering Technology, subscribe to biomedical engineering journals such as the *Annals of Biomedical Engineering*, and go to conferences listed there. You can search medical databases such as MEDLINE at www.ncbi.nlm.nih.gov/PubMed/. You can search patents at www.uspto.gov/. Commercial databases available at universities include ISI, OVID, and INSPEC. Perform general searches using search engines such as www.google.com/.

1.7 Problems

1.1   Compare the scientific method with that of making a diagnosis. List three things these processes have in common and three differences.

1.2   Give an example of a problem requiring different areas of biomedical engineering to be integrated in order to solve it.

1.3   Describe why a differential diagnosis is important.

1.4   Apply the steps of diagnosing diseases to diabetes.

1.5   For a system with a continuous, constant drift, explain what type of calibration would work best.

1.6   Your samples from a population are 1, 1, 3, 5, 5. Estimate the mean $\bar{x}$ and standard deviation s.

1.7   Your samples from a population are 1, 1, 3, 5, 5. Estimate the standard deviation of the mean $s_{\bar{x}}$.

1.8   For the Poisson probability density distribution, calculate and plot $p(K;m)$ for $K = 0, 1, 2, 3, 4$, and $m = 2$.

1.9   An X-ray image has 1000×1000 picture elements (pixels). Calculate the number of X-ray photons required so that the signal-to-noise ratio (SNR) = (average number of photons)/(standard deviation of the number of photons) = 8.

1.10 The results of test X are based on a threshold; positive results are greater than the threshold and negative results are less than the threshold. If the threshold is increased, explain how the sensitivity and specificity of test X change.

1.11 Find the prevalence, sensitivity, specificity, positive predictive value, and negative predictive value for a test where TP = 80, FN = 20, TN = 30, FP =10.

1.12 A test had 90% sensitivity and 80% specificity. If there are 5 people actually affected by the condition and 60 not affected, then find the number of true positives, true negatives, false positives, and false negatives.

1.13 A test has 83.3% sensitivity and 91.7% specificity. If the tests report that 15 people are positive and 57 people are negative, calculate the number of TP, TN, FP, and FN.

1.14 Express positive predictive value and negative predictive value in terms of the prevalence, sensitivity, and specificity.

1.15 For the blood pressure of adults, the mean $\mu = 120$ mmHg and the standard deviation $\sigma = 5$ mmHg. Calculate the number N of adults we must test to obtain a probability $p = 0.95$ that our sample population average blood pressure $\bar{x}_s$ does not differ from μ by more than $\delta = 3$ mmHg.

1.16 Find a magazine or journal article about a new technology in biomedical engineering. Provide the complete reference. Write a one-paragraph summary of what it may be used for and how it works.

1.17 Find web information about a new technology in biomedical engineering. Provide the complete reference. Write a one-paragraph summary of what it may be used for and how it works.

1.8 References

Biomedical Engineering: IEEE EMBS [Online] www.eng.unsw.edu.au/embs/index.html.

Dawson-Saunders, B. and Trapp, R. G. 1990. *Basic and Clinical Biostatistics*. Norwalk, CT: Appleton & Lange.

Doebelin, E. O. 1990. *Measurement Systems: Application and Design*. 4th ed. New York: McGraw-Hill.

Kanji, G. K. 1993. *100 Statistical Tests*. Newbury Park, CA: SAGE Publications.

Nathan, D. M. 1993. Long-term complications of diabetes mellitus. *N. Engl. J. Med.,* **328** (23): 1676–85.

Distribution Tables 2002. [Online] www.statsoft.com/textbook/stathome.html

The Diabetes Control and Complications Trial Research Group. 1993. The effect of intensive treatment of diabetes on the development and progression of long-term complications in insulin-dependent diabetes mellitus. *N. Engl. J. Med.,* **329** (14): 977–86.

Selwyn, M. R. 1996. Principles of Experimental Design for the Life Sciences. Boca Raton: CRC Press.

Webster, J. G. (ed.) 1998. *Medical Instrumentation: Application and Design*. 3rd ed. New York: John Wiley & Sons.

The Whitaker Foundation [Online] www.whitaker.org.

Chapter 2

Basic Concepts of Electronics

Hong Cao

Medical instruments are widely used in clinical diagnosis, monitoring, therapy, and medical research. They provide a quick and precise means by which physicians can augment their five senses in diagnosing disease. These instruments contain electric components such as sensors, circuits, and integrated circuit (IC) chips. Modern electronics technology, which includes transistors, ICs, and computers, has revolutionized the design of medical instruments.

Biomedical engineers should have a fundamental understanding of their operations and a basic knowledge of their component electric and electronic systems. Using this knowledge provides a better understanding of the principles of various measurements, as well as developing new measurements and instruments.

Electrical engineering is too large a topic to cover completely in one chapter. Thus, this chapter presents some very basic concepts in several fields of electrical engineering. It discusses analog components such as resistors, capacitors, and inductors. It then goes on to basic circuit analysis, amplifiers, and filters. From this it moves to the digital domain, which includes converters, sampling theory, and digital signal processing. It then discusses the basic principles of microcomputers, programming languages, algorithms, database systems, display components, and recorders.

2.1 Electronic Components and Circuit Analysis

2.1.1 Current

An atom contains a nucleus surrounded by electrons. Most of the electrons are tightly bound to the atom, while some electrons in the outer orbits are loosely bound and can move from one atom to another. In conductors, there are many free electrons. This process of electron transfer occurs in random directions in materials.

Suppose there is an imaginary plane in a conductor (Figure 2.1(a)). The loosely bound outer orbital electrons continuously cross this plane. Due to the random direction of the electron movement, the number of electrons that cross the plane from left to right equals the number that cross from right to left. Thus, the net flow of electrons is zero.

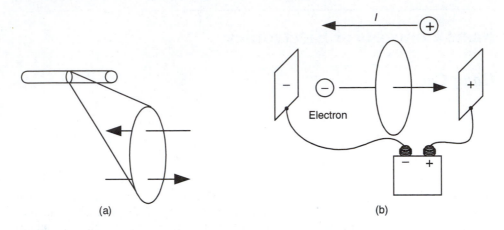

Figure 2.1 Electric current within a conductor. (a) Random movement of electrons generates no current. (b) A net flow of electrons generated by an external force.

When an electric field is applied across a conductor, it causes a net flow of electrons in the conductor because the electrons are attracted to the positive side of the electric field (Figure 2.1(b)). The rate of flow of electrons through a region is called electric current.

If ΔQ is the amount of charge that passes perpendicularly through a surface with area A, in time interval Δt, the average current, I_{av}, is equal to the charge that passes through A, per unit of time.

$$I_{av} = \Delta Q / \Delta t \tag{2.1}$$

The rate at which current flows varies with time, as does charge. We therefore can define the instantaneous current, I, as the differential limit of Eq. (2.1).

$$I = dQ/dt \tag{2.2}$$

The unit of current is the ampere (A), which represents a net flow of 1 coulomb (1 C), of charge or 6.242×10^{18} electrons across the plane per second (s).

The electron has a negative charge. When a negative charge moves in one direction, it yields the same result as a positive charge moving in the opposite direction. Conventionally, we define the direction of positive charge movement to be the direction of the electric current. Figure 2.1(b) shows that the direction of current is opposite to that of the flow of electrons.

Current can be generated in a circuit by a current source or by a voltage source and resistor in series.

2.1.2 Voltage and Potential

In moving a charge (+ or –) from point A to point B in an electric field, the potential energy of the charge changes. That change in energy is the work, W, done by the electric field on the charge. The amount of work is measured in Joules, J. If we let the electric potential, V, be equal to the potential energy, U, per unit charge, q_0, then we can define a potential difference, or voltage, as

$$\Delta V = V_B - V_A = \frac{W_{AB}}{q_0} \tag{2.3}$$

where V_B and V_A are the potentials at points B and A, respectively. The unit of potential is the volt (V), where 1 J/C = 1 V. If we choose the potential at infinity to be zero, the absolute potential of a point in an electric field can be defined as the total work per unit charge that has been done to move the charge from infinity to the point.

It is important that potential difference not be confused with difference in potential energy. The potential difference is *proportional* to the change in potential energy, where the two are related by $\Delta U = q_0 \Delta V$.

From Eq. (2.3), we can determine the work needed to move a charge from A to B if we know the potential difference between the two points. We are more interested in potential difference than the absolute potential for a single point. Notice that the potential is a property of the electric field, whether there is a charge in the field or not.

Voltage can be generated in a circuit by a voltage source or by a current source and resistor in parallel. Even though they exist, current sources are much less common than voltage sources in real circuits.

Example 2.1 How much work is needed to move an electron (a charge of 1.6×10^{-19} C) from 0 V to 4 V?

Rearranging Eq. (2.3)

$$W_{AB} = q_0 (V_B - V_A) = 1.6 \times 10^{-19} \, \text{C} (4 \, \text{V} - 0 \, \text{V}) = 6.4 \times 10^{-19} \, \text{J}$$

2.1.3 Resistors and Ohm's Law

When free electrons move in a conductor, they tend to bump into atoms. We call this property resistance and use a resistor as the electric component implementation. The unit of resistance is the Ohm (Ω), in honor of Georg Simon Ohm, who discovered what is now known as Ohm's law. Ohm's law states that for many materials, when a potential difference is maintained across the conductor, the ratio of the current density J (current per unit area) to the electric field that is producing the current, E, is a constant, σ, that is independent of the electric field. This constant is called the conductivity of the conductor.

$$J = \sigma E \tag{2.4}$$

It is useful to consider Ohm's law in the case of a straight wire with cross-sectional area a and length l, as in Figure 2.2. $V = V_b - V_a$ is the potential difference maintained across the wire, which creates an electric field and current in the wire. The potential difference can be related to the electric field by the relationship

$$V = El \tag{2.5}$$

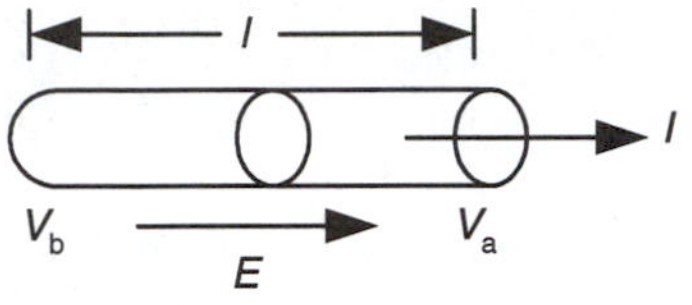

Figure 2.2 A model of a straight wire of length l and cross-sectional area a. A potential difference of $V_b - V_a$ is maintained across the conductor, setting up an electric field E. This electric field produces a current that is proportional to the potential difference.

Since the magnitude of the current density is related as $J = \sigma E,$ and current density J is defined as I/a, the potential difference can be rewritten as

$$V = (l/\sigma)J = (l/\sigma a)I \tag{2.6}$$

The quantity $(l/\sigma a)$ is defined as the resistance, R, of the conductor. The relationship can also be defined as a ratio of the potential difference to the current in the conductor

$$R = (l/\sigma a) = V/I \tag{2.7}$$

Experimental results show that the resistance, R, of a conductor can be rewritten as

$$R = \rho \frac{l}{a} \tag{2.8}$$

where ρ, the reciprocal of conductivity σ, is the resistivity ($\Omega \cdot$ m), l is the length of the conductor in meters (m), and a is the cross-sectional area of the conductor (m^2).

Ohm's law is written as

$$V = IR \tag{2.9}$$

where V is the voltage across the resistor and I is the current through the resistor.

Resistance depends on geometry as well as resistivity. Good electrical conductors have very low resistivity, where good insulators have very high resistivity. Resistivity is important in biomedical applications regarding not only electrical component resistances, but also resistances caused by body substances such as red blood cells, that may act as an insulator in counting procedures. Common carbon composition resistors used in electronic circuits have values between 10 Ω and 22 MΩ.

Example 2.2 What is the resistance of a 12.0 cm long piece of copper wire, if the wire has a resistivity of 1.7×10^{-8} $\Omega \cdot$ m and a radius of 0.321 mm?

The cross-sectional area of the wire is

$$a = \pi r^2 = \pi \left(0.321 \times 10^{-3} \text{ m}\right)^2 = 3.24 \times 10^{-7} \text{ m}^2$$

The resistance can be found using Eq. (2.8).

$$R = \rho \frac{l}{a} = 1.7 \times 10^{-8} \ \Omega \cdot \text{m} \left(\frac{0.12 \text{ m}}{3.24 \times 10^{-7} \text{ m}^2} \right) = 6.30 \times 10^{-3} \ \Omega$$

Most electric circuits make use of resistors to control the currents within the circuit. Resistors often have their resistance value in ohms color coded. This code serves as a quick guide to constructing electric circuits with the proper resistance. The first two colors give the first two digits in the resistance value. The third value, called the multiplier, represents the power of 10 that is multiplied by the first two digits. The last color is the tolerance value of the resistor, which is usually 5%, 10%, or 20% (Figure 2.3). In equation form, the value of a resistor's resistance can be calculated using the following

$$AB \times 10^C \pm D\% \tag{2.10}$$

where A is the first color representing the tens digit, B is the second color representing the ones digit, C is the third color or multiplier, and D is the fourth color or tolerance. Table 2.1 gives the color code for each of these four categories, and their corresponding digit, multiplier, or tolerance value.

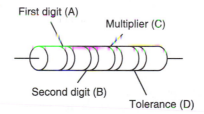

Figure 2.3 The colored bands that are found on a resistor can be used to determine its resistance. The first and second bands of the resistor give the first two digits of the resistance, and the third band is the multiplier, which represents the power of ten of the resistance value. The final band indicates what tolerance value (in %) the resistor possesses. The resistance value written in equation form is $AB \times 10^C \pm D\%$.

Example 2.3 What is the resistance value of a resistor where the first band is orange, second band is blue, third band is yellow, and fourth band is gold?

Using Table 2.1, the first band, orange, represents the first digit, 3. The second band gives the second digit, 6. The third band is yellow, 4. The fourth band, or tolerance, is gold, 5%. Therefore, the resistance value is AB $\times$ 10^C $\pm$ D% or $36 \times 10^4 = 360$ kΩ, with a tolerance value of 5% or 18 kΩ.

Table 2.1 The color code for resistors. Each color can indicate a first or second digit, a multiplier, or, in a few cases, a tolerance value.

Color	Number	Tolerance (%)
Black	0	
Brown	1	
Red	2	
Orange	3	
Yellow	4	
Green	5	
Blue	6	
Violet	7	
Gray	8	
White	9	
Gold	−1	5%
Silver	−2	10%
Colorless		20%

We define power, the rate at which work is done, as

$$P = VI \tag{2.11}$$

The unit of power is the watt (W). V is the voltage between the two terminals of the resistor and I is the current through the resistor.

Using Eq. (2.7) we can rewrite Eq. (2.11) as

$$P = \frac{V^2}{R} = I^2 R \tag{2.12}$$

Although power is most often used to describe the rate at which work can be done by mechanical objects, in electronics it represents the amount of energy dissipated by a component. The power lost as heat in a conductor, for example, is called joule heat and is referred to as a loss of I^2R.

Example 2.4 Assume the chest is a cylinder 10 cm in diameter and 40 cm long with a resistivity of 0.8 $\Omega \cdot$ m. For a voltage of 2 kV during defibrillation, calculate the current and power dissipated by the chest.

Calculate the current and power by using Eqs. (2.8), (2.9), and (2.11).

$$R = \rho \frac{l}{a} = 0.8 \frac{0.4}{\frac{\pi}{4} \times (0.1)^2} = 40.8 \, \Omega$$

$$I = \frac{V}{R} = \frac{2000}{40.8} = 49 \text{ A}$$

$$P = VI = 2000 \times 49 = 98 \text{ kW}$$

2.1.4 Basic Circuit Analysis

An electric system generally contains many components such as voltage and current sources, resistors, capacitors, inductors, transistors, and others. Some IC chips, such as the Intel Pentium, contain more than one million components.

Electric systems composed of components are called networks, or circuits. We can analyze the performance of simple or complex circuits using one theorem and two laws of linear circuit analysis.

Superposition theorem: The current in an element of a linear network is the sum of the currents produced by each energy source.

In other words, if there is more than one source in the network, we can perform circuit analysis with one source at a time and then sum the results. The superposition theorem is useful because it simplifies the analysis of complex circuits.

One concept we need to introduce is the polarity of a voltage drop. Conventionally, if the direction of current flow in a circuit is known or assumed, then the direction of current flow across a circuit element is from + to –. In other words, by current flowing through an element, a voltage drop is created (as shown by Ohm's law) by that element with the polarity of + to – in the direction of current flow (Figure 2.4(a)).

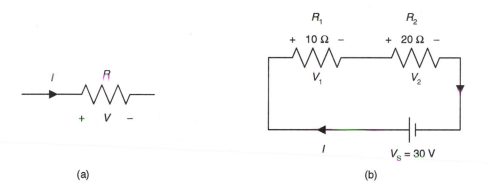

(a)   (b)

Figure 2.4 (a) The voltage drop created by an element has the polarity of + to – in the direction of current flow. **(b)** Kirchhoff's voltage law.

In a circuit, a loop is defined as a closed path containing circuit elements. Kirchhoff's voltage law (KVL) states

$$\sum V = 0 \qquad\qquad (2.13)$$

or the sum of voltage drops around a loop is zero. In other words, starting at any point on the closed loop, we can use the voltage law to form an equation by adding each voltage drop across every element (resistors and sources). We can assume the direction of current, but if our calculations give a negative result then the actual current flows in the opposite direction. This law follows from conservation of energy. A charge that moves in a closed loop must gain as much energy as it loses if a potential can be found at each point in the circuit.

Figure 2.4(b) shows a circuit to which we can apply Kirchhoff's voltage law. If we sum the voltages counterclockwise around the loop we get

$$-V_s + V_1 + V_2 = 0$$

$$V_s = V_1 + V_2$$

Notice that the voltage source, V_s, has a polarity opposite that of the other voltage drops.

A resistor that is traversed in the direction of the current gives a drop in potential across the resistor equal to $-IR$, while a resistor that is traversed in the direction opposite of a current gives an increase in potential across the resistor equal to $+IR$. This is directly related to Eq. (2.9), where the direction of current flow across a resistor affects the potential change across that resistor. Also a voltage source that is traversed from the + to − terminal gives a drop in potential of $-V$, while a voltage source that is traversed from the − to + terminal gives a increase in potential of $+V$.

In circuit analysis, a node is a junction of two or more branches. Kirchhoff's current law (KCL) states that at any node

$$\sum I = 0 \tag{2.14}$$

or the sum of currents entering or leaving any node is zero, this follows the law of conservation of charge. In other words, the currents entering a node must equal the currents leaving the node (Figure 2.5).

When we use Kirchhoff's current law, we can arbitrarily label the current entering the node + and leaving the node −. The sum of the currents entering the node in Figure 2.5(b) is

$$+3 + 6 - I = 0$$

$$I = 9 \text{ A}$$

Kirchhoff's voltage law and current law are basic laws for circuit analysis. There are two types of circuit analysis, each based on one of Kirchhoff's laws.

We assume unknown currents in the loops and set up the equations using Kirchhoff's voltage law and then solve these equations simultaneously.

We can use Figure 2.4(b) to illustrate loop analysis. We assume a current flows through the circuit in the direction already indicated in the figure. If we assumed the current in the other direction, our result would just be negative. Recalling Ohm's law, we have for the sum of the voltages through the clockwise loop

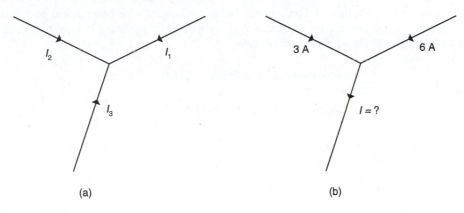

Figure 2.5 (a) Kirchhoff's current law states that the sum of the currents entering a node is 0. (b) Two currents entering and one "negative entering," or leaving.

$$V_S = R_1 I + R_2 I$$

$$30 = 10I + 20I$$

$$I = 1\,\text{A}$$

Now that we know the current through the loop, we can find the voltage drop through either resistor using Ohm's law. In this example, the voltage drop for R_1 is 10 V.

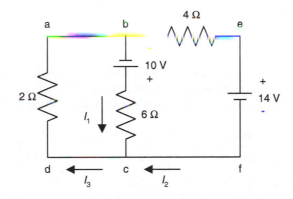

Figure 2.6 Kirchhoff's current law for Example 2.5.

Example 2.5 Find the currents I_1, I_2, and I_3 in the circuit shown in Figure 2.6.
 Applying Kirchhoff's current law to the current junction at node c, we have

$$I_1 + I_2 = I_3 \qquad (1)$$

All together there are three loops, *abcda*, *befcb*, and *aefda* (the outer loop) to which Kirchhoff's voltage law can be applied. Applying Kirchhoff's voltage law to the two inside loops in the clockwise direction, we have

$$\text{Loop } abcda: \; 10\,\text{V} - (6\,\Omega)I_1 - (2\,\Omega)I_3 = 0 \qquad (2)$$

$$\text{Loop } befcb: \; -14\,\text{V} - 10\,\text{V} + (6\,\Omega)I_1 - (4\,\Omega)I_2 = 0 \qquad (3)$$

Expressions (1), (2), and (3) represent three equations with three unknowns, and we can solve this system of three equations, first by substituting (1) into (2), giving

$$10 = 8I_1 + 2I_2 \qquad (4)$$

Second by dividing each term in (3) by 2, we get

$$-12 = -3I_1 + 2I_2 \qquad (5)$$

Finally, subtracting (5) from (4), I_2 is eliminated, giving

$$22 = 11I_1$$

$$I_1 = 2\,\text{A}$$

Substituting this value into (5), we find $I_2 = -3$ A., and substituting both these values in (1), we find $I_3 = -1$ A.

The other type of analysis is nodal analysis. We assume unknown voltages at each node and write the equations using Kirchhoff's current law. We solve these equations to yield the voltage at each node. Figure 2.7 shows an example using nodal analysis.

In Figure 2.7, we are given the voltage at A as 30 V and at B as 20 V. We assume the voltage at C is V. We apply Kirchhoff's current law at node C

$$\frac{30-V}{20} + \frac{20-V}{20} - \frac{V-0}{20} = \frac{50-3V}{20} = 0$$

$$V = 16.7\,\text{V}$$

When solving for voltages or currents in a circuit using the loop or nodal method, it is important to note that the number of independent equations needed is equal to the number of unknowns in the circuit problem. With an equal number of equations and unknowns, large numbers of unknowns can be solved using a matrix and linear algebra.

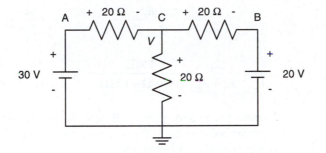

Figure 2.7 Example of nodal analysis.

2.1.5 Attenuators

When we amplify the 1 mV electrocardiogram signal from the heart, the signal may be undesirably reduced (attenuated) by the input resistance of the amplifier. For example, Figure 2.8 shows that the resistance of the skin, R_s, may be 100 kΩ, and the input resistance, R_i (input resistance of the oscilloscope used as an amplifier), is 1 MΩ.

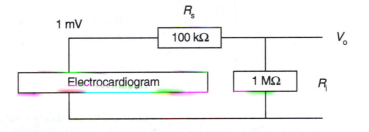

Figure 2.8 The 1 mV signal from the electrocardiogram is attenuated by the resistive divider formed by the 100 kΩ skin resistance and the 1 MΩ input resistance of the oscilloscope.

The current through the resistive divider is

$$i = \frac{V}{R} = \frac{1\,\text{mV}}{100\,\text{k}\Omega + 1\,\text{M}\Omega} = 0.91\,\text{nA}$$

The output voltage V_o is

$$V_o = iR_i = (0.91\,\text{nA})(1\,\text{M}\Omega) = 0.91\,\text{mV}$$

Thus the input voltage (ECG) has been undesirably attenuated by an amount that we can calculate from the voltage divider equation

$$\frac{V_o}{V_i} = \frac{R_i}{R_s + R_i}$$

(2.15)

$$\frac{V_o}{V_i} = \frac{R_i}{R_s + R_i} = \frac{1\,M\Omega}{100\,k\Omega + 1\,M\Omega} = 0.91$$

Figure 2.8 is a typical example of voltage divider. With two resistors in series, the output voltage is part of the input voltage. By adjusting the values of R_i and R_s, one can flexibly obtain any percentage of the input voltage.

A potentiometer is a three-terminal resistor with an adjustable sliding contact that functions as an adjustable voltage divider or attenuator. Figure 2.9 shows that if the slider is at the top, $v_o = v_i$. If the slider is at the bottom, $v_o = 0$. If the slider is in the middle, $v_o = 0.5v_o$. Potentiometers are usually circular with a rotatable shaft that can be turned by hand or a screwdriver. The potentiometer is useful to provide a variable gain for an amplifier or for the volume control on a radio. Alternatively, a potentiometer can be used as a two-terminal variable resistor by using the variable resistance between the slider and only one end of the potentiometer.

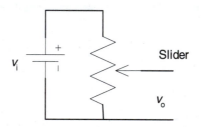

Figure 2.9 A potentiometer is a three-terminal resistor with an adjustable sliding contact shown by the arrow. The input signal v_i is attenuated by the potentiometer to yield an adjustable smaller voltage v_o.

2.1.6 Common Electrical Instruments

Galvanometer

The galvanometer is a main component that is used in creating ammeters (devices that measure current in a circuit) and voltmeters (devices that measure potential difference in a circuit), as the galvanometer can be used in conjunction with other circuit elements to measure current or potential differences of an electronic signal.

A common type of galvanometer consists of a coil of wire mounted in such a way that it is free to deflect or rotate in the presence of a magnetic field provided by a permanent magnet. The main principal behind the galvanometer makes use of a torque that acts on the loop of current when a magnetic field is applied. The torque is proportional to the amount of current that passes through the galvanometer, such that the larger the current, the greater amount of deflection or rotation of the coiled wire.

Typically, an off-the-shelf galvanometer is not directly used as an ammeter since it has a large resistance (about 60 Ω) that would considerably reduce the amount of current in the circuit in which the galvanometer is placed. Also, the fact that the galvanometer gives a full-scale deflection for low currents (1 mA or less) makes it unusable for currents greater in magnitude. If a galvanometer is placed in parallel with a *shunt resistor*, R_p, with a relatively small resistance value compared with that of the galvanometer, the device can be used effectively as an ammeter. Most of the current that is measured will pass through this resistor (Figure 2.10(a)).

If an external resistor, R_s, is placed in series with the galvanometer, such that its resistance value is relatively larger than that of the galvanometer, the device can be used as a voltmeter. This ensures that the potential drop across the galvanometer doesn't alter the voltage in the circuit to which it is connected (Figure 2.10(b)).

Older ammeters and voltmeters that used galvanometers with moving coils have been largely replaced by digital multimeters with no moving parts.

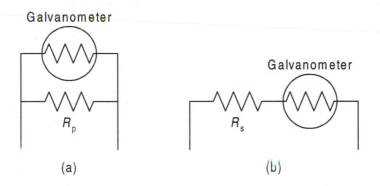

Figure 2.10 (a) When a shunt resistor, R_p, is placed in parallel with a galvanometer, the device can be used as an ammeter. (b) When a resistor, R_s, is connected in series with the galvanometer, it can be used as a voltmeter.

Wheatstone bridge

Often in biomedical instruments, unknown resistances or changes in resistance are measured within a circuit. Many times, an electric element known as Wheatstone bridge is used to measure these unknown resistances. The Wheatstone bridge has several applications and is widely used in electronics.

One application of the Wheatstone bridge consists of a common source of electric current (such as a battery) and a galvanometer (or other device that measures current) that connects two parallel branches, containing four resistors, three of which are known. One parallel branch contains one known resistance (R_1) and an unknown resistance (R_x) as shown in Figure 2.11. The other parallel branch contains resistors of known resistances, R_2 and R_3.

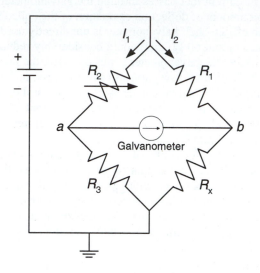

Figure 2.11 A circuit diagram for a Wheatstone bridge. The circuit is often used to measure an unknown resistance R_x, when the three other resistances are known. When the bridge is balanced, no current passes from node a to node b.

In order to determine the resistance of the unknown resistor, the resistances of the other three are adjusted and balanced until the current passing through the decreases to zero. Since the voltage potentials at a and b are equal, the potential difference across R_2 must equal R_1. Similarly, the potential difference across R_3 must also equal that across R_x. With this we have

$$I_1 R_2 = I_2 R_1 \tag{1}$$

$$I_1 R_3 = I_2 R_x \tag{2}$$

Dividing (2) by (1), we can eliminate the current, and solve for R_x

$$R_x = \frac{R_1 R_3}{R_2} \tag{2.16}$$

From Eq. (2.16), we can calculate the value of the unknown resistor.

The fact that a Wheatstone bridge is valuable for measuring small changes of a resistance makes it also suitable to measure the resistance change in a device called a *strain gage.* This device, commonly used in biomedical instruments to measure experimental stresses, often consists of a thin wire matrix attached to a flexible plastic backing and glued to the stretched metal. Stresses are measured by detecting changes in resistance of the strain gage as it bends. The resistance measurement is made with the strain gage as one or more elements in the Wheatstone bridge. Strain gages diffused into the diaphragm of a silicon integrated circuit blood pressure sensor are formed into a Wheatstone bridge.

2.1.7 Capacitors

From Ohm's law we know the voltage across a resistor is related to the current through the resistor by the resistance value. Now we investigate two elements with different relationships between voltage and current.

Conventionally, we use capital letters to denote variables that do not change with time and small letters to denote variables that change with time.

A capacitor is a two-terminal element in which the current, i, is proportional to the change of voltage with respect to time, dv/dt, across the element (Figure 2.12), or

$$i = C\frac{dv}{dt} \tag{2.19}$$

where C is capacitance and is measured in farads (F). One farad is quite large, so in practice we often see µF (10^{-6} F), nF (10^{-9} F), and pF (10^{-12} F). Common capacitor values are 10 pF to 1 µF. Capacitors are commonly used in a variety of electric circuits and are a main component of electronic filtering systems (introduced in 2.3) in biomedical instruments.

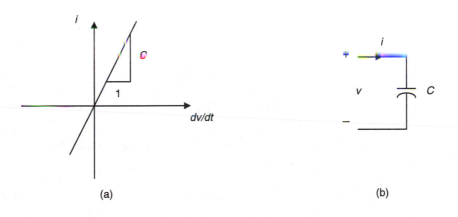

Figure 2.12 (a) Capacitor current changes as the derivative of the voltage. (b) Symbol of the capacitor.

We can also represent v as a function of i and t

$$v(t) = v(t_0) + \frac{1}{C}\int_{t_0}^{t} i\,dt \tag{2.20}$$

where t_0 and t are the beginning and ending times over which we observe the current, $v(t_0)$ is the initial voltage across the capacitor. We usually choose a t_0 of zero.

A capacitor usually consists of two conductors that are separated by an insulator. The capacitance of a device depends on its geometry and the dielectric material. A parallel–plate capacitor consists of two parallel plates of equal area A, separated by a distance d (Figure 2.13). One plate has a charge $+Q$ and the other $-Q$, where the charge per unit area on either plate is $\sigma = Q/A$. If the charged plates are very close together, compared to their length and width, we can assume that the electric field is uniform between the plates and zero everywhere else.

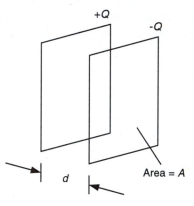

Figure 2.13 Diagram of a parallel–plate capacitor. The component consists of two parallel plates of area A separated by a distance d. When charged, the plates carry equal charges of opposite sign.

We can then see that the electric field between the plates is

$$E = \sigma/\varepsilon_0 = Q/\varepsilon_0 A \tag{2.21}$$

where ε_0 is the permittivity of free space (8.85 pF/m). Since the potential difference between the plates is equal to Ed, we find that the capacitance is

$$C = \varepsilon_0 A/d \tag{2.22}$$

Capacitances can be found in many different configurations within a circuit. Two capacitances, or several capacitances, can be reduced to one capacitance value by simple algebraic expressions depending on the situation. For the series capacitances in Figure 2.14(a), the equivalent capacitance, C_e, can be calculated by

$$C_e = \left(\frac{1}{C_1} + \frac{1}{C_2}\right)^{-1} = \frac{C_1 C_2}{C_1 + C_2} \qquad \text{2 capacitors}$$

$$\frac{1}{C_e} = \frac{1}{C_1} + \frac{1}{C_2} + \frac{1}{C_3} + \Lambda\,\Lambda \qquad \text{multiple capacitors} \qquad (2.23)$$

For the parallel capacitances in Figure 2.14(b), the equivalent capacitance, C_e, can be calculated by

$$C_e = C_1 + C_2 \qquad \text{2 capacitors}$$

$$C_e = C_1 + C_2 + C_3 + \Lambda\,\Lambda \qquad \text{multiple capacitors} \qquad (2.24)$$

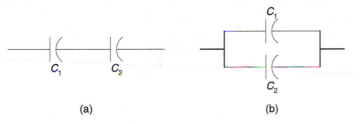

Figure 2.14 (a) A series combination of two capacitors. (b) A parallel combination of two capacitors.

2.1.8 Inductors

An inductor is a two-terminal element in which the voltage, v, is proportional to the change of current with respect to time, di/dt, across the element (Figure 2.15), or

$$v = L\frac{di}{dt} \qquad (2.25)$$

where L is the inductance. Its unit is the henry (H). Like the farad, this unit is large so we use more convenient units such as mH (10^{-3} H) and µH (10^{-6} H). This is usually taken to be the defining equation for the inductance of any inductor, regardless of its shape, size, or material characteristics. Just as resistance is a measure of opposition to a current, inductance is a measure of the opposition to any change in current. The inductance of an element often depends on its geometry.

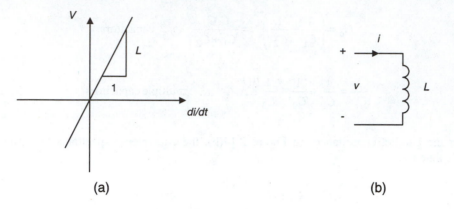

Figure 2.15 (a) Inductor voltage changes as the derivative of the current. (b) Symbol of the inductor.

We can also represent i in terms of v and t

$$i(t) = i(t_0) + \frac{1}{L}\int_{t_0}^{t} v\,dt \qquad (2.26)$$

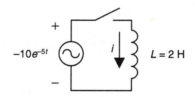

Figure 2.16 Simple inductor circuit.

Example 2.6 In Figure 2.16, an inductor is connected, at time $t = 0$ by a switch, to a voltage source whose signal can be described as $-10e^{-5t}$ V. Derive an equation that describes the current through the inductor as a function of time.

Using Eq. (2.26) we can solve for $i(t)$. There is no current through the inductor at $t = 0$, so $i(t_0) = 0$.

$$i(t) = i(t_0) + \frac{1}{L}\int_{t_0}^{t} v\,dt$$

$$i(t) = 0 + \frac{1}{2\,\mathrm{H}} \int_0^t -10e^{-5t} \text{ V } dt = \frac{1}{2}\left(\frac{-10}{-5}\right)e^{-5t}$$

$$i(t) = e^{-5t} \text{ A}$$

For multiple series inductors, the equivalent inductance L_e, can be calculated by

$$L_e = L_1 + L_2 + L_3 + \Lambda\,\Lambda \tag{2.27}$$

For multiple parallel inductors, L_e can be calculated by

$$\frac{1}{L_e} = \frac{1}{L_1} + \frac{1}{L_2} + \frac{1}{L_3} + \Lambda\,\Lambda \tag{2.28}$$

2.1.9 First-Order Systems

Consider the simple resistor and capacitor (RC) circuit shown in Figure 2.17(a).

Suppose that at time $t = 0$ the voltage across the capacitor is $v_C(0)$. For $v_C(t)$ when $t \geq 0$, we have

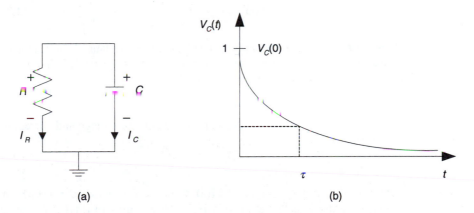

(a)   (b)

Figure 2.17 (a) Simple RC circuit with $v(0)$ on capacitor at time $t = 0$. (b) Normalized voltage across the capacitor for $t \geq 0$ (normalized means the largest value is 1).

$$v_R = v_C \text{ and } i_R + i_C = 0 \quad \Rightarrow \quad \frac{v_C}{R} + C\frac{dv_C}{dt} = 0 \tag{2.29}$$

Upon rearranging, we get

$$\frac{dv_C}{dt} + \frac{1}{RC}v_C = 0 \qquad (2.30)$$

This equation contains the variable v_C and its first derivative dv_C/dt. This is a first-order differential equation. Many complex systems can be reduced to multiple first-order systems, which are more easily solved than high-order systems.

The solution of this differential equation, assuming at $t = 0$, $v_C(t) = v_C(0)$, is

$$v_C(t) = v_C(0)e^{-\frac{t}{RC}} \qquad (2.31)$$

Figure 2.17(b) shows $v_C(t)$ versus t for $t \geq 0$. Also, we have

$$i_C(t) = -i_R(t) = \frac{v_C(t)}{R} = -\frac{v_C(0)}{R}e^{-\frac{t}{RC}} \qquad (2.32)$$

From these results we can see that voltage and current in the circuit decrease as time increases and approach zero as time goes to infinity. The voltage and current in this circuit can be described by equations of the form

$$f(t) = Ke^{-\frac{t}{\tau}} \qquad (2.33)$$

where K and τ are constants. The constant τ is called the time constant of the first-order system. In this case, $\tau = RC$. When time t reaches τ, $v_C(t = \tau)$ is $v_C(0)e^{-1}$. When $t = 3\tau$, the value of $v_C(t)$ is down to only 5% of the value of $v_C(0)$. We can assume the value of $v_C(t)$ is practically zero after 5 time constants.

In the circuit shown in Figure 2.17(a), there was no independent source (or input), only an initial state of the circuit. For this reason, we call the voltage and current we obtained the natural response.

Example 2.7 Consider the circuit in Figure 2.17(a). Suppose $R = 1$ kΩ, $C = 40$ μF and the initial voltage across the capacitor at $t = 0$ is 10 V. How long will it take for the capacitor to discharge to half of its original voltage?

This simple RC circuit can be solved with a first-order differential equation. $v_C(0) = 10$ V, so half of that voltage would give us $v_C(t) = 5$ V. Using Eq. (2.31) and algebra we can solve for t.

$$v_C(t) = v_C(0)e^{-\frac{t}{RC}}$$

$$\ln\left(\frac{v_C(t)}{v_C(0)}\right) = -\frac{t}{RC}$$

$$t = -RC\ln\left(\frac{v_C(t)}{v_C(0)}\right) = -(1\,\text{k}\Omega)(40\,\mu\text{F})\ln(0.5) = 27.7\,\text{ms}$$

Now let us consider the case of nonzero input using the circuit shown in Figure 2.18(a). From $t = -\infty$ to just before $t = 0$ ($t = 0^-$), the switch has been open. Therefore, no current has been flowing and no charge is in the capacitor. At $t = 0$, we close the switch which causes current to flow and charge to accumulate in the capacitor.

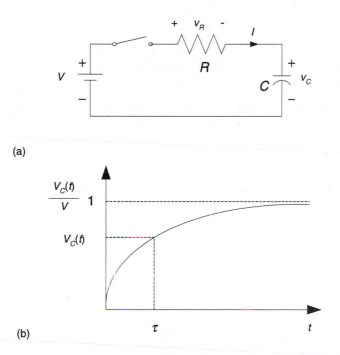

(a)

(b)

Figure 2.18 (a) Series RC circuit with voltage step input at time 0. (b) Normalized voltage across the capacitor.

Using KVL, the equation for this current is

$$V = C\frac{dv_C}{dt} + \frac{v_C}{R} \quad \text{for } t \geq 0$$

or   (2.34)

$$\frac{dv_C}{dt} + \frac{1}{RC}v_C = \frac{V}{RC} \quad \text{for } t \geq 0$$

The solution of the equation is

$$v_C = V - Ve^{-\frac{t}{RC}} \tag{2.35}$$

The voltage across the resistor, v_R, and current, i, are

$$v_R = Ve^{-\frac{t}{RC}}$$

$$i = \frac{V}{R}e^{-\frac{t}{RC}} \tag{2.36}$$

The voltage $v_C(t)$ is shown in Figure 2.18(b). The voltage across the resistor decreases exponentially and the voltage across the capacitor increases exponentially. From the definition of a capacitor, we know the voltage across it cannot change abruptly or the current will become infinite. When the switch is closed, the voltage on the capacitor is still zero and the voltage drop across the resistor is V, the source voltage. As time increases, the capacitor becomes charged and as time approaches infinity, all the source voltage is across the capacitor while the voltage drop across the resistor is zero.

Example 2.8 Consider the circuit in Figure 2.18(a). Suppose that $V = 12$ V, $R = 1$ kΩ and $C = 10$ μF. At time $t = 0$ the switch closes. Sketch the voltage across the capacitor versus time. How long will it take for the capacitor to become 90% of the voltage source?

Using Eq. (2.35):

$$v_C = V - Ve^{-\frac{t}{RC}}$$

$$v_C = 12 - 12e^{-\frac{t}{0.01}} \text{ V}$$

90% of the voltage source is 10.8 V. Substituting this value in for v_C in the above equation and solving for t yields a time of 23 ms with a time constant of 10 ms.

We can also use a resistor and an inductor to make a first-order circuit where $\tau = L/R$.

2.1.10 Frequency

Sinusoidal waves are widely used in electric circuits. Figure 2.20 shows two sinusoidal waveforms, which can be represented as

$$A\sin(2\pi\,ft+\theta) = A\sin(\omega t+\theta) = A\sin\left(\frac{2\pi}{T}t+\theta\right) \qquad (2.37)$$

where: A = amplitude of sine wave
f = frequency of sine wave in hertz (Hz)
$\omega = 2\pi f$ = angular frequency of sine wave (radians per second)
θ = phase angle of sine wave (radians)
$T = 1/f$ = period (seconds)

Note: The sine function lags 90° in phase behind the cosine function.

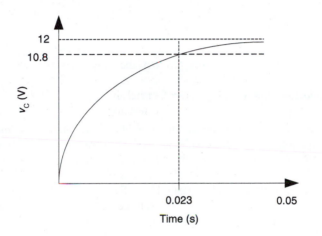

Figure 2.19 Plot of v_C for Example 2.8.

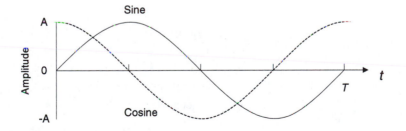

Figure 2.20 One period, T, of the sine and cosine waveforms.

Example 2.9 Figures 2.21 and 2.22 show a comparison of a sinusoidal waveform with different frequency and different phase angle.

In Figure 2.21 the solid line shows the function $y = \sin(\omega t)$, the dashed line shows the function with the frequency doubled, $y' = \sin(2\omega t)$. By doubling the frequency the period is reduced by half by $T = 1/f$.

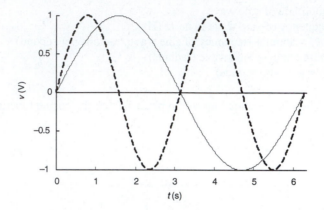

Figure 2.21 Sinusoidal waveforms with different frequencies.

In Figure 2.22 the solid line again shows the function $y = \sin(\omega t)$, the dashed line shows the function with a phase shift of 180° (π in radians), $y' = \sin(\omega t - \pi)$. The minus sign in front of the phase angle shifts this waveform to the right and can be said that it is leading the original waveform. Also, with a phase shift of 180°, the two waveforms are said to be completely out of phase.

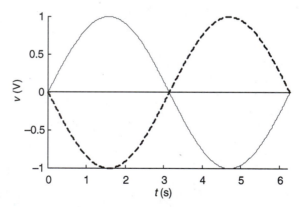

Figure 2.22 Sinusoidal waveforms with 0° phase angle (solid) and 180° phase angle (dashed).

When a voltage or current source is a sinusoidal wave, Kirchhoff's laws still apply in differential form to include capacitors and inductors. Since sinusoidal sources change with time and its differential is still sinusoidal, a more efficient way to solve the equations is to use complex sinusoidal forms.

For example, we can write $A\cos(\omega t + \phi)$ in complex sinusoidal form as

$$Ae^{j(\omega t+\phi)} = Ae^{j\phi}e^{j\omega t} \tag{2.38}$$

This representation comes from Euler's identity and has the same amplitude, frequency, and phase angle as the noncomplex form. Let us denote the complex number

$$\mathbf{A} = Ae^{j\phi} \tag{2.39}$$

by

$$\mathbf{A} = A\angle\phi \tag{2.40}$$

which is known as a phasor, with bold type meaning it is a vector. Equation 2.40 can be read as: "The vector $\mathbf{A}$ has a magnitude of A and an angle of ϕ."

Let us consider a resistor. If the current and voltage of the resistor are all complex sinusoids, say

$$i = Ie^{j(\omega t+\theta)}$$

$$v = Ve^{j(\omega t+\phi)} \tag{2.41}$$

Then, by Ohm's law

$$v = Ri$$

$$Ve^{j(\omega t+\theta)} = RIe^{j(\omega t+\phi)}$$

$$Ve^{j\theta} = RIe^{j\phi} \quad \text{or} \quad V\angle\theta = RI\angle\phi \tag{2.42}$$

Eq. (2.42) shows that $\theta = \phi$, which means the voltage and the current are in phase across the resistor (Figure 2.23(a)). We can rewrite this as the phasor

$$\mathbf{V} = R\mathbf{I} \tag{2.43}$$

As to an inductor, the voltage across it is given by

$$v = L\frac{di}{dt} \tag{2.44}$$

The V–I relationship can be written as

$$Ve^{j\theta} = j\omega LIe^{j\phi} \quad \text{or} \quad V\angle\theta = j\omega LI\angle\phi \tag{2.45}$$

Therefore,

$$V = j\omega L I \tag{2.46}$$

and $\theta = \phi + 90°$, which shows that current lags the voltage by 90° (Figure 2.23(b)). Using the methods presented in Eq. (2.44) through (2.46) yields

$$V = \frac{1}{j\omega C} I \tag{2.47}$$

and $\theta = \phi - 90°$, which shows that current leads the voltage by 90° (Figure 2.23(c)).

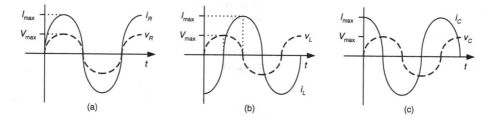

Figure 2.23 Plots of the current and voltage as a function of time. (a) With a resistor both the current and the voltage vary as sin(ωt), the current is in phase with the voltage, meaning that when the current is at a maximum, the voltage is also. (b) For an inductor, the current lags behind the voltage 90°. (c) For a capacitor, the current leads the voltage by 90°.

From the equations above, we can see a general equation in phasor notation for the resistor, capacitor, and inductor. That equation is

$$V = ZI \tag{2.48}$$

where Z is the impedance of the element. For the three elements,

$$Z_R = R \qquad Z_C = \frac{1}{j\omega C} \qquad Z_L = j\omega L \tag{2.49}$$

With impedance form of Eq. (2.49) and the method introduced in Section 2.1.4, the relationship of voltage and current is expressed in algebraic form instead of differential form. Phasor equations simplify the circuit analysis and can be used by computers to solve the problem.

Note that the impedance of capacitors and inductors changes with frequency. Therefore, the performance of the circuit will also change with frequency. This change in circuit performance is called the frequency response of the circuit.

Example 2.10 The current through a 50 mH inductor is $100\angle 0°$ mA. If $\omega = 1000$ rad/s, what is the voltage across the inductor?

Using the phasor equation in Eq. (2.46), we can solve for **V**.

$$\mathbf{V} = j\omega L \mathbf{I}$$

$$\mathbf{V} = j1000(0.050)(100\angle 0° \text{ mA}) = 5\angle 90° \text{ V}$$

2.1.11 Series and Parallel Impedances

Now that the impedances for the three passive circuit elements are defined (Eq. 2.49), it is time to study two types of configurations for them in a circuit. Figure 2.24(a) shows two impedances in series. These two impedances, and in general infinitely many impedances, can be combined into one equivalent impedance. For the series impedances in Figure 2.24(a), the equivalent impedance, Z_e, can be calculated by

$$Z_e = Z_1 + Z_2 \qquad\qquad (2.50a)$$

For multiple impedances in series, the equivalent impedance Z_e is

$$Z_e = Z_1 + Z_2 + Z_3 + \Lambda\,\Lambda \qquad\qquad (2.50b)$$

For the parallel impedances in Figure 2.24(b), the equivalent impedance, Z_e, can be calculated by

$$Z_e = \frac{1}{\dfrac{1}{Z_1} + \dfrac{1}{Z_2}} = \frac{Z_1 Z_2}{Z_1 + Z_2} \qquad\qquad (2.51a)$$

For multiple impedances in parallel, the equivalent impedance Z_e is

$$\frac{1}{Z_e} = \frac{1}{Z_1} + \frac{1}{Z_2} + \frac{1}{Z_3} + \Lambda\,\Lambda \qquad\qquad (2.51b)$$

A shorthand notation for indicating parallel circuit elements as in Eq. (2.51) is to use a double vertical bar between the parallel elements, such as $Z_e = Z_1 \| Z_2$.

Using these definitions, it is possible to reduce large complicated circuits down to (ideally) one or two impedances and a voltage source. Figure 2.24(c) shows the equivalent impedance of the combined series or combined parallel circuit.

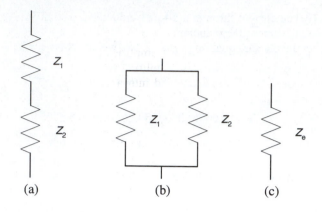

Figure 2.24 (a) Series circuit. (b) Parallel circuit. (c) Single impedance equivalent.

2.1.12 Electrical Safety

Because safety is paramount in considering biomedical devices in the real world setting, great emphasis is placed on currents and voltages that exist in an apparatus and what dangers these may pose to patients and operators who use the device.

Electric shock may result in fatal burns and can cause muscles and even the heart to malfunction. Often the degree of damage depends on the magnitude of current that is applied, how long the current acts, and through which point on the body the current passes. Typically, skin currents of 5 mA or less may cause a light sensation of shock, but usually do no damage. Currents larger than 10 mA tend to cause the muscles to contract to the point a person is unable to let go of a live wire. If currents of 100 mA pass through the body, even for a few seconds, respiratory muscles can become paralyzed and breathing stops. The heart can go into ventricular fibrillation, which is fatal (macroshock). If an electrode or catheter is within the heart, current must not exceed 10 μA because the current density is high (microshock). Contact with live wires or voltages above 24 V is not recommended.

Often in biomedical engineering, currents and voltage outputs must be amplified, attenuated, or filtered safely in order to achieve the best possible electronic signal with minimal harm to the patient or operator. In the following sections, we will explore ways in which this can be done.

2.2 Amplifiers

Most bioelectric signals have a very small magnitude (on the order of millivolts or microvolts) and therefore require amplification so that users can process them. This section emphasizes the operational amplifier (op amp) for use in amplifier design.

2.2.1 Basic Op Amp

An op amp is a high-gain dc differential amplifier (differential meaning that any signals not the same at both inputs are greatly amplified). An op amp is made of many transistors and other components that are integrated into a single chip. We need not consider the components within the chip, but simply the terminal characteristics. Focusing on the terminal behavior of op amps allows us to see many applications for this circuit in bioinstrumentation.

Figure 2.25(a) shows the equivalent circuit of an op amp. Figure 2.25(b) shows the symbol of the op amp. v_1 and v_2 are the two input terminals, v_o is the output terminal and V^+ and V^- are the positive and negative power supplies. Gain, also called amplification, is defined as an increase in voltage by an amplifier, expressed as the ratio of output to input (A in Figure 2.25).

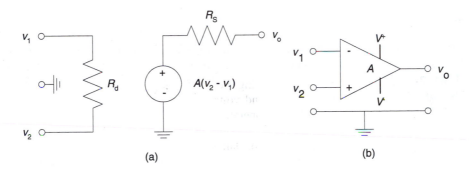

(a)   (b)

Figure 2.25 (a) Equivalent circuit for op amp. (b) Symbol of op amp. Many times V^+ and V^- are omitted in the op amp symbol, but it is understood that they are present.

2.2.2 Ideal Op Amp

When designing a circuit, we assume that we are using an ideal op amp. Then we investigate whether or not the nonideal characteristics affect the performance. If they do, we revise the initial design to accommodate the nonideal characteristics.

For an ideal op amp, we make the following assumptions:

1. $A = \infty$ (gain, or amplification of the input, is infinity)
2. $v_o = 0$ when $v_1 = v_2$ (no offset voltage)
3. $R_d = \infty$ (differential input impedance is infinity)
4. $R_s = 0$ (output impedance is zero)
5. Bandwidth $= \infty$. (Bandwidth will be discussed in greater detail in Sections 2.27 and 2.3.)

Two basic rules should be stated before we discuss designing circuits with op amps.

Rule 1: The two input terminals are at the same voltage.

If the two input terminals were not at the same voltage, the differential input voltage would be amplified by infinite gain and yield infinite output voltage. This is not realistic. For an op amp with a power supply of ±15 V, most manufactures guarantee a linear output range of ±10 V. Also the op amp saturates about ±13 V, which means that when v_o exceeds ±13 V, it saturates and further increases in v_i will not produce change in the output.

Rule 2: No current flows into either of the two input terminals.

This is true because we assume the input impedance of an op amp is infinity and no current flows into an infinite impedance.

2.2.3 Inverter

Figure 2.26 shows an inverting amplifier, or inverter. It is the most widely used circuit in instrumentation. Note that R_f is connected from the output to the negative input terminal and thus provides negative feedback. The negative feedback increases the bandwidth and lowers the output impedance of the op amp circuit.

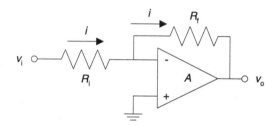

Figure 2.26 An inverting amplifier. The gain of the circuit is $-R_f/R_i$.

In Figure 2.26, the positive input terminal of the op amp is connected to ground and is always at 0 V. By rule 1, the negative input of the op amp is also at 0 V and remains at 0 V no matter what happens to the other components of the circuit. This clamping of the negative input to 0 V is called a virtual ground.

By rule 2, we know no current flows into the op amp. Therefore, the current flows through R_i and then through R_f to the output terminal, then through the output circuit of the op amp to ground.

Using KCL at the "–" terminal

$$i_{in} = i_{out} \Rightarrow \frac{v_i - 0}{R_i} = \frac{0 - v_o}{R_f} \quad \Rightarrow \quad v_o = -\frac{R_f}{R_i}v_i \quad \Rightarrow \quad \frac{v_o}{v_i} = -\frac{R_f}{R_i} \tag{2.52}$$

The output v_o has the opposite sign of the input v_i and is amplified by an amount R_f/R_i.

When using op amps in circuits it is important to consider the output impedance of any source that they are connected to, as this value will affect the total input resistance and the gain of the op amp.

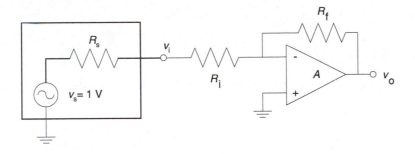

Figure 2.27 Inverter circuit attached to a generator that contains an internal resistance.

Example 2.11 Suppose in Figure 2.27 that $v_i = 1$ V, $R_f = 4$ kΩ and $R_i = 1$ kΩ, we would have, from Eq. (2.52), a gain of –4 and therefore, an output voltage of –4 V. However if the signal generator shown in Figure 2.27 were connected and had output impedance R_s, this impedance would add to R_i of the op amp and the new input impedance would be $R_s + R_i$. R_s would therefore change v_o as well as the gain of the circuit. If $R_s = 150$ Ω, then using Ohm's law we obtain the gain

$$\frac{v_o}{v_s} = -\frac{R_f}{R_i + R_s} = -3.48$$

which is much less than our desired value.

2.2.4 Noninverting Amplifier

Figure 2.28(a) shows the circuit for a noninverting amplifier. From rule 1, we know the voltage of the negative input always equals the voltage of the positive input terminal, which in this case is v_i. From Figure 2.28(a) we have

$$\frac{v_i - 0}{R_i} = \frac{v_o - v_i}{R_f}$$

$$\frac{v_o}{v_i} = \frac{R_f + R_i}{R_i} \quad \text{or} \quad v_o = \frac{R_f + R_i}{R_i} v_i = v_i\left(1 + \frac{R_f}{R_i}\right)$$

(2.53)

Notice the output v_o has the same sign as the input voltage v_i. The amplifier gain is determined by the ratio $(R_f + R_i)/R_i$.

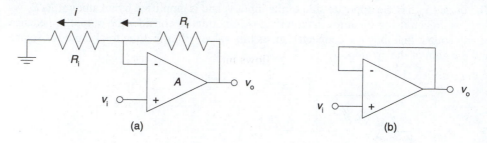

Figure 2.28 (a) A noninverting amplifier has a gain of $(R_f + R_i)/R_i$. (b) A follower, or buffer, with unity gain.

If we choose $R_i = \infty$ (open circuit), the circuit can be simplified as Figure 2.28(b). The gain, $(R_f + R_i)/R_i$, would be unity. Therefore $v_o = v_i$, or the output voltage follows the input voltage. At first glance, this circuit does nothing, since the output equals the input. However, this circuit is very useful as a buffer, to prevent a high source resistance from being loaded by a low-resistance load.

Loading can be expressed by the voltage divider equation (Eq. 2.15). In Figure 2.8, if R_s is a significant fraction of R_i, we would get an attenuated signal for V_o. We say that the source resistance, R_s, is being loaded by R_i. This can be seen mathematically in Eq. (2.15)—as R_s increases to infinity, V_o/V_i approaches zero. In general, we like to have $R_i \gg R_s$, which is what the buffer circuit does. It provides an extremely high input impedance and an extremely low output impedance.

Example 2.12 Suppose the signal generator in Figure 2.27 were connected in a noninverting configuration, as in Figure 2.29, will the generator's output impedance affect the gain? No, since no current will flow into the positive input terminal, $v_i = 1$ V, the gain would remain $(R_f + R_i)/R_i$. If $R_f = 3$ kΩ and $R_i = 1$ kΩ, the gain would be 4, independent of the value of R_s.

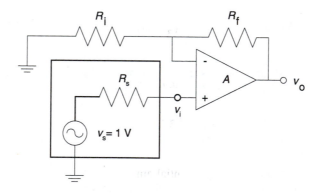

Figure 2.29 The gain of the noninverting amplifier is not affected by the addition of the impedance R_s due to the generator.

2.2.5 Differential Amplifiers

Figure 2.30 is a simple one-op-amp differential amplifier. Current flows from v_2 through R_1 and R_2 to the ground, and no current flows into the op amp. Thus, R_1 and R_2 act as a voltage divider.

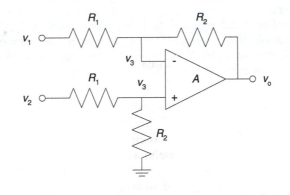

Figure 2.30 A differential amplifier uses two active inputs and a common connection.

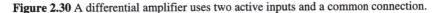

The voltage that appears at the positive input terminal determines the voltage at the negative terminal. The top part of the amplifier works like an inverter.

The voltage at the positive input terminal is

$$v_3 = \frac{R_2}{R_1 + R_2} v_2 \tag{2.54}$$

Since there is no flow of current into the op amp,

$$i = \frac{v_1 - v_3}{R_1} = \frac{v_3 - v_o}{R_2} \tag{2.55}$$

Combining these two equations yields

$$v_o = \frac{R_2}{R_1}(v_2 - v_1) \tag{2.56}$$

This is the equation for a differential amplifier. If the two inputs are hooked together and driven by a common source, then the common-mode voltage is equal to V_c ($v_1 = v_2$). The differential amplifier circuit gives an output of 0 and the differential amplifier common-mode gain (G_c) is 0. When $v_1 \neq v_2$, the differential voltage ($v_2 - v_1$) produces a differential gain (G_d), which equals R_2/R_1.

An important application of differential amplifier is to remove interference from the input signal. Take the ECG for example. The wires from patient to the ECG machine are about 1 to 2 m in length and are exposed to different interferences such as power line interference. Since their wires have almost the same amount of noise because they are exposed to the same common noise sources, this common noise is rejected by the differential amplifier according to Eq. (2.56). The output of the amplifier solely depends on the input from the patient.

In some instances, the signal on both input lines (such as ECG input terminals) is corrupted by a surrounding interference (such as power line interference) and is not acceptable for us to process. From Eq. (2.56) we know that only different signals input into the differential amplifier will be amplified. Therefore, the common interference on both inputs will not be amplified.

No differential amplifier perfectly rejects the common-mode voltage. To quantify the performance of a differential amplifier, we use the term common-mode rejection ratio (CMRR), which is defined as

$$CMRR = \frac{G_d}{G_c} \tag{2.57}$$

In real applications a CMRR greater than 100 may be acceptable, depending on the situation. A high-quality differential amplifier may have a CMRR greater than 1,000,000 (120 dB).

To measure CMRR, we have to know both G_d and G_c. We can apply the same voltage to the two different inputs and measure the output to obtain G_c. Then we can connect one input to ground and apply a voltage to the other one, and obtain G_d from the output. With these two numbers, we can calculate the CMRR from Eq. (2.57).

As with the inverting amplifier, we must be careful that resistances external to the op amp circuit do not affect the circuit. A differential voltage connected to a differential amplifier can have two output impedances affecting the op amp input impedance. This is a problem when the added impedance fluctuates (as skin resistance fluctuates, for example). One solution to the problem is to use buffer amplifiers as in Figure 2.31. Because no current flows into the input terminal, there is no current across the resistor. Thus, there is no loss of voltage.

Example 2.13 For the differential amplifier in Figure 2.32, solve for v_o.

Using KCL, we can solve for v_o. First, the voltage at the positive terminal of the amplifier is part of a voltage divider.

$$v_3 = \frac{R_3}{(R_1 \parallel R_2) + R_3} 8\,V = \frac{1\,k\Omega}{(1\,k\Omega \parallel 500\,\Omega) + 1\,k\Omega} 8\,V = 6\,V$$

Because there is no flow of current into the op amp, we can solve for v_o using Kirchhoff's current law.

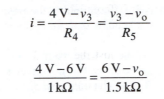

$$i = \frac{4\,\mathrm{V} - v_3}{R_4} = \frac{v_3 - v_o}{R_5}$$

$$\frac{4\,\mathrm{V} - 6\,\mathrm{V}}{1\,\mathrm{k}\Omega} = \frac{6\,\mathrm{V} - v_o}{1.5\,\mathrm{k}\Omega}$$

$$v_o = 9\,\mathrm{V}$$

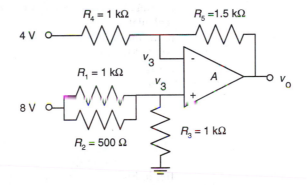

Figure 2.31 Differential amplifier attached to a differential voltage that contains varying imped-ances. Adding buffers ensure that fluctuations in R_s do not affect the gain.

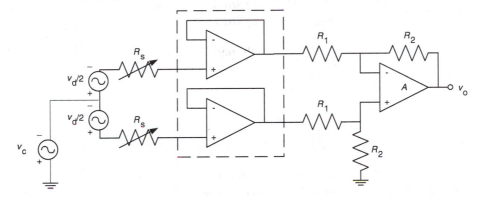

Figure 2.32 Differential amplifier for Example 2.13.

2.2.6 Comparators

A comparator is a circuit that compares the input voltage with some reference voltage. The voltage output flips from one saturation limit to the other as the negative input of the op amp passes through 0 V. If the input is greater than the reference, the output is the negative saturation voltage. Otherwise, the output is the positive saturation voltage. Note that the saturation voltage is always smaller than the voltage of the power supply for the op amp.

The simplest comparator is the op amp itself. If a reference voltage is connected to the positive input and v_i is connected to the negative input, the circuit is complete. However, the circuit in Figure 2.33 is more widely used.

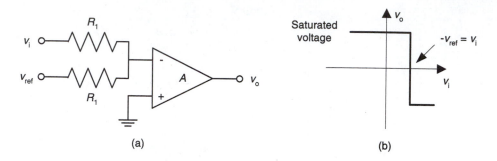

(a)   (b)

Figure 2.33 (a) A comparator. (b) The input–output characteristic of the comparator in (a).

The positive input is always at 0 V since it is grounded. Because the current flows from v_i to v_{ref} through two equal resistors, the voltage at the negative input is their average value $(v_i + v_{ref})/2$. If $v_i > -v_{ref}$, the voltage at the negative input is larger than the voltage at the positive input, which is 0 V. Because the op amp is open loop and its gain is infinite, the output voltage v_o saturates at the negative saturation value, which is usually about −13 V. On the other hand, if $v_i < -v_{ref}$, the output is the positive saturation value, which is about 13 V.

An application of the comparator is detecting heartbeats from the ECG signal, as shown in Figure 2.34. The normal ECG signal contains a P wave, a QRS complex, and a T wave. The QRS complex normally has a much larger amplitude than the other two waves. A threshold is set, which is larger than the P and T waves but smaller than the R waves. Every time a QRS complex passes through the comparator, the comparator gives out a pulse that corresponds to the heart beat. These pulses are sent to a counter that calculates heart rate.

2.2.7 Frequency Response

An ideal op amp has unlimited bandwidth; its gain is same for low frequency and high frequency. However, an op amp consists of several stages of circuitry in order to obtain very high gain. Each of the stages has some stray capacitance or junction capacitance associated with it. This capacitance reduces the high frequency gain of the op amp. Figure 2.35 shows the typical frequency response of an open loop op amp.

It would appear that an open-loop op amp has very poor frequency response, since its gain is reduced when the frequency exceeds 10 Hz. Because of this, we never use an open loop op amp to build an amplifier. Instead, we introduce large negative feedback to control the gain of the amplifier circuit. For example, if we have an op amp with a circuit gain of 10 (Figure 2.35), the frequency response is flat up to 100 kHz and is reduced above that frequency.

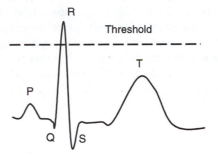

Figure 2.34 Heart beat detector uses a comparator to determine when the R wave exceeds a threshold.

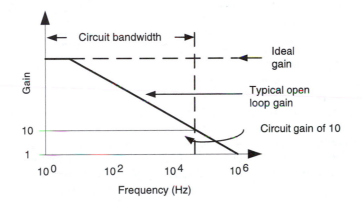

Figure 2.35 The typical op amp open loop gain is much larger, but less constant, than the circuit gain. However the in circuit bandwidth is larger than the open loop bandwidth.

We use the unity gain–bandwidth product to quantify the frequency performance of an op amp. The gain–bandwidth product is the product of gain and bandwidth at a particular frequency. Figure 2.35 shows that the unity gain bandwidth is 1 MHz. Along the entire curve with a slope of –1, the gain–bandwidth product is constant at 1 MHz. Thus, for any amplifier circuit, we can obtain the bandwidth by dividing the unity gain–bandwidth product by the amplifier circuit gain.

2.3 Filters

To obtain a specific output from a device, it is sometimes necessary to electronically sort through and pick out exactly what is desired from the incoming mass of signals. Filters can simplify this process by altering the amplitude and phase of the input signal in a desired way. In most applications, the signal we obtain contains more frequency components than we want, so we use a filter to reject undesired components. Take a stereo for example; it has bass (low-pass filter) and treble (high-pass filter) controls to adjust the pitch of the sound. It also has oscillator circuitry (bandpass filter) to tune to different radio stations. This is the basic idea behind electronic filters. More specifically, a filter is an electronic circuit that, by design, attenuates sinusoidal voltage signals at undesired frequencies and passes signals at preferred frequencies. In other words, it rejects undesired components of a signal. Similarly, in the case of biomedical applications, filters are useful when recording physiological signals such as the ECG (electrocardiogram) and EEG (electroencephalogram).

2.3.1 Input–Output Relationship of Filters

The changes made to a signal depend on the frequencies contained in the signal and the filter design. Generally a filter is inserted between an input source and an output. Suppose the input, v_i, is a signal with a frequency of ω.

$$v_i = V_m \sin(\omega t + \phi)$$

$$v_o = V_m |T(\omega)| \sin(\omega t + \phi + \theta(\omega)) \tag{2.58}$$

where V_m is the amplitude of the input signal, $|T(\omega)|$ is the amplitude of the transfer function, or filter, and $\theta(\omega)$ is the angle of $T(\omega)$.

Thus the filter multiplies the amplitude of the input sine wave by $|T(\omega)|$ and adds to the phase of the sine wave $\theta(\omega)$. Since $|T(\omega)|$ and $\theta(\omega)$ are functions of frequency, they are named the amplitude response and phase response, respectively, of the filter.

For example, the ideal low-pass filter has a magnitude response of

$$T(f) = \begin{cases} 1 & \text{if } f < f_c \\ 0 & \text{if } f \geq f_c \end{cases} \tag{2.59}$$

where $f = \omega/2\pi$.

The low-pass filter passes all frequency components lower than f_c and rejects all frequency components higher than f_c. Realistically, building a filter with such a sharp transition is not possible since the transition from passband ($T(f) = 1$), to stopband ($T(f) = 0$) takes some finite number of frequencies.

Figure 2.36 shows the magnitude characteristics of four widely used filters: low-pass, high-pass, bandpass and bandstop filter. All these filters have a passband, where the filter passes the frequency components, and a stopband, where the filter rejects or attenuates the frequency components.

Circuits can be used to approximate ideal filters. The closer the actual circuit characteristics approach the ideal response (i.e. the steeper the transition band), the better the approximation. However, the complexity of the circuit increases as the ideal response is more closely approximated.

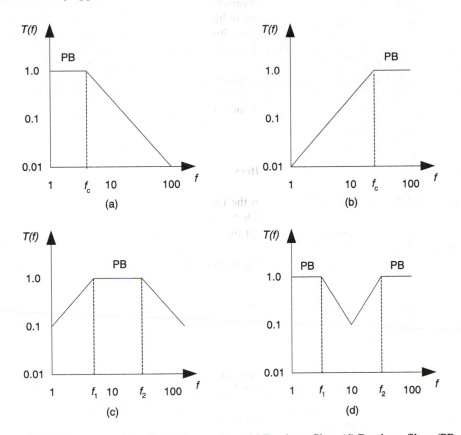

Figure 2.36 (a) Low-pass filter. (b) High-pass filter. (c) Bandpass filter. (d) Bandstop filter. (PB denotes passband)

2.3.2 Low-pass Filter

Low-pass filters pass low-frequency signals and attenuate high-frequency signals. The corner frequency, ω_c, distinguishes the passband from the stopband.

The first-order RC and RL networks are approximations of a low-pass filter. Figure 2.37 shows the RC and RL filter.

For the RC filter in Figure 2.37(a), the frequency response is calculated using the voltage divider principle for complex impedance

$$T(\omega) = \frac{1/j\omega C}{R + 1/j\omega C} = \frac{1}{1 + j\omega RC} = \frac{1}{1 + j\omega/\omega_c} \qquad (2.60)$$

where

$$\omega_c = \frac{1}{RC} \tag{2.61}$$

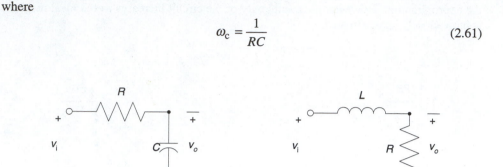

(a)   (b)

Figure 2.37 Low-pass filter. (a) *RC* circuit. (b) *RL* circuit.

For the *RL* circuit in Figure 2.37(b), we have

$$T(\omega) = \frac{R}{R + j\omega L} = \frac{1}{1 + j\omega/\omega_c} \tag{2.62}$$

where

$$\omega_c = \frac{R}{L} \tag{2.63}$$

The magnitude and phase responses of these filters are similar to those in Figure 2.36(a). At $\omega = \omega_c$ ($f = f_c$), the magnitude is $1/\sqrt{2}$ times the magnitude at $\omega = 0$. We designate ω_c the corner frequency. This is also known as the half power frequency because at $\omega = \omega_c$ the power delivered to the load is half of what it is at $\omega = 0$.

A typical use of a low-pass filter in bioinstrumentation is to remove the radio frequency (RF) interference from an ECG signal. A combination of unshielded ECG electrode leads and long electrode-to-amplifier connecting wires causes RF signals from other medical equipment to couple to the amplifier. The RF signal is rectified by the transistor junctions and causes interference below 150 Hz. However, frequency components of the ECG also lie below 150 Hz. Therefore, the ECG is always sent through a low-pass filter before it is amplified in order to filter out radio-frequency signals. For example, a 10 kΩ resistor and 300 pF capacitor can make the circuit in Figure 2.37(a), which will minimize RF interference.

2.3.3 High-Pass Filter

High-pass filters can also be realized with first-order *RC* or *RL* circuits (Figure 2.38).

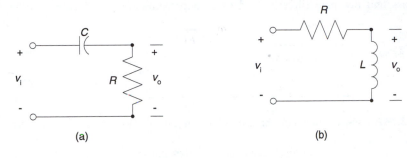

Figure 2.38 High pass filter. (a) *RC* circuit. (b) *RL* circuit.

The frequency responses of these two filters have a similar form:

$$T(\omega) = \frac{1}{1 - j\omega_c/\omega} \tag{2.64}$$

where

$$\omega_c = \frac{1}{RC} \quad \text{for the } RC \text{ circuit}$$
$$\omega_c = \frac{R}{L} \quad \text{for the } RL \text{ circuit} \tag{2.65}$$

The frequency response of the first order high-pass filter is similar to that shown in Figure 2.36(b). The corner frequency of the high pass filter is ω_c.

The ECG signal always has some dc offset and low-frequency artifacts due to the electrodes. A high-pass filter with a corner frequency of 0.05 Hz is used to filter the signal. A resistor of 3.2 MΩ and a capacitor of 1 μF can be used to make a high-pass filter that passes frequencies higher than 0.05 Hz.

2.3.4 Other Filters

More complex circuits, such as second, third, or higher order circuits, can be used to make the changes in magnitude from passband to stopband more abrupt.

Second order circuits, such as *RLC* circuits, can be used to realize bandpass and bandstop filters. Alternatively, a combination of a low-pass circuit and a high-pass circuit can be used together to achieve the same filtering effect.

For example, a device requires a bandpass filter with corner frequencies ω_1 and ω_2 ($\omega_1 < \omega_2$). In this case, a low-pass filter with a corner frequency of ω_2 and a high-pass filter with corner frequency of ω_1 are suitable. These two filters can be cascaded so the signal will first go through the low-pass filter and then through the high-pass filter. The high-pass filter rejects the frequency components smaller than ω_1 while the low-pass filter rejects frequency components greater than ω_2. As a result, only the frequency components between ω_1 and ω_2 will pass through the cascaded filter. This combination of a high-pass and low-pass filter is used as an approximation of a bandpass filter (Figure

high-pass and low-pass filter is used as an approximation of a bandpass filter (Figure 2.39).

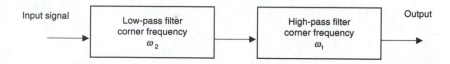

Figure 2.39 A low-pass filter and a high-pass filter are cascaded to make a bandpass filter.

Similarly, if a single filter cannot satisfy the requirement of signal processing, several filters can be cascaded together to form a multistage filter until they satisfy the need for application. However, keep in mind that the response of cascaded filters is more complex than that of a single filter.

2.3.5 Timers

In electronic design, there is often a need to generate signals which repeat at regular intervals. One type of signal is the square wave, shown in Figure 2.40.

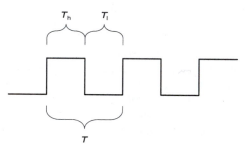

Figure 2.40 A square wave of period T oscillates between two values.

The voltage of a square wave is high for a fixed amount of time, T_h, then it drops to a lower voltage for a length of time T_l. This pattern of alternating high and low cycles continuously repeats. The total period of the square wave, the time it takes to repeat, is thus

$$T = T_h + T_l \tag{2.66}$$

The duty cycle of a square wave is defined as the percentage of the time that the square wave is at its higher output voltage. Thus

$$\text{Duty cycle} = \frac{T_h}{T} \times (100\%) \tag{2.67}$$

For example, a square wave in which $T_h = T_l$ is said to have a 50% duty cycle.

There are many ways to generate square waves. Digital systems use square waves with 50% duty cycles as clocks to synchronize digital logic; thus, there are many commercially available clock generator chips which yield square waves with 50% duty cycles.

Many times, however, we want to generate square waves with duty cycles other than 50%. A popular means of doing this is with a 555 timer. The 555 timer is an 8-pin integrated circuit, as shown in Figure 2.41(a). 555 timers form the core of many different kinds of timing circuits. One popular configuration is shown in Figure 2.41(b). When powered, this circuit oscillates internally, alternately charging and discharging capacitor C. Figure 2.41(c) shows the output of the circuit. Note that the duty cycle of this circuit is always greater than 50% because R_a must be nonzero. To get square waves with duty cycles less than 50%, the output of this circuit may be fed into an inverting amplifier or logic inverter.

This method of generating square waves is simple and requires only a small IC and four external components. The circuit of Fig. 2.41(b), however, is not very useful for precision timing applications, because of the difficulty of creating precision capacitors. Using typical off-the-shelf components, the period may vary by as much as 25% from the nominal values. This can be minimized by using variable resistances for R_a and R_b, which allows fine-tuning of the time constants.

2.4 Analog-to-Digital (ADC) and Digital-to-Analog (DAC) Conversion

The past sections discuss analog signals, which are usually continuous in amplitude, phase, and time. Now we will discuss digital circuits, in which signals take on a limited number of values over discrete time. Binary systems employ two number values and are the most common digital system. High or low voltages and the symbols 1 and 0 represent the two possible binary system levels.

The simplest binary system is a switch to control light. When a switch is closed and the light goes on, we denote that state as 1. When the switch is open and the light is turned off, the state is 0. There are many examples of binary systems.

The main advantage of digital signals is their high error tolerance. In digital systems, we often define voltages in some range as high or 1 and in other ranges as low or 0. For example, in TTL logic circuits, we define a voltage of 4 V to 5 V to be high and 0 V to 1 V to be low. The voltages between 1.0 V and 4.0 V are undefined. If a high state (5 V) signal is corrupted to 4.5 V during transmission due to interference, it is still assumed as high state at the receiver end and can be restored to 5.0 V. For an analog signal, the corruption is more difficult to recover because of the random signal corruptions, such as transmission loss, interference, and signal deterioration.

Digital circuits play a very important role in today's electronic systems, and they are employed in almost every aspect of electronics, including communications, control, instrumentation, and computing. This widespread usage is due to the development of inexpensive integrated circuit (IC) packages. The circuit complexity of a digital IC ranges from a small number of transistors (~100 to 10,000) to a complete computer chip (42 million transistors for a Pentium 4 chip).

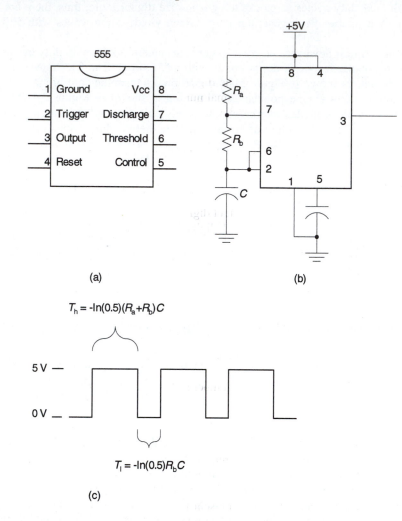

(a)

(b)

$$T_h = -\ln(0.5)(R_a + R_b)C$$

$$T_l = -\ln(0.5)R_bC$$

(c)

Figure 2.41 The 555 timer. (a) Pinout for the 555 timer IC. (b) A popular circuit that utilizes a 555 timer and four external components creates a square wave with duty cycle > 50%. (c) The output from the 555 timer circuit shown in (b).

2.4.1 Number Systems

The most commonly used numbering system is the decimal system, which was devised by Hindu mathematicians in India around 400 A.D. Although other numbering systems have been developed, the decimal system is the most natural and widely accepted.

A decimal number is represented by a string of digits. Depending on its position in the string, each digit has an associated value of an integer raised to the power of 10. For example, the decimal number of 215, can be computed as

$$215 = 2\times10^2 + 1\times10^1 + 5\times10^0$$

Conventionally, we only write the digits and infer the corresponding powers of 10 from their positions. In general, a decimal number can be written as a string of digits as

$$A_n A_{n-1} A_{n-2}...A_1 A_0 \qquad (2.68)$$

Each coefficient A_i is one of the 10 digits (0, 1, 2, 3, 4, 5, 6, 7, 8, 9).

The decimal number system is base 10 because the coefficients are multiplied by powers of 10 and the system uses 10 distinct digits. In general, a number in a base r number system contains digits 0, 1, 2,..., $r - 1$ digits and is expressed as a power series in r with the form

$$A_n r^n + A_{n-1} r^{n-1} + A_{n-2} r^{n-2} +...+ A_1 r^1 + A_0 r^0 \qquad (2.69)$$

or in position form

$$(A_n A_{n-1} A_{n-2}...A_1 A_0)_r \qquad (2.70)$$

For the number 314 in a base-5 number system, we have

$$(314)_5 = 3\times5^2 + 1\times5^1 + 4\times5^0 = 75+5+4 = (84)_{10}$$

In addition to decimal, we use binary, octal and hexadecimal number systems when dealing with computers. These are base-2, 8, and 16 number systems, respectively.

The binary number system is a base-2 system that uses the digits 0 and 1. Two numbers are chosen because they can represent the two states of the digital system (e.g. voltage high and low, or switch closed and open). A binary number can be converted to a decimal number using Eq. (2.69). For example

$$(100101)_2 = 1\times2^5 + 0\times2^4 + 0\times2^3 + 1\times2^2 + 0\times2^1 + 1\times2^0$$

$$= 32+4+1 = (37)_{10}$$

To change a decimal number to a binary number, the method of dividing the integer number by the base successively until 0 results is most often used. The remainders at every division form the binary number. The remainder of the last division is the most significant bit (MSB), and the remainder of the first division is the least significant bit (LSB). For example

$$
\begin{array}{lll}
11 \div 2 = 5 & \text{with remainder} & 1 \text{ (LSB)} \\
5 \div 2 = 2 & & 1 \\
2 \div 2 = 1 & & 0 \\
1 \div 2 = 0 & & 1 \text{ (MSB)}
\end{array}
$$

so

$$
(11)_{10} = (1011)_2
$$

Binary digital impulses, alone, appear as long strings of ones and zeros, and have no apparent meaning to a human observer. However, if a digital-to-analog converter (DAC) is used to decode the binary digital signals, meaningful output appears. Examples of this output might be a computerized voice, a picture, or a voltage to control mechanical motion.

2.4.2 Digital-to-Analog Converters (DAC)

The digital representation of a continuous analog signal is discrete both in time and amplitude. Often an analog signal must be sampled discretely with an analog-to-digital converter (ADC). Digital information is stored in a memory device for real time or future processing. A digital signal needs to be converted to an analog signal with a digital-to-analog converter (DAC) since many devices need an analog signal for controlling, monitoring, or feedback. For example, modems use ADCs and DACs in converting digital information in computers to analog signals for phone lines and back again to communicate with the Internet.

The DAC changes digital input to analog output. Figure 2.42 shows the static behavior of a 3-bit DAC. All the combinations of the digital input word are on the horizontal axis, while the corresponding analog outputs are on the vertical axis. The maximum analog output is 7/8 of the reference voltage V_{ref}. For each digital input, there is a unique analog output. The difference between the consecutive analog outputs is

$$
\frac{1}{8} V_{ref} = \frac{1}{2^3} V_{ref}
$$

We call this value the resolution of the DAC. In general, for an n-bit DAC or ADC, the resolution is

$$
\text{resolution} = \frac{1}{2^n} V_{ref} \tag{2.71}
$$

Equation (2.71) shows that if an application calls for high resolution, the DAC needs many bits. DACs are available in 4, 8, 10, 12, 16, 20 or more bits. The more bits a DAC has, the more complex and expensive it is.

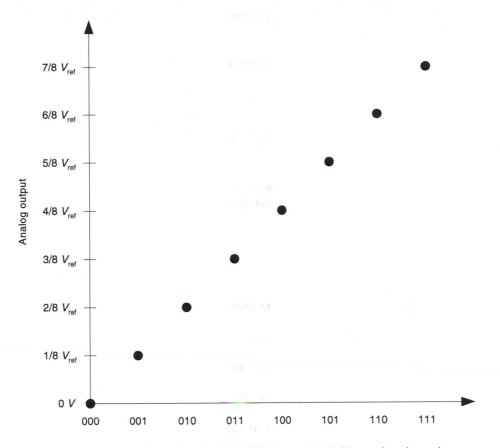

Figure 2.42 The ideal static behavior of a 3-bit DAC. For each digital input, there is a unique analog output.

A variety of circuit configurations are suitable for a DAC. We present the simplest one: a voltage scaling DAC. Figure 2.43 shows a simple 3-bit voltage scaling DAC. The V_{ref} in Figure 2.43 is the reference voltage of the DAC. The analog output is derived from this reference. The accuracy and stability of the reference should be carefully considered in the circuit design.

The reference voltage is divided by the series resistor network to obtain the desired discrete voltages on each node. The digital inputs are decoded by the decoder and the corresponding switch is closed. This voltage is then sent out as the analog output.

Voltage scaling is suited to IC technology because this technology is optimized for making many copies of the same structure. The value of R may have an error up to 50%, but as long as the difference between two resistors is still below 1%, the discrete output retains the accuracy of 1%.

The output of the DAC is discrete and noncontinuous. Interpolation should be used to make the output voltage smoother. Interpolation could be implemented by an electronic circuit with a voltage holder, linear interpolation, or low-pass filter.

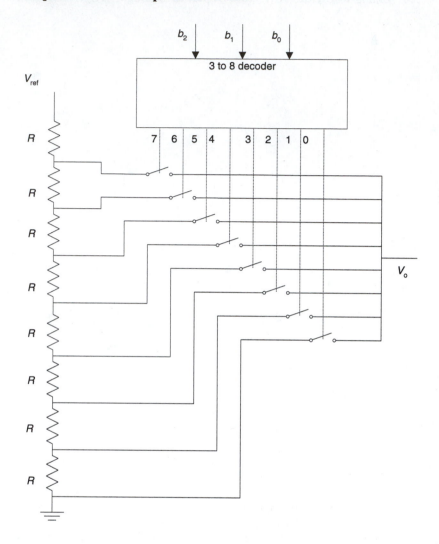

Figure 2.43 A 3-bit voltage scaling DAC converter.

2.4.3 Analog-to-Digital Converters (ADC)

Many signals used in biomedical engineering such as voltage, temperature, pressure, strain, flow, and speed, will originally be in analog form. These signals need to be converted to a digital signal before being transferred to the computer. The objective of an ADC is to convert analog input data to an equivalent output digital word. Since we are going to map a continuous value (infinite number set) to a limited number set, there will be overlap for different input continuous values. Figure 2.44 shows the converting characteristic of a 3-bit ADC. Figure 2.44 shows that the resolution of an ADC depends on the number of bits of the ADC. We can determine resolution using the same Eq. (2.71) as the DAC.

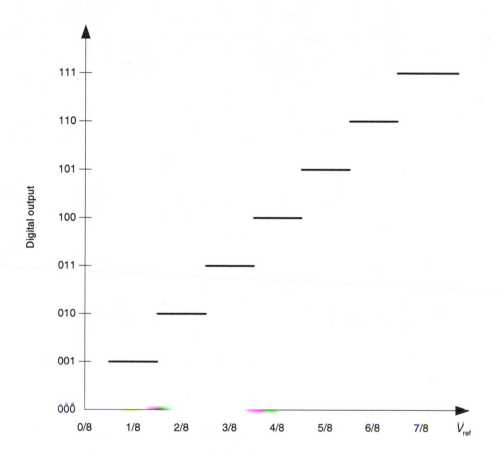

Figure 2.44 Converting characteristic of 3-bit ADC.

Many circuit configurations could be employed in an ADC, such as a counter ADC, dual slope ADC, successive approximation ADC, or parallel ADC. Here we present the successive approximation ADC.

Figure 2.45 is the block diagram for a successive approximation ADC. It consists of a comparator, a DAC, and digital control logic. When the conversion begins, it assumes the MSB (most significant bit) to be 1 and other bits to be 0. This digital value is sent to the DAC and generates an analog value of 0.5 V_{ref}. The comparator compares this generated value with the input analog signal. If the input is higher than the generated value, the MSB is indeed 1. Otherwise, the MSB is changed to 0. This process is repeated for the remaining bits.

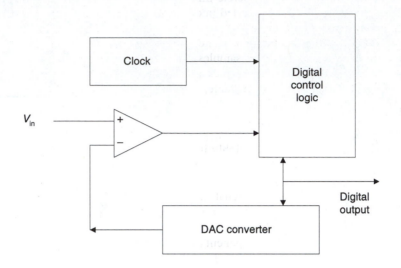

Figure 2.45 Block diagram of a typical successive approximation ADC.

Figure 2.46 shows the possible conversion paths for a 3-bit converter. The total number of clock cycles required to convert an n-bit word is n.

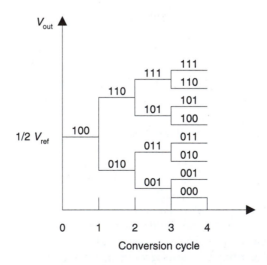

Figure 2.46 The possible conversion paths of a 3-bit successive approximation ADC.

2.5 Digital Signal Processing

Digital signal processing uses discrete-time technology to process continuous-time signals. A continuous signal is converted into a sequence of samples by the ADC. This sequence is processed by algorithms to yield an output sequence. After discrete-time processing, a DAC converts the output sequence back to a continuous-time signal.

In some cases, the input samples can be stored in memory and the processing will be implemented after all the samples are obtained. But in other digital signal processing, real-time operation is often desirable, meaning that the system is implemented so that samples of the output are computed at the same rate at which the continuous signal is sampled.

Compared with analog signal processing, digital signal processing has some advantages: (1) A digital filter is highly immune to interference because of the way it is implemented (digital signals are far less susceptible to noise than analog signals). (2) Accuracy is dependent on round-off error, which is directly determined by the number of bits used to represent the digital signal. (3) With the help of modern computers, digital signal processing is much more flexible and powerful than analog signal processing. (4) Performance of a digital signal system is minimally affected by environmental factors such as temperature variation, component aging, and power supply fluctuations.

Because of these advantages, digital signal processing is widely used in communication systems, radar and sonar, speech and video coding and enhancement, and biomedical engineering.

There are some limitations of digital signal processing. The greatest drawback is its relatively slow speed, which is affected by two factors. One is the sampling speed of the ADC and DAC. The fastest ADCs and DACs work at frequencies in the range of tens of megahertz, while analog signal processing can work at frequencies in the range of a gigahertz. The other is the speed of the signal processing hardware. The speed of analog signal processing is limited by the delay of the circuits. In digital signal processing, it depends on the complexity of the algorithms and the clock speed of the processor.

2.5.1 Digital Signals

Analog signals are continuous in time and amplitude. Digital signals are discrete in time and amplitude. They are represented by a sequence of numbers, x, in which the nth number in the sequence is denoted $x[n]$, where n is an integer. Such sequences are usually obtained by the periodic sampling of an analog signal $x_a(t)$. In the sampled sequence, the value of the nth sample in the sequence is equal to the value of $x_a(t)$ (rounded to the precision of the ADC) at time nT, or

$$x[n] = x_a(nT) \tag{2.72}$$

where T is the sampling period and its reciprocal is the sampling frequency. As an example, Figure 2.47(a) shows a segment of a sine wave and Figure 2.47(b) is the sampled sequence of the sine wave.

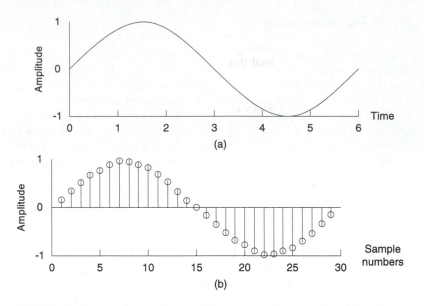

Figure 2.47 (a) Continuous signal. (b) Sampled sequence of the signal in (a) with a sampling period of 0.2 s.

2.5.2 Sampling Theorem

To convert a continuous-time signal into a discrete-time signal and process it with a microprocessor, we assume that we can represent a continuous time signal by its instantaneous amplitude values taken at periodic points in time. We also assume that we will be able to reconstruct the original signal perfectly with just these sampled points.

The sampling theorem, developed by Shannon, guarantees that the original signal can be reconstructed from its samples without any loss of information. It states that, for a continuous band-limited signal that contains no frequency components higher than f_c, the original signal can be completely recovered without distortion if it is sampled at a frequency of at least $2f_c$. A sampling frequency f_s, which is twice the highest frequency in a signal, is called the Nyquist frequency.

Suppose the original signal can be represented by function $X_c(f)$ in the frequency domain and its highest frequency component is f_c. When it is sampled periodically at frequency f_s, then the sampled signal can be represented as (Oppenheim and Schafer, 1989)

$$X_s(f) = f_s \sum_{k=-\infty}^{+\infty} X_c(f - kf_s) \tag{2.73}$$

where $X_s(f)$ is the frequency representation of the sampled sequence. Figure 2.47 is a sketch of the sampling process. We know the sampled signal consists of periodically repeating copies of $X_c(f)$. These copies are shifted by integer multiples of the sampling frequency and then superimposed to produce the periodic samples. Figure 2.48 shows that if we use a low-pass filter in which f_n is the corner frequency, we can reconstruct the original by passing the sampled signal through the low-pass filter (f_n is between f_c and $f_s - f_c$).

In Figure 2.48, if the frequency $f_s - f_c$ is less than f_c, the copies of the original signal overlap for some of the frequency components. This overlap will corrupt the original signal and we will not be able to reconstruct the original signal from the sampled signal. To guarantee that we can reconstruct the original signal, we must meet the condition of the sampling theorem

$$f_s - f_c > f_c \quad \text{or} \quad f_s > 2f_c \tag{2.74}$$

Here we present an example of a moving window digital filter for an ECG signal.

In some ECG signals, there is the higher frequency electromyogram (EMG) which interferes with the ECG signal. We can use a low-pass digital filter to attenuate the EMG using the following equation

$$y[n] = \frac{1}{4}(x[n] + 2x[n-1] + x[n-2]) \tag{2.75}$$

We can also change the coefficients of the digital filter to get different results or use it for other applications. This is more flexible than the analog filter, which would require a change in components or even construction of a new circuit.

2.6 Microcomputers

The computer is one of the most important inventions of the twentieth century. In the last twenty years, computer technology, especially microcomputer technology, has had a significant impact on bioinstrumentation.

Computers are divided, by their size and use, into several categories: supercomputers and mainframe computers, workstations, microcomputers (or personal computers), and microprocessor-based (or microcontroller) systems. The PC and microprocessor-based systems are generally chosen for bioinstrumentation.

2.6.1 Structure of a Microcomputer

A microcomputer is composed of its CPU (central processing unit), memory, I/O devices, and buses, as shown in Figure 2.49.

The CPU plays an important role in a computer. It fetches, decodes and executes instructions and controls the performance of other modules. The most commonly used

CPU in microcomputers is the Intel Pentium 4, which contains 42 million transistors on one chip.

Read-only memory (ROM) can only be read by the CPU and random-access memory (RAM) can be read and written to by the CPU. The core programs, such as start-up instructions and interface instructions for the CPU, are generally stored in ROM. ROM retains these programs even if there is no power. Because it holds basic programs, a modern microcomputer contains tens of kilobytes of ROM.

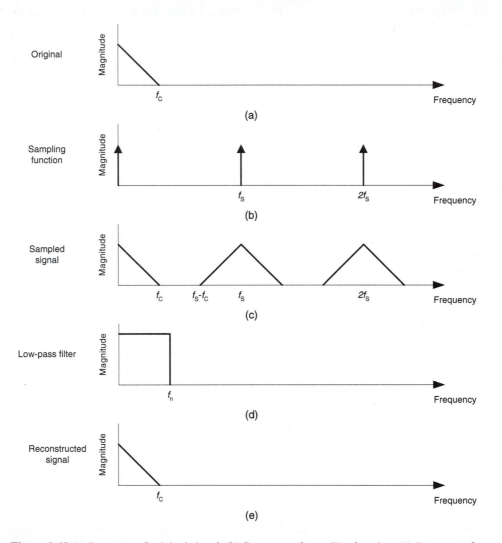

Figure 2.48 (a) Spectrum of original signal. (b) Spectrum of sampling function. (c) Spectrum of sampled signal. (d) Low-pass filter for reconstruction. (e) Reconstructed signal, which is the same as the original signal.

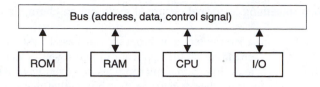

Figure 2.49 General block diagram of a microcomputer system (arrows represent the data flow direction).

The system software and applications are loaded to RAM for processing. Data and instructions are also stored in RAM. With advances in VLSI technology, the capacity of RAM becomes greater while the price keeps decreasing. In 2003, most computers have more than 256 megabytes (MB) of RAM, with some having several gigabytes (GB) of memory.

The bus is a communication path used to transfer information (such as instructions, data, addresses or control signals) between the modules that make up the system. One or more functionally separate buses may link the various microcomputer modules.

The clock is an oscillation signal generated to synchronize the performance of the modules in the system (which is not shown in Figure 2.49). The faster the clock speed, the faster the computer system. The first IBM PC had an Intel 8088 with a clock speed of 4.8 MHz. In 2002, the Intel Pentium 4 has a clock speed of 2.2 GHz. Intel or other companies will manufacture much faster chips in the future.

I/O (input/output) devices are used by the computer to input and output information. These devices can be divided into several categories.

The most common input devices for a microcomputer are the keyboard and mouse. A user gives the computer instructions or data through these devices. The computer can display the result on the output devices (monitors or printers) for the user.

Mass storage devices, like floppy disk drives, hard disk drives, CD-ROM drives, and tape drives are used to permanently store software and data on mass storage media.

A computer can be connected to other computers via a network using devices such as a modem or network card. When connected to the network, computers can share resources, such as devices, software, and databases, which improves the performance of connected computers.

In bioinstrumentation, ADC and DAC devices are widely used to collect the data for processing and send out signals to control other equipment.

2.6.2 Microprocessor-Based Systems

The structure of microprocessor-based systems is similar to that of the microcomputer except that it contains fewer modules. For example, a clinical ECG monitoring system contains a microprocessor, ROM for a program, RAM for data and an I/O port for ECG signals. It does not have mass storage (a floppy disk or hard disk), keyboard, or monitor for display, but may have some keys to choose functions or a small printer to print out the

ECG signals. When something abnormal occurs, it gives out an alarm via a speaker and light signals.

Compared to a microcomputer, a microprocessor-based system contains less memory (tens of kilobytes, depending on the application) and fewer devices. The advantages are its small size and flexibility. Some microprocessors even have ADC/DAC circuits, communication circuits and other devices on the chip so that one single chip satisfies all the needs of the application.

2.7 Software and Programming Languages

Section 2.6 discussed the components of a computer system. The computer and its peripherals (printer, keyboard, monitor, etc.) describe the hardware of the computer system.

Besides the hardware for bioinstrumentation, software plays a critical role to the performance of the system. The computer itself can do nothing without software. The types of software are limited by the hardware system. Figure 2.50 shows three levels of software between the hardware and the real world: the operating system (UNIX, DOS, Windows, and Macintosh), the support software, and the application software. The application software gives the computer its bioinstrumentation abilities.

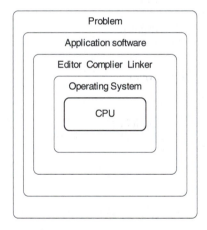

Figure 2.50 Three levels of software separate the hardware of microcomputer from the real problem.

Two aspects should be considered for developing a biomedical system: (1) the choice of operating system to support the task, and (2) the choice of the programming language to implement the application. There are many combinations of operating systems and programming languages that are able to address the problem, but some choices are better than others for certain applications.

2.7.1 Operating System

The operating system (OS) is a set of procedures that handle the resources of the computer system, which include the CPU, file system, display, printer and so on. The OS manages the file system control, CPU control, memory control, and system security. It provides a virtual machine so that the user just deals with the operating system instead of the computer hardware. The operating system also provides some frequently-used functions so that users can work more efficiently.

An example of the type of task an OS handles is file system management. In a computer, there are several storage media such as a floppy disk, hard disk, or tape. When saving files, we must consider the following: (1) the mechanisms of these devices are different, so we should use different procedures to write our files to each one, (2) for a device, we have to decide the format and location of the file, (3) we have to determine a method for file indexing so that we can obtain access the file. The OS provides an interface between user and media. When a user commands the OS to save a file, it automatically executes the task. Without the OS, computer users have to communicate with the computer hardware directly by writing some code, which would take a lot of effort and time.

Below is a brief discussion of three common operating systems: Unix, DOS/Windows, and Mac.

Unix is widely used on mainframe computers and workstations, although some versions are available on microcomputers. It is very flexible, provides the capability to maximally manipulate a computer system, and gives excellent control of input/output and other facilities. It also allows multiple users to execute multiple tasks (i.e. it is a multi-user, multitasking operating system). Because of this great flexibility, Unix requires considerable expertise to use all of its capabilities. Different hardware companies (such as HP, IBM and others) provide their own versions of Unix for their products, however, all versions are very similar.

The Macintosh is a computer hardware/software system by Apple designed for ease of use and graphics-oriented applications. Macintosh provides a user friendly graphical user interface (GUI) by sacrificing a great deal of direct user control of the hardware system. The concept was to produce a personal computer that would be optimal for running applications instead of writing new ones. Without training, an individual can sit down and quickly learn to use the Macintosh because of its intuitive OS.

MS-DOS was designed by Microsoft for the IBM PC and compatibles, and became the most popular operating system because of the wide use of the IBM PC. In the 90s, MS-DOS evolved to Windows system (3.1, NT, 2000, XP), which has a graphical user interface. In DOS/Windows, a user can still control the hardware of the system, but it is not as flexible as the Unix system.

A typical need for bioinstrumentation is the ability for real-time processing. In other words, we need the results when we collect the data. Sometimes this is critical, especially in the intensive care unit of a hospital. In most cases, biomedical applications require a single-user system that provides maximal performance from the computing power, and the DOS/Windows system satisfies this requirement. The Unix system is not desirable because of its multiuser/multitask property. The Macintosh system is not desirable, as the user cannot control the hardware structure.

2.7.2 Programming Languages

The OS takes care of talking to the hardware but the user still has to give the computer instructions on how to implement the solution (i.e. program the computer).

The instructions for the computer are in a binary format and are machine dependent, meaning that binary codes can be used on only one kind of computer. For example, the binary code for PC machines that use DOS cannot be used on Macintosh machines.

Since CPU instructions are binary, the final codes for the computer should be in binary format. When the computer was first invented, engineers had to remember instructions in binary code and write their own programs in binary format. After 40 years of development, there are now many programming languages, some of which are similar to our native languages. The codes developed in the programming language are called source codes. A complier is used to translate the source codes to binary codes that computers can understand. One advantage of programming languages is ease of use. Another advantage is transportability, which means user can use the source codes on different machines with different operating system provided that the language compiler is available in the different operating systems.

There are numerous programming languages, but all languages can be classified in one of two categories.

One category is low-level languages, which includes assembly language. Every instruction in assembly language corresponds to a binary instruction code for the CPU. Therefore, it is known as a low-level language. Assembly language can achieve the greatest run-time performance because it provides the direct manipulation of the processor. However it is also the most difficult language to write since we have to think about the problem in the way the processor does. Another drawback is that the program is machine dependent. If we want to use another system, we have to write the code again from scratch.

The other category of programming languages is high-level languages. The instructions are similar to the way we think of a problem and programs are relatively easy to write. High-level languages make developing programs much easier. The tradeoff is that the programs usually are not as efficient as those in assembly language, especially for real-time processing.

The most widely used high-level languages are FORTRAN, Pascal and C/C++. FORTRAN was developed for scientific calculation (FORmula TRANslation). Pascal was designed for teaching students structural programming techniques.

The C language is the most popular language and is used to develop Unix systems. It has the characteristics of a high-level language, but also contains functions and interfaces for low-level programming. C++ is a modern version of the C language and allows object-oriented structure. C/C++ is the current language of choice for real-time programming. Some parts of the code may be written in assembly language for increased efficiency.

There are two other special languages worth mentioning. Optimized for laboratory applications, LabVIEW is a visual computing language available for many operating systems. Programming is accomplished by interconnecting functional blocks that represent processes such as data acquisition and analysis, or virtual instruments. Thus, unlike traditional programming by typing command statements, LabVIEW is a purely graphical, block diagram language. In most cases, LabVIEW is used to acquire and process data in the laboratory and in some industrial applications.

MATLAB is general purpose software for matrix manipulation. It is used for matrix calculation and also has many packages containing various functions such as signal processing, image processing, data analysis, and so on. MATLAB is very useful for digital signal processing, which in turn is very important for biomedical instrumentation.

2.7.3 Algorithms

An algorithm is a list of steps that solve a given problem. Such a description should be precise enough to be used in a completely automatic manner (i.e. by a computer). It is the programmer's responsibility to ensure the correctness of the algorithm.

There are several aspects of the correctness of an algorithm. For every satisfying input, the results of the algorithm should be correct. Also for every satisfying input, the algorithm should execute to the end and not abort in the middle. If these conditions are not met, the system will produce incorrect results, especially for real-time processing. The algorithm should also terminate in a finite number of steps. An algorithm that needs an infinite number of steps is useless.

Another consideration for an algorithm is its efficiency. Execution of an algorithm on a computer requires a certain amount of computer resources as well as a certain amount of computing time. Before implementing an algorithm on a computer, we have to investigate whether it is feasible to use the algorithm for problem solving. A lot of known algorithms grow too computationally expensive as the size of the input increases. For many real-time processing applications in bioinstrumentation, execution time and memory requirement are important considerations in system design.

2.7.4 Database systems

A database is a collection of stored operational data used by applications. Database systems are composed of three parts. First, the database itself is a collection of data stored on some storage media. Second, a set of ordinary programs enables users to access the data for performing functions such as retrieving, updating, inserting, and deleting. Third, the database is *integrated*, meaning that it contains data for many different users, even though an individual user may be concerned with only a small portion of the data.

Why are databases important? One general answer is that they provide users with centralized control of the operational data. If every user in a group or each application has its own private file, then the operational data would be widely dispersed. In hospitals, all the information about a patient is stored in computers from which doctors can retrieve the data. Even if it is the first time that patient is admitted to a certain hospital, doctors can obtain his or her former information from the other hospitals using a database system.

There are several advantages of having centralized control of data. First, the amount of data redundancy in the stored data can be reduced. If each application has its own private files, this often leads to considerable redundancy in the stored data. For example, a patient's files for different doctors all contain his or her name, social security number, address, age, gender and so on. With central control, this data can be stored only once and shared by different doctors. Second, the data can be shared. This means any doctor can use the database if he or she is granted access. Third, data security restrictions can be applied. The database administrator can assign access to users for certain data. This can protect data from abuse. Another important advantage of centralization is the data independence of databases. Most applications are data-dependent, which means the way in which the data are stored and the way in which they are accessed are both dictated by the application.

With data independence, the application does not need to know the storage structure of the data and access strategy. By using the same commands, the application can get access to different databases without the knowledge of the physical storage and access format. Different applications can have different views of the same data and the data administrator responds to different requirements automatically.

The architecture of the database determines the application performance. The data model is the heart of any database system. There are different ways to implement a database. The three best-known approaches are the relational approach, the hierarchical approach, and the network approach.

2.8 Display Devices

For improved understanding of the processes of the instrument or of an object, it is important to display the results as either characters or graphically.

The most common device for displaying electric signals is the oscilloscope. The main part of an oscilloscope is a cathode ray tube (CRT). Figure 2.51 is a sketch of a CRT.

The cathode emits electrons, which are accelerated by the axial high voltage, hit the screen, and light is emitted due to fluorescence. The electrons pass through a transverse electric field before hitting the screen and deflect at a certain angle, controlled by an electric signal. The deflection control unit controls the location of the spot on the screen. For an oscilloscope, the horizontal deflection control signal is a time swept signal and the vertical deflection control signal is the signal we display.

The heart of a TV monitor is also a CRT, but it is a little different from the CRT for the oscilloscope. In a TV monitor, the signal received from the antenna controls the number of electrons emitted from the cathode and the brightness of the spot on the screen. If many electrons hit the spot, it will emit more light and we observe that the location is white. If no electrons hit the spot, it emits no light and we observe it is black. The deflection control unit controls the location of the spot (scanning) and we can see an image on the screen. For a color TV monitor, there are three colors (red, blue, green) of fluorescent material on the screen. These three colors compose the different colors we observe.

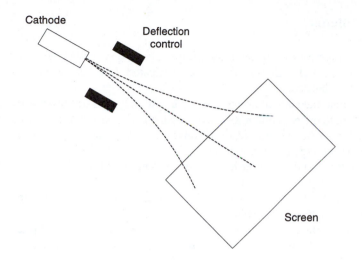

Figure 2.51 Sketch for cathode ray tube (CRT). There are two pairs of electrodes to control the deflection of the electron, but only one pair is shown.

2.9 Recording Devices

In many applications, we want to save the data permanently as a record. A strip chart recorder is a common electromechanical device that records an electric signal versus time. A strip of paper travels at a constant speed (i.e., 5 cm/s, 10 cm/s) and an ink pen is in contact with the paper. The pen arm moves freely normal to the direction of paper movement, and its movement is controlled by the input electric signal. It records (plots) the electric signal on the paper. A chart recorder is suitable for low-frequency signals.

A tape recorder is also a common device for recording analog signals, and is similar to the tape recorder we use in everyday life. The main difference being that we record the input electric signal instead of music. We can retrieve the signal from the tape and send it to a device to display the signal or make a measurement.

In recent years, technological advances have made digital techniques more widely used in recording signals.

The digital oscilloscope can sample the signal at a high sampling speed and display the signal. It also has an interface that can be used to save data on a floppy disk or send data to a computer. The computer printer output replaces the strip chart recorder in many biomedical instruments such as ECG, pulse oximeter, and spirometer.

Another commonly used device is a PC data acquisition system, such as Lab-VIEW Biobench. A personal computer uses an ADC unit to sample the signal and save the signal as a file. It can also display the signal on the screen for manual or automatic measurement, in which case it is similar to the oscilloscope. The advantage of this kind of system is the power of the computer can be employed to process the data, making the system more flexible than an oscilloscope.

2.10 Problems

2.1 One ampere flows through an electrosurgical dispersive electrode that has a resistance of 50 Ω. Calculate the power dissipated. Discuss the possibility of burns under the electrode.

2.2 Calculate the equivalent resistance for a 400 Ω and 1 kΩ resistor in series.

2.3 Calculate the equivalent resistance for a 400 Ω and 1 kΩ resistor in parallel.

2.4 Design a resistive attenuator to convert a 20 V voltage from a signal generator to a 10 mV voltage for an amplifier.

2.5 An operational amplifier has limited current output of 20 mA. Calculate the time to charge a 1 μF capacitor from 0 to 10 V.

2.6 From a 2 kV source in series with a 20 kΩ resistor, calculate the time required to charge a 100 μF defibrillator capacitor to 1.9 kV.

2.7 Calculate the time response of a 10 kΩ, 30 mH high-pass and low-pass RL circuits (similar to the two cases of RC circuit, but $\tau = L/R$) to a 1 V step input.

2.8 Design an inverting amplifier with a gain of -10. If a voltage of $2\sin(2\omega t)$ is applied, sketch the waveform of the output and label axes.

2.9 Design a noninverting amplifier with a gain of $+10$. If a voltage of $2\sin(2\omega t)$ is applied, sketch the waveform of the output and label axes.

2.10 In Figure 2.30, $R1 = 1$ kΩ, the two $R2$ are 10 kΩ and 9 kΩ respectively (the one connected to ground is 9 kΩ). Calculate the CMRR.

2.11 Design a comparator to detect an amplified electrocardiogram whose Q wave is 0 V and R wave is 4 V.

2.12 A low-pass filter with $R = 10$ kΩ and $C = 300$ pF is placed before an ECG amplifier. Calculate by what factor it reduces 1 MHz interference from an electrosurgical unit.

2.13 Calculate the number of bits to obtain a resolution of 1/1000 V_{ref} for an ADC.

2.14 Explain the advantages and disadvantages of digital signals versus analog signals.

2.15 Define Nyquist frequency. For a signal that contains frequency components up to 1 kHz, calculate the minimum sampling frequency.

2.16 Several students are attempting to digitally process a sinusoidal signal that repeats every 250 μs. The students randomly select a DSP that takes 360,000 samples per minute. Will they be successful in processing the signal correctly? Why or why not?

2.17 Calculate the minimum bit size and sampling speed for a DSP to handle an incoming signal with a range from 0 to 5 V, frequency components between 25 and 250 Hz and sensitive to ±1 mV.

2.18 Explain the function of the following devices in a microcomputer: CPU, RAM, ROM, I/O port, data bus.

2.19 Give two examples for each category of software: application software package, high-level programming language, low-level programming language. Also state on which platform each operates.

2.20 Describe how a CRT functions.

2.21 Go to www.bmenet.org/BMEnet/. Click on jobs. Find an industrial (not university) job and describe in a few sentences.

2.22 A radio-frequency (RF) catheter is introduced into the heart to inject power at 500 kHz to ablate (destroy) tissue that causes tachycardia. This causes interference in the electrocardiogram (ECG). Design a filter that reduces this interference to 0.0005 of its former level.

2.11 References

Alexandridis, N. A. 1984. *Microprocessor System Design Concepts.* Rockville, MD: Computer Science Press.

Banachowski, L., Kreczmar, A. and Rytter, W. 1991. *Analysis of Algorithms and Data Structure.* Reading, MA: Addison-Wesley.

Bobrow, L. S. 1987. *Elementary Linear Circuit Analysis.* New York: Holt, Rinehart and Winston.

Date, C. J. 1977. *An Introduction to Database Systems.* Reading MA: Addison-Wesley.

Hansen, P. B. 1973. *Operating System Principles.* Englewood Cliffs, NJ: Prentice-Hall.

Hilburn, J. L. and Julich, P. M. 1979. *Microcomputers/Microprocessors: Hardware, Software and Applications.* Englewood Cliffs, NJ: Prentice-Hall.

Lurch, E. N. 1963. *Electric Circuits.* New York: John Wiley & Sons.

Oppenheim, A. V. and Schafer, R. W. 1989. *Discrete-time Signal Processing.* Englewood Cliffs, NJ: Prentice-Hall.

Tompkins, W. J. (ed.). 1993. *Biomedical Digital Signal Processing.* Englewood Cliffs, NJ: Prentice-Hall.

Webster, J. G. (ed.) 1998. *Medical Instrumentation: Application and Design.* 3rd ed. New York: John Wiley & Sons.

Chapter 3

Analysis of Molecules in Clinical Medicine

Mat Klein

Methods used to measure molecules in clinical chemistry, toxicology, and pulmonary medicine have a tremendous impact on the fields of bioengineering and medicine. The ability to measure certain molecules in the body enables us to better understand the biological and chemical processes occurring in the body. This chapter presents various techniques used to measure different molecules in the body, along with the importance and impact of their measurement.

3.1 Spectrophotometry

Photometry is based on measurements of electromagnetic energy emitted, absorbed, or transmitted under controlled conditions. Spectrophotometry and flame photometry are two different types of photometry commonly used to determine the concentration of light absorbing molecules in a solution. Based on the use of light intensity measurements at a particular wavelength or a discrete portion of wavelengths, spectrophotometry determines the concentration of desired molecules in solution by the amount of radiant energy these molecules absorb.

3.1.1 Components

The principles of spectrophotometry can be understood by examining the single beam spectrophotometer. Figure 3.1 shows that the components of the single beam spectrophotometer include a light source, monochromator, cuvette, detector, and readout device.

Radiant Energy Sources

The purpose of the light source in a spectrophotometer is to provide incident light of sufficient intensity to the sample. The wavelength and intensity of incident light desired determines what light source to use. The most common source for substances that absorb radiation in the visible, near infrared, and near ultraviolet regions, is a glass enclosed tungsten filament. These distinctions between the spectrum are made because silica, used to make cuvettes, transmits light effectively at wavelengths greater than 220 nm. Low-pressure iodine or bromine vapor is usually contained in the bulb to increase the lifetime of the tungsten filament. The tungsten light bulb does not supply enough radiant energy for measurements below 320 nm. High-pressure hydrogen or deuterium discharge lamps are sufficient for measurements in the near ultraviolet region. At wavelengths below this, the emission is no longer continuous. Two advantages deuterium lamps have is that they produce about three times the light intensity of hydrogen lamps and have a longer lifetime. Xenon arc or high-pressure mercury vapor lamps provide high levels of continuous ultraviolet illumination. Since they become extremely hot during operation, these lamps may require thermal insulation with or without auxiliary cooling to protect surrounding components.

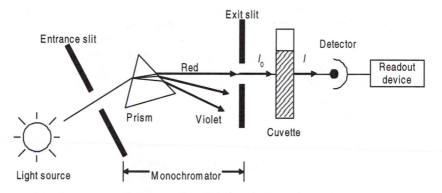

Figure 3.1 Block diagram of a single beam spectrophotometer. The prism serves as the dispersing device while the monochromator refers to the dispersing device (prism), entrance slit, and exit slit. The exit slit is moveable in the vertical direction so that those portions of the power spectrum produced by the power source (light source) that are to be used can be selected.

Monochromator

A monochromator is a system for isolating radiant energy of a desired wavelength. The term monochromator refers to the dispersing device and associated slits and components used to isolate the desired wavelength. A monochromator commonly used in spectrophotometers uses prisms or diffraction gratings. Both components separate white light into a spectrum from which the desired wavelength may be chosen. A prism separates white light into a continuous spectrum by refraction. As white light passes through the prism, shorter wavelengths are refracted, or bent, more than longer wavelengths. Although longer wavelengths are closer together than shorter wavelengths, since refraction is nonlinear, with the proper components, the desired spectrum (a narrow band) can be isolated. A diffraction grating separates white light, such as that produced by a tungsten filament, into a continuous linear spectrum. Diffraction gratings used in spectrophotometers consist of many closely spaced parallel lines on a substrate such as glass covered with a polished aluminum or aluminum–copper alloy. When radiation strikes the grating, light rays bend around the sharp edges of the closely spaced parallel lines. The amount of bending is dependent upon the light's wavelength. A tiny spectra is produced for each line of the grating. Wavefronts are formed as the light waves move past the corners. Wavefronts that are in phase when they cross reinforce each other, while wavefronts that are out of phase cancel out, leaving a complete spectrum from which to choose a narrow band of wavelengths or particular wavelength.

Cuvette

Spectrophotometry ascertains the absorption of the desired molecules (the solute) in a solvent. A cuvette, also referred to as an absorption cell, holds the solute and solvent. Cuvettes can be round, square, or rectangular, and have a light path of constant length, commonly 1 cm. For measurements made in the visual range (above 340 nm), cylindrical test tubes are sufficiently accurate. They are made from glass tubing, are not perfectly round or polished, and contain surface aberrations. For measurements below 340 nm, square or rectangular cuvettes made of quartz or silica, which are free of optical aberrations, are required.

Detector and Readout Device

There are two requirements for a detector of radiant energy in a spectrophotometer. The photosensitive detector must have a linear response and be sensitive enough in the part of the spectrum in which it is used. The most common devices used to detect the amount of radiant energy leaving the cuvette include barrier layer cells, photodiode arrays, and photomultiplier tubes. These devices convert electromagnetic energy to electric energy, which can then be measured.

   A readout device displays the electric energy from the detector onto some type of scale such as absorbance or transmittance. The two types of readout devices are the direct reading system and the null point system. In a direct readout meter, there is a linear relationship between milliamperes and %T (percent transmittance) and a log relationship between millivolts and absorbance. This characteristic makes direct readout systems fast and simple. In a null point system, the meter is calibrated (zeroed) by a potentiometer, and the absorbance, transmittance, or any other arbitrary scale is fitted to the potentiometer scale.

3.1.2 Theory

The determination of the concentration of a light-absorbing substance in a solution using a spectrophotometer is based on the discoveries of four individuals. Bouguer and Lambert noted that transmittance of light through an absorbing material, such as that contained in a cuvette, decreases exponentially with an increase in the light path through the cuvette. Beer and Bernard observed that the concentration of a substance in solution [less than 10^{-2} M (molar)] is directly related to its absorbance. An amalgamation of these two discoveries is known as Beer's Law, as stated by Bouguer: "Equal thickness of an absorbing material will absorb a constant fraction of the energy incident upon it" (Wheeler, 1998). This relationship is

$$I = I_0 10^{-aLc} \tag{3.1}$$

where

I_0= radiant power arriving at the cuvette
I = radiant power leaving the cuvette
a = absorptivity of the sample (extinction coefficient)
L = length of the path through the sample
c = concentration of the absorbing substance

Transmittance is the ratio of light intensity leaving the cuvette I to light intensity entering the cuvette I_0

$$T = \frac{I}{I_0} \tag{3.2}$$

Absorbance (A) is exponentially related to the reciprocal of transmittance.

$$A = \log\left(\frac{I_0}{I}\right) = \log\left(\frac{100}{\%T}\right) = 2 - \log(\%T) \tag{3.3}$$

As the concentration of the substance in solution increases (or decreases), the transmittance varies logarithmically and inversely. The fraction of incident light absorbed by a substance in solution at a particular wavelength is a constant characteristic of the substance called absorptivity, a. Let the length of the light path (usually the outer diameter of the cuvette) be a constant, L, and the concentration of the substance in solution be c. The relation of these parameters to the total absorbance is expressed as:

$$A = aLc \tag{3.4}$$

where a is in liters per gram times centimeters, L is in cm, and c is in grams per liter. Absorptivity is often replaced by molar absorptivity, ε, expressed in liters per mole times centimeters. Epsilon is a constant corresponding to a 1 molar solution of the absorbing substance, with a light path of $L = 1$ cm and a given wavelength. Hence, absorbance can be expressed as

$$A = \varepsilon Lc \tag{3.5}$$

If the concentration and absorbance of a substance is known at a specific wavelength and this concentration is changed, we can calculate the unknown concentration of the substance using the absorbance of the unknown concentration and the equation

$$\frac{c_u}{c_s} = \frac{A_u}{A_s}$$

(3.6)

where A_u is the absorbance of the unknown concentration and c_u is the unknown concentration.

3.1.3 Calibration

The purpose of calibrating a spectrophotometer is to make sure that the wavelength of radiant energy emitted from the exit slit of the monochromator is within certain tolerances of the radiant energy wavelength that appears on the wavelength selector of the instrument. This ensures that measurements made with the spectrophotometer are accurate and reproducible.

3.1.4 Clinical Laboratory Tests

Table 3.1 lists common molecules measured in the clinical laboratory and their normal and toxic levels

3.1.5 Microdialysis

Microdialysis is a technique for sampling the chemistry of the individual tissues and organs of the body, and applies to both animal and human studies (de la Peña, 2000). The basic principle is to mimic the function of a capillary blood vessel by perfusing a thin dialysis tube implanted into the tissue with a physiological liquid. The perfusate is analyzed chemically and reflects the composition of the extracellular fluid with time due to the diffusion of substances back and forth over the membrane. Microdialysis is thus a technique whereby substances may be both recovered from and supplied to a tissue. The most important features of microdialysis are: sampling the extracellular fluid, the origin of all blood chemistry; continuously sampling for hours or days without withdrawing blood; and sample purification, simplified chemical analysis by excluding large molecules from the perfusate. However, the latter feature renders the technique unsuitable for sampling large molecules such as proteins. The technique has been extensively used in the neurosciences to monitor neurotransmitter release, and is now finding application in monitoring the chemistry of peripheral tissues in both animal and human studies.

Table 3.1 Normal and toxic levels of various molecules in the body given in both U.S. and European (SI) units. NA denotes "not available."

Molecule	Type of test	Normal levels US units (mg/dL)	Normal levels SI units (mmol/L)	Toxic levels (mg/dL)
Total bilirubin	Blood (serum)	0.2–1.0	$3.4–17.1 \times 10^{-3}$	High levels result in jaundice.
Indirect		0.2–0.8	$3.4–12 \times 10^{-3}$	
Direct		0.1–0.3	$1.7–5.1 \times 10^{-3}$	
Lactate	Blood (serum)	5–12	0.5–1.3	>45
Creatinine	Blood (serum)			>4
Female		0.6–1.1	$53–97 \times 10^{-3}$	
Male		0.7–1.2	$62–115 \times 10^{-3}$	
Urea	Blood (serum)	10–20	3.6–7.1	>100
Glucose	Blood	70–105	3.9–5.8	Adult male: <50 or >400 Adult female: <40 or >400
Sodium	Blood (serum)	136–145 mEq/L	136–145	<120 or >160 mEq/L
Potassium	Blood (serum)		3.5–5.1	<3 or >7.5 mmol/L
Lithium	Blood (serum)	0.8–1.2 mEq/L	NA	>2.0 mEq/L

3.2 Oxygen Saturation

Hemoglobin is a protein and the main component of red blood cells. Hemoglobin transports oxygen from the lungs, where oxygen tension (partial pressure of oxygen) PO_2 is high, to the tissues where oxygen tension is low (see Chapter 6). Oxygen saturation, SO_2, is defined as the ratio of bound oxygen to the total oxygen capacity.

$$HbO_2SAT = \frac{[HbO_2]}{[RHb] + [HbO_2]} \tag{3.7}$$

where $[HbO_2]$ is the concentration of oxyhemoglobin, and $[RHb]$ is the concentration of deoxyhemoglobin (hemoglobin that is not carrying O_2). Oxygen saturation is calculated as a percent or fraction.

$$HbO_2SAT(\%) = HbO_2SAT(\text{fraction}) \times 100 \tag{3.8}$$

Oxygen saturation measurements are made in order to determine the response to therapeutic intervention (e.g. supplemental oxygen administration or mechanical ventilation) and/or diagnostic evaluation (e.g. the effect of exercise on O_2 levels). Oxygen saturation measurements are also needed to monitor the severity and progression of some diseases. SO_2 can be determined by using spectrophotometry on a venous whole blood sample. The extinction coefficients of both hemoglobin species are equal at the isosbestic wavelength 805 nm, while reduced hemoglobin [RHb] is more transparent (smaller extinction coefficient) than oxyhemoglobin [HbO$_2$] in the infrared region. An oximeter uses a spectrophotometer to measure light transmission at two wavelengths to calculate SO_2. By using four wavelengths, the CO-oximeter also measures carboxyhemoglobin ([COHb]) and methemoglobin ([MetHb]).

To measure SO_2 noninvasively, we shine light through the finger or earlobe, which measures a mixed sample of venous and arterial blood. To measure the arterial blood alone, we use the pulse oximeter (Webster, 1997). Light emitting diodes (LEDs) transmit light at 660 nm and 940 nm and a photodiode measures the transmission at the minimum blood pressure (diastole) and the maximum blood pressure (systole). By subtracting the measurements, we calculate the SO_2 for only the additional blood added to the expanded artery, and thus, all arterial blood.

The OxiFirst™ Fetal Oxygen Saturation Monitoring System is inserted into the mother's uterus and placed against the fetus's temple or cheek when fetal heart rate indicates that the baby may be in distress. This allows labor to progress when SO_2 indicates it is safe to do so and thus minimizes the cesarean section rate (Mallinckrodt, 2003).

3.3 Bilirubin

Red blood cells are replaced approximately every 100 days, which means that every day one percent of the body's red blood cells, produced in bone marrow, are replaced. Bilirubin is waste resulting from the removal of old red blood cells.

Hemoglobin consists of four subunits. Each subunit has one chain of protein, known as globin, and one molecule of heme. Heme is made up of a single iron atom attached to porphyrin, a ring shaped molecule. When a red blood cell is destroyed, the body recycles the iron. The ring shaped molecule is toxic and consequently broken down into bilirubin. Unconjugated bilirubin is produced in the spleen when the porphyrin is broken down. The unconjugated bilirubin enters the blood stream and travels to the liver where it is converted to conjugated bilirubin and excreted.

Unconjugated bilirubin is produced when red blood cells are destroyed. Abnormally high levels of conjugated bilirubin in the bloodstream, known as jaundice, result from liver disease and can turn skin and the whites of a person's eyes yellow. Neonatal jaundice, a common problem that occurs after birth, results from the mother's antibodies attacking the baby's red blood cells. In general, blood samples can be taken and used to measure bilirubin concentration when diagnosing liver and/or biliary disease.

The most common method to quantitatively measure bilirubin is based on the diazo reaction and spectrophotometry. This technique is used to measure total and conjugated bilirubin in blood serum or plasma. A detailed discussion of this method, also known as the Jendrassik and Grof Technique, is referenced in (Burtis and Ashwood, 1994).

3.4 Lactate

Lactate, also known as blood lactate in medicine, is the anionic form of lactic acid present in the blood. Lactic acid is a metabolic intermediate involved in many biochemical processes including glycolysis and gluconeogenesis (the formation of new glucose from noncarbohydrate sources). Lactate measurement is important because it discovers any disease that causes decreased tissue oxygen consumption. This condition is manifested by increased amounts of lactate in the blood. Some of these diseases include diabetes mellitus, neoplasia (tumor growth), and liver disease. Lactate measurement is important in industry for the regulation and control of food products such as milk (Canh, 1993). Lactic acidosis (very high levels of lactate in the blood) is caused by hypoperfusion (a decreased blood flow through an organ). Prolonged lactic acidosis may result in permanent cellular dysfunction and death. Lactic acid is measured from a sample of whole blood or plasma.

Lactate is most commonly measured by attaching the enzyme lactate oxidase to a PO_2 electrode (see Section 3.8.1) to detect oxygen consumption, or on a platinum electrode to detect the amount of hydrogen peroxide formed. This latter reaction is

$$\text{L-lactate} + O_2 \xrightarrow{\text{lactate oxidase}} \text{pyruvate} + H_2O_2 \tag{3.9}$$

The enzyme lactate dehydrogenase (LDH) recycles pyruvate into lactate

$$\text{Pyruvate} + \text{NADH} + H^+ \xrightarrow{\text{LDH}} \text{L-lactate} + \text{NAD}^+ \tag{3.10}$$

Coupling these two enzymes enables lactate measurement as low as 80 nM (Canh, 1993). This is the method of choice for measuring lactate due to its simplicity and high specificity (see Chapter 1).

3.5 Creatinine

Found in muscle cells, phosphocreatine converts adenosine diphosphate (ADP) back to adenosine triphosphate (ATP), replenishing the muscle cell's energy. During this conversion, creatine is produced. Creatine is usually converted back to phosphocreatine and the cycle starts over. However, when creatine needs to be excreted, it is dehydrated and converted to creatinine. Creatinine is not reabsorbed when going through the kidneys during urine production. Therefore, it is an ideal candidate for measuring the condition of the kidneys.

Measured in blood (serum) and urine, elevated levels of creatinine result from muscle damage or strenuous physical activity. In a person with kidney disease, creatinine builds up in the blood since it is being produced faster than it is being eliminated. Therefore, measurement of creatinine in blood (serum) is a rough estimate of the health of the kidneys. Measurement of creatinine from urine is a much better measure of kidney function (see Chapter 9) and is the most prevalent clinical test for approximating glomerular filtration rate (the rate of filtration by the kidneys).

The most common method used to measure creatinine is based on the Jaffe reaction

$$\text{Creatinine} + \text{Picric Acid} \xrightarrow{\text{alkaline}} \text{Picrate–creatinine Complex} \qquad (3.11)$$

In an alkaline (basic) medium, creatinine and picric acid form a red–orange compound whose structure has been postulated but not confirmed. The hypothesized picrate–creatinine complex forms from a creatinine to picric acid ratio of 1:1. The complex can be measured spectrophotometrically at wavelengths between 505 to 520 nm. The concentration of the hydroxyl ion (alkaline medium) affects the rate of the complex formation. Most methods use a 0.5 M concentration of sodium hydroxide and picric acid in excess of stochiometric amounts so that picric acid is not the limiting reagent. One of the main problems with the Jaffe reaction is that it is nonspecific when used to measure creatinine in plasma. Molecules interfering with Jaffe reaction specificity include glucose, protein, ascorbic acid, acetone, and pyruvate. As a result, several modifications exist which increase the specificity of the reaction (Burtis and Ashwood, 1994). However, many interference problems are still unresolved.

3.6 Urea

Urea, NH_2–CO–NH_2, a nitrogen containing molecule, is a metabolic product of breaking down proteins. Figure 3.2 shows how urea is formed in the liver.

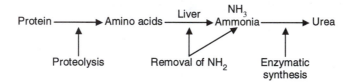

Figure 3.2 Origin of urea in the body (Burtis and Ashwood, 1994). The liver produces urea by first breaking proteins down into their building blocks, amino acids, by a process that breaks the peptide bonds between the amino acids (proteolysis). The amino group (NH_2) of amino acids is removed and ultimately used to form ammonia (NH_3) and urea.

Over 90% of urea is excreted through the kidneys to the urine. The body produces urea to rid itself of excess nitrogen.

The liver produces ammonia and converts it to urea as a waste product of gluconeogenesis. Urea is transported in the blood to the kidneys as blood urea nitrogen (BUN). Although the urea nitrogen measurement is often referred to as BUN, it is never measured from whole blood. Urea nitrogen is most often measured from blood serum (watery fluid separated from coagulated blood) and sometimes plasma. An above normal amount of urea in the blood indicates decreased kidney function, and therefore possible kidney disease.

The two primary methods of measuring urea nitrogen are spectrophotometric. The first method measures urea indirectly by quantifying the concentration of the ammonium ion spectrophotometrically. Urea is hydrolyzed by water in the presence of urease (an enzyme) and the resulting ammonium ion is quantified

$$NH_2-CO-NH_2+H_2O \xrightarrow{\text{urease}} 2NH_2+CO_2 \longrightarrow 2NH_4^+ +CO_3^{2-} \quad (3.12)$$

Two ways of quantifying the ammonium ion concentration are Berthoelot's reaction and the enzymatic assay with glutamate dehydrogenase. An electrochemical approach exists but is too slow to be used repeatedly (Burtis and Ashwood, 1994).

The second method measures urea directly. It is based on the Fearon reaction, and measures urea in urine and plasma. Used in several automated systems, such as the Dupont Dimension 380, the reaction is

$$\text{Urea} + \text{Diacetyl} \xrightarrow{H^+, \text{Heat}} \text{Diazine} + H_2O \quad (3.13)$$

Diazine can be measured using spectrophotometry and absorbs electromagnetic energy at 540 nm. Since diacetyl monoxime is unstable in the presence of water, diacetyl can be generated by the condensation reaction

$$\text{Diacetyl Monoxime} + H_2O \xrightarrow{H^+} \text{Diacetyl} + \text{Hydroxylamine} \quad (3.14)$$

3.7 Glucose

Glucose is the main source of energy for all organisms. Diabetes mellitus is a group of metabolic carbohydrate metabolism disorders in which glucose is underutilized, producing hyperglycemia (high blood sugar levels). The two types of diabetes mellitus are insulin dependent diabetes mellitus (type I) and noninsulin dependent diabetes mellitus (type II). Normally, sugar is not present in the urine, however, when a person has high blood glucose, glucose shows up in the urine.

Insulin is one of the hormones that controls whether glucose is taken out of storage and put into the blood, or vice versa. In type I diabetes, the body does not produce insulin because there is some destruction of the pancreatic islets that produce it. Type II patients produce insulin, but cells do not recognize it (i.e. they have defective insulin receptors). Type I, generally seen in children, is a more severe form of diabetes mellitus, and must be treated with insulin. Type II is seen in older people, and for many, careful control of diet and exercise are adequate treatment. In more severe cases, insulin is taken.

Self-monitoring of blood glucose is required for diabetic patients, especially insulin dependent, in order to maintain normal blood glucose levels (glycemia). When an abnormal amount of glucose is present in the blood, the individual needs to correct the abnormality to avoid short and long term health complications. By regulating blood glucose levels, patients are mimicking the body by providing themselves with the correct amount of insulin. Insulin dependent patients need to measure their blood glucose levels about three times a day.

One reason blood glucose levels need tight regulation is that glucose is the only source of energy neurons can consume (they do not have the enzymes to consume anything else). Low blood glucose levels result in hypoglycemia. When this occurs, an individual's neurons have no source of energy and if low enough, may induce a coma and/or death. High blood glucose levels result in hyperglycemia. When blood glucose becomes too high, the glucose molecules denature (alter the shape of) proteins, such as collagen and hemoglobin, throughout the body. Collagen attaches to the lining of blood vessels (the basement membrane), and when glucose denatures collagen, blood vessels are destroyed. This leads to decreased blood flow in the arms and legs (lower perfusion). When glucose denatures proteins associated with neurons, nerve damage occurs and results in a person's inability to feel. Diabetes mellitus is the leading cause of amputation because decreased blood flow causes tissue to die and damaged nerves hinder the sensation thus making widespread damage much more likely.

The most common techniques for measuring blood glucose are enzymatic. The glucose oxidase method is a very popular manual procedure used for self-monitoring. The Hexokinase method is widely used in laboratories since the procedures for it are carried out by automated equipment.

3.7.1 Glucose Oxidase Method

The glucose oxidase method used in a large number of commercially available simple strip tests allows quick and easy blood glucose measurements. A strip test product, One Touch II (Lifescan, Milpitas, CA), depends on the glucose oxidase–peroxidase chromogenic reaction. After a drop of blood is combined with reagents on the test strip, the reaction shown in Eq. (3.15) occurs.

$$\text{Glucose} + 2H_2O + O_2 \xrightarrow{\text{glucose oxidase}} \text{Gluconic Acid} + 2H_2O_2 \qquad (3.15)$$

Adding the enzymes peroxidase and o-dianiside, a chromogenic oxygen, results in the formation of a colored compound that can be evaluated visually.

$$\text{o-dianisine} + H_2O_2 \xrightarrow{\text{peroxidase}} \text{oxidized o-dianisine} + H_2O \qquad (3.16)$$

Glucose oxidase chemistry in conjunction with reflectance photometry produces a system for monitoring blood glucose levels (Burtis and Ashwood, 1994). In the Lifescan system (Figure 3.3), a drop of blood is applied to the reagent pad, the strip is inserted into the meter, and a digital screen displays the results 45 s later. The device automatically starts and times the test and has a range of 0 to 600 mg/dL.

3.7.2 Hexokinase method

Much of the automated equipment is based on the hexokinase method. The general reactions are

$$\text{Glucose} + \text{ATP} \xrightarrow{\text{hexokinase}} \text{glucose–6–phosphate} + \text{ADP} \qquad (3.17)$$

$$\text{Glucose–6–phosphate} + \text{NAD}^+ \xrightarrow{\text{G–6–PD}} \text{6–phosphogluconate} + \text{NADH} + \text{H}^+ \quad (3.18)$$

Glucose is phosphorylated to glucose-6-phosphate in the presence of ATP, Mg^{2+}, and the enzyme hexokinase. The glucose-6-phosphate formed in Eq. (3.17) oxidizes to 6-phosphogluconate by the enzyme glucose-6-phosphate dehydrogenase (G-6-PD) in the presence of nicotinamide-adenine dinucleotide (NAD$^+$) (Eq. 3.18). The NADH produced absorbs electromagnetic energy at 340 nm and is proportional to the glucose concentration. This method detects glucose in both serum and plasma. Several other techniques based on the hexokinase method exist (Burtis and Ashwood, 1994).

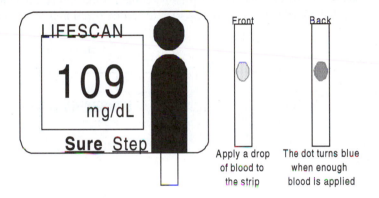

Figure 3.3 LifeScan, Inc., a system by Johnson and Johnson for self-monitoring glucose levels.

3.8 Amperometric Biosensors for Oxygen and Glucose

3.8.1 Oxygen (PO_2)

Amperometry measures the amount of current flowing through an electrochemical cell when a constant potential is applied to the electrodes (Burtis and Ashwood, 1994). The Clark electrode, an amperometric sensor that measures the partial pressure of O_2 (PO_2), is actually an electrochemical cell consisting of two electrodes: a glass covered platinum cathode and a silver/silver chloride anode. The silver/silver chloride anode is in a phosphate buffer containing potassium chloride to buffer the hydroxyl ions produced at the cathode (Eq. (3.20)). Figure 3.4 shows the main components of the Clark (PO_2) electrode.

At the anode, the following reaction occurs

$$4Ag + 4Cl^- \longrightarrow 4AgCl + 4e^- \tag{3.19}$$

At the cathode, the following reaction occurs:

$$O_2 + 2H_2O + 4e^- \longrightarrow 2H_2O_2 + 4e^- \longrightarrow 4OH^- \tag{3.20}$$

$$4OH^- + 4KCl \longrightarrow 4KOH + 4Cl^-$$

The electrode at which the reduction–oxidation (redox) reaction involving the molecule of interest occurs is referred to as the working electrode. According to the Nernst equation (Eq. 3.21), a cell reaction spontaneously proceeds to the right if the potential of the cell is greater than zero ($E > 0$) because the O_2 concentration is greater than H_2O. A cell reaction spontaneously proceeds to the left if the potential for the cell less than zero ($E < 0$). When the potential for the cell is zero ($E = 0$), the redox reaction is at equilibrium.

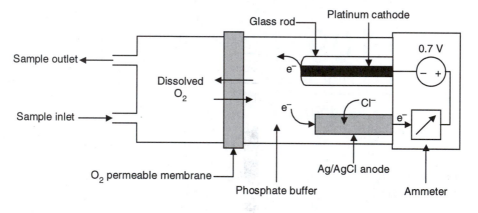

Figure 3.4 In the PO_2 electrode, O_2 dissolved in the blood diffuses through a permeable membrane with current proportional to PO_2 (Hicks et al., 1987).

$$E = E_0 + \frac{0.059}{n} \log \frac{C_{H_2O}}{C_{O_2}} \tag{3.21}$$

where n is the valence of the electrode (cathode) material and E_0 is the standard half-cell potential. E_0 is the half-cell potential when $C_{O_2} = C_{H_2O}$. A dc voltage between 600 and 800 mV allows the electron transfer reaction between O_2 and H_2O to occur. The entire reaction at the working electrode consists of several steps, each having its own inherent rate. These steps include the transport of O_2 from the solution to the working electrode, the transfer of electrons from O_2 to H_2O, interactions between O_2 and H_2O (the oxidized and reduced forms of the species of interest), and the transport of H_2O back to the solution. The amount of current produced is controlled by the slowest of these steps. Because O_2 is consumed at the cathode, there is a concentration gradient from the dissolved O_2 to the cathode. It takes about 60 s to establish this concentration gradient for a stable measurement. The cathode may be vibrated to ensure that fresh O_2 is available at the cathode.

The current produced by this reaction is directly proportional to the rate of electrons transferred from O_2 to H_2O and the rate of electrons transferred from O_2 to H_2O is directly proportional to the concentration of O_2.

3.8.2 Glucose

The first enzyme electrode to measure glucose was developed by Clark and Lyons in 1962 (Figure 3.5). These biosensors still serve as the classical example for measuring glucose because their operation explains the basic concepts for measuring glucose with an amperometric enzymatic electrode (Hall, 1991).

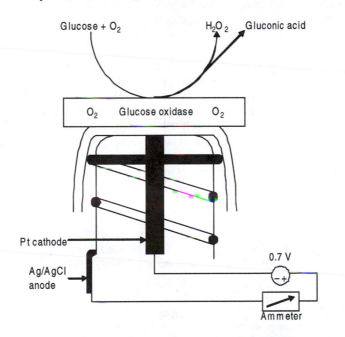

Figure 3.5 When glucose is present in the glucose enzyme electrode, it combines with O_2, thus decreasing the O_2 that reaches the cathode.

It is often referred to as a first generation biosensor due to its structure and the level at which its components are integrated (Taylor and Schultz, 1996). This glucose sensor uses the PO_2 electrode and the enzyme glucose oxidase immobilized on a membrane. The reaction for the basic glucose sensor is

$$\text{Glucose} + O_2 + 2H_2O \xrightarrow{\text{glucose oxidase}} \text{Gluconic acid} + 2H_2O_2 \qquad (3.22)$$

Glucose and oxygen react in the presence of the enzyme glucose oxidase. The amount of O_2 consumed in the reaction, indicated by a decrease in PO_2, is proportional to the concentration of glucose in the solution. (The concentration of glucose is indicated by the consumption of O_2 by the enzyme glucose oxidase.) The disadvantage of the single oxygen electrode glucose sensor is that it is sensitive to variations of PO_2 in the blood.

A dual oxygen electrode glucose sensor (Peura, 1998) eliminates the problem of variations in PO_2 that the single oxygen electrode is sensitive to. As the name implies, the dual oxygen electrode uses two oxygen electrodes to compensate for the PO_2 sensitivity. An enzyme within a gel covers one electrode, thus depleting oxygen when glucose is present. The other electrode, not covered by the enzyme, senses only oxygen. The amount of glucose is determined as a function of the difference between the electrode responses. The disadvantage of enzymatic glucose sensors is that they can only be used in vivo for a few months, and the enzyme is unstable.

A more recent approach to amperometric enzyme glucose electrodes involves the use of the electron transfer mediator dimethyl ferrocene. Electron transfer mediators, such as dimethyl ferrocene, are low molecular weight species, which shuttle electrons between the working electrode and the oxidation–reduction center of the enzyme (Taylor and Schultz, 1996). There are several advantages in using mediators. First, mediators are much more readily reduced than the substrate cofactor (in this case O_2). Second, they allow glucose measurements that are independent of the sample variations in PO_2 (Burtis and Ashwood, 1994). In the series of reactions shown in Figure 3.6, a reduced form of the mediator dimethyl ferrocene reduces the oxidized form of glucose oxidase. Measurement using this reaction can be made using whole blood and the current produced is directly proportional to the glucose concentration. One disadvantage of this technique is that dimethyl ferrocene (and all mediators) is not very soluble and adsorbs to the electrode surface.

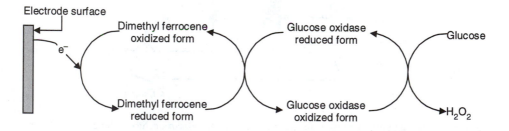

Figure 3.6 The sequence of reactions involved in the mediated reaction of a glucose sensor (Taylor and Schultz, 1996). Dimethyl ferrocene expedites the transfer of electrons from the electrode surface to the redox center of the enzyme glucose oxidase. The use of the mediator dimethyl ferrocene reduces the sensor's dependence on oxygen tension.

Cygnus, Inc. has developed a GlucoseWatch™. An electric current is used to extract glucose across the skin into a hydrogel using iontophoresis. Within the hydrogel, the extracted glucose undergoes a reaction with the enzyme glucose oxidase to produce gluconic acid and hydrogen peroxide in the presence of oxygen. The hydrogen peroxide further diffuses to and reacts on a platinum electrode, producing two electrons, water, and oxygen (Kurnik et al., 1998).

3.9 Ion-selective Electrodes for pH and CO_2

Ion-selective electrodes are used to measure the activity of a specific ion in a cell or solution. An ion-selective electrode contains a membrane that is permeable to a specific cation or anion. In the presence of the specific ion, a membrane potential results which indicates the amount of ion activity. pH is the negative logarithm to the base 10 of the hydrogen ion concentration [H+] in mol/L.

$$pH = -\log_{10}[H^+] \tag{3.23}$$

The measurement of pH is obtained by using an ion-selective electrode (ISE) with a glass membrane. To prevent pH measurement errors occurring due to other ions, a highly selective H+ glass membrane is desired. One such glass composite consists of silicon dioxide, lithium oxide, and calcium oxide in the ratio 68:25:7 (Burtis and Ashwood, 1994).

$$1 \mid \text{glass membrane} \mid 2 \tag{3.24}$$

The glass membrane separates the test solution on side 2 from the reference solution, usually hydrochloric acid of known pH, on side 1 (as seen in Eq. (3.24)). When the pH of the solutions on side 1 and 2 differ, a potential across the membrane develops, $V = V_1 - V_2$. The potentials of solution 1 (hydrochloric acid) are recorded with a reference electrode such as a silver/silver chloride electrode. The potential of solution 2 can be recorded with a calomel electrode. Ideally, the pH electrode obeys the Nernst equation

$$E = K + \left(2.303 \frac{RT}{ZF} \right) \log(a_i) \tag{3.25}$$

where E is the potential, R is the universal gas constant, F is the Faraday constant (= 96487 C/mol, C = coulomb), T is the absolute temperature, Z is the ionic charge, a_i is the activity of the ion, and K is a constant containing contributions from various sources. The activity of an ion, a_i, equals the activity coefficient λ_i times its concentration c.

$$a_i = \lambda_i c \tag{3.26}$$

Factors such as other ions in the solution and the strength of H+ in the solution influence λ_i, which decreases as c increases. λ_i for plasma is about 0.75 for Na+, 0.74 for

K^+, and 0.31 for Ca^{2+}. In solutions where the primary ion (H^+) is of very low concentration (μM), the membrane potential is not directly proportional to the logarithm of the activity of the diffusible ion H^+ in solution 1 or 2. The relationship is better expressed by the Nikolski–Eisenman equation.

The potential can also be calculated from

$$E = -0.0615 \log_{10} \frac{[H^+]_i}{[H^+]_o} \tag{3.27}$$

where $[H^+]_i$ is the concentration of H^+ inside the cell, and $[H^+]_o$ is the concentration of H^+ ions outside the cell.

IQ Scientific Instruments (2001) offers pH measurement without glass utilizing a stainless steel probe with a silicon chip sensor that can be stored dry. The ISFET consists of a silicon semiconductor substrate with two electrical contacts (source and drain) a small distance apart. Deposited on the substrate between the source and drain is a silicon electrical insulator. Hydrogen ions at or near the surface of the insulator cause a variable voltage potential between the insulator and the underlying semiconductor material between the source and drain. The variable voltage potential is proportional to the relative concentration of hydrogen ions in the sample solution. The pH is derived from this voltage to a very high level of accuracy.

The pH measures the concentration of H^+ protons. The pH level in the body is more strictly regulated than glucose. The normal range of pH is 7.36 to 7.44. A pH of less than 7.36 is called acidosis and a pH less than 7.0 causes a person to go into a coma and die. A pH greater than 7.44 is called alkalosis, and a pH greater than 7.8 causes the body's muscles to seize up, a condition known as tetany. Explanations for these phenomena lie in acid–base chemistry. Molecules in the body, particularly proteins, have sites where protons can associate or dissociate, and when the body's pH is altered, the charge of that particular portion of the molecule is changed. The electrostatic charges on the parts of the molecule may attract or repel in different ways than they should. As a result, proteins do not fold properly and DNA does not form as it should, among other things.

The metabolism of substances in the body tends to produce protons. With the exception of vomiting, protons leave the body via the kidney. The kidney excretes acids directly or produces bicarbonate, which combines with a proton to form carbonic acid. Carbonic acid spontaneously breaks down into water and carbon dioxide. The body also eliminates acid by exhaling carbon dioxide through the lungs. An individual can exhale too much or too little carbon dioxide, thus changing the acidity of the blood. Or, if the kidney is prevented from eliminating acid, acid buildup can occur, causing a condition known as acidosis. Both the lungs and the kidneys regulate the body's pH level and having two ways to eliminate acid allows one method to compensate if the other becomes impaired. Three pH measurements determine abnormal kidney and lung function: carbon dioxide concentration, pH, and bicarbonate concentration. If the problem is abnormal CO_2 concentration, then we know the problem is with the lungs. If the bicarbonate concentration is abnormal, then a condition called metabolic acidosis exists, and the cause might be either the kidney or something within the metabolic pathway. Further studies are required for abnormal bicarbonate concentration, including sodium concentration or chloride concentration measurements of the blood.

In the CO_2 electrode, blood comes in contact with a plastic membrane. Holes in the membrane permit gas to diffuse through, but block everything else. The CO_2 diffuses to a chamber containing a pH electrode, forms carbonic acid, and increases the acidity. The reaction for this process is

$$CO_2 + H_2O \leftrightarrow H_2CO_3 \leftrightarrow H^+ + HCO_3^-$$

Since CO_2 concentration is proportional to H^+ concentration; measurements from the pH electrode yield PCO_2.

3.10 Flame Photometry

Flame photometry, a subset of emission photometry, is a technique commonly used to quantify sodium, potassium, and lithium concentrations in bodily fluids. It is a suitable technique for quantifying these elements, because compared to calcium and magnesium, they have low excitation potentials. Figure 3.7 shows a block diagram of a flame photometer. The purpose of the flame is to provide energy to the elements such that their valence electrons will be excited from ground states to higher excited states. Electrons in the excited states are unstable and emit energy as photons of a particular wavelength as they return to their ground states. The wavelength(s) of the light emitted by an element is characteristic of that element. When sodium, potassium, and lithium are present as cations in an aqueous solution and exposed to a flame, sodium produces yellow, potassium violet, and lithium red colors. The intensity of the light given off is directly proportional to the number of photons being emitted, which in turn is directly proportional to the number of atoms, or concentration of cations, in the solution.

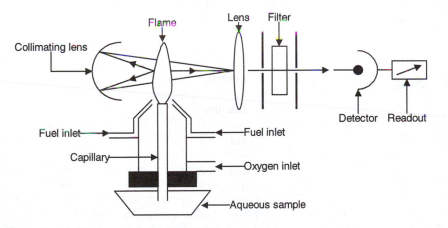

Figure 3.7 The flame photometer aspirates a sample containing metal ions and heats it to incandescence. Detector output is proportional to concentration.

In flame photometry, the monochromator and detector are similar to those found in the spectrophotometer, although the detector measures emission as opposed to absorption. The distinguishing features of the system are the flame and the atomizer. The atomizer draws the sample solution containing the cations through an aspirator and into the flame. This is achieved by passing a gas over the upper outlet of a capillary tube at a high velocity, while the lower end of the capillary tube is inserted into the sample. The fuel used to generate the flame can be either propane, natural gas, or acetylene mixed with compressed air or oxygen. Lithium can be used in calibration because it does not naturally occur in the body (provided the patient is not taking lithium medication). A known concentration of lithium is added to the sample. Then a ratiometric technique can be used to compare the ratios of sodium and potassium to lithium.

3.10.1 Lithium

Lithium is not produced nor does it naturally appear in the body. The purpose of lithium measurement is to measure lithium carbonate, which is used in the treatment of the psychiatric disorder manic depression (also known as bipolar disorder).

3.10.2 Sodium and Potassium

Sodium and potassium are used by the body to maintain concentration gradients in nerve and muscle cells, which allows them to conduct action potentials. To keep these cells working properly, the amount of sodium and potassium in the body must be regulated. It is the responsibility of individual cells to do the regulation. Sodium and potassium enter the body via eating. For people who do not eat properly, it is the kidneys' job to regulate and compensate the levels of sodium and potassium after eating. For example, if an individual eats a lot of salt, the excess salt in their blood passes in the urine. Potassium chloride injected into the blood stream raises the level of potassium in the body, although it is important to be extremely careful with the dosage. Too much potassium can kill a person by stopping their heart as a result of a decreased concentration gradient, and hence the ability to generate an action potential in the heart muscle cells.

3.11 Mass Spectrometry

Mass spectrometry is a powerful analytical technique used to identify unknown compounds, quantify known materials, and elucidate the chemical and physical properties of molecules. Mass spectrometry provides valuable information to a wide range of professionals including physicians, biologists, chemists, and engineers. A mass spectrometer separates a material according to its atomic or molecular mass and the sample of molecules can be a gas, liquid, or solid, as long as the specimen is not too volatile. Mass spectrometry involves ionization of a gas–phase sample, fragmentation of the ionized species, sorting of the fragments according to their mass and charge, and measurement of ion abundance at each mass. As far as the scope of mass spectrometry is concerned, many different techniques and types of instrumentation have been devised to accomplish each of the above processes. This discussion will be limited to those most useful in the toxicology laboratory.

3.11.1 Sample Inlet

Figure 3.8 shows that the specimen containing known or unknown molecules enters the mass spectrometer through the sample inlet, where it is vaporized under a high vacuum. Several types of sample inlet systems exist that accommodate certain aspects of the sample such as its temperature, vapor pressure, and volatility. An important inlet system, the gas chromatograph interface, allows the components of a mixture to enter the mass spectrometer continuously as they emerge from the chromatograph.

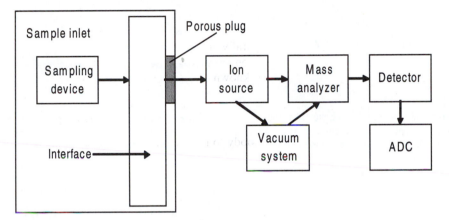

Figure 3.8 The mass spectrometer separates molecules in a high vacuum after they have passed through a porous plug.

3.11.2 Ion Source

Figure 3.9 shows the ion source where the sample gets ionized. Many different methods to ionize the sample exist, one of which is the heated filament. A filament, (rhenium or tungsten) when heated to greater than 2000 °C by electric current, emits electrons under the influence of an electric field. Typical current flow is about 500 µA. The electrons are focused in a magnetic field, which causes them to spiral toward the positively charged target. Electron energy is typically 70 to 80 eV. The ionization process is initiated when a sample is bombarded by the high-energy electrons.

By bombarding molecules of the vaporized sample with electrons from a heated filament, electrons can be removed from the molecules. When an electron is removed from a molecule, it becomes positively charged or ionized (+). Since the mass of an electron is negligible relative to the mass of a molecule, the mass of the molecular ion remains essentially the same

$$ABC + e^- \rightarrow ABC_\bullet^+ + 2e^-$$
$$ABC + e^- \rightarrow ABC^-$$

(3.28)

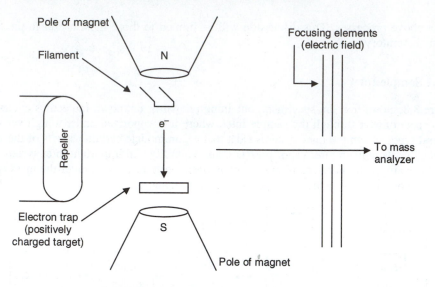

Figure 3.9 An electron impact ion source bombards the molecules and carrier gas from a gas chromatography column with electrons. The electron collisions with ion ABC (Eq. (3.28)) produce both positive and negative ions. The positive ions are directed by focusing elements to the mass analyzer.

For a time span of 0.1 to 10 ns after the collision of the electron with the ion, several molecular ions disintegrate, forming radicals (•) and positively charged fragment ions.

$$ABC_\bullet^+ \rightarrow AB^+ + C^+$$
$$A^+ + BC^+$$
$$AB_\bullet^+ + C \text{ (Loss of neutral)}$$
$$AC^+ + B^\bullet \text{(Rearrangement)}$$
$$etc.$$

(3.29)

Not all of the positively charged molecular ions undergo fragmentation. These unfragmented positively charged molecular ions are drawn out of the ion source by an electric field while the vacuum system pumps the negative and neutral fragments away. An electric field then appropriately focuses and accelerates them into the mass analyzer. The intensity of the electron beam and size of the sample determine the number of fragments and molecular ions, whereas the chemical stability of the molecule determines the relative amounts of molecular ions and fragment ions. Figure 3.10 is an example of a mass spectrum for a sample and its resulting parts. A mass spectrum is a plot that represents the relative ion abundance at each value of *m/z* (mass to charge ratio).

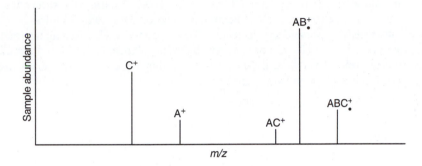

Figure 3.10 The mass spectrum displays the relative abundance of charged molecular ions and fragments.

3.11.3 Mass Analyzer

The function of the mass analyzer is to separate the stream of ions produced from the ion source by their mass to charge ratio (*m/z*), typically with a magnetic field. Figure 3.11 shows the main components of a quadrupole mass analyzer.

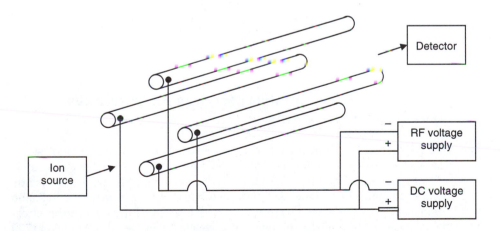

Figure 3.11 A quadrupole mass analyzer accelerates ions from the ion source. The dc and RF voltages that are applied to the four ceramic rods stabilize the paths of ions with selected *m/z* values. Ions that do not stabilize collide with the rods and do not reach the detector.

3.11.4 Detector

The three most commonly used detection schemes are the electron multiplier, photographic emulsions, and Faraday cups (Haven et al., 1995). Figure 3.12 shows the main components of an electron multiplier. Positively charged ions collide violently into the cathode, thus releasing secondary electrons. These secondary electrons are attracted to the first dynode, which is 100 V more positive than the cathode. A larger number of tertiary electrons are released and attracted to the second dynode, etc. to build up a current of about 1 µA, which is measured at the anode.

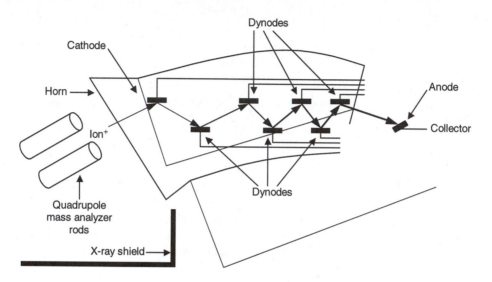

Figure 3.12 Block diagram of an electron multiplier. An electron multiplier is sensitive enough to detect the presence of a single ion. A 10- to 12-dynode electron multiplier produces a signal amplification of 10^5 to 10^6 (Haven et al., 1995).

3.12 Carbon Dioxide Concentration Measurement by Infrared Transmission Spectroscopy

A spectrum usually shows the intensity of electromagnetic waves as a function of wavelength. The intensities depend on the medium(s) the waves pass through. Spectroscopy is the analysis of spectra (components of electromagnetic radiation). Spectra are formed by a spectroscope, an optical instrument that physically breaks up electromagnetic radiation into its component parts, usually with a prism or grating. Infrared transmission spectroscopy is the analysis of spectra over the range of wavelengths in the infrared region (770 to 1000 nm) of the electromagnetic spectrum and can be used to measure the concentration of carbon dioxide in a gaseous mixture.

The majority of carbon dioxide in the body is expired through the lungs as an end product of the respiration cycle. Measuring the amount of CO_2 expired by the lungs helps diagnose disease. Gases with dipole moments such as CO_2, N_2O, or water vapor (H_2O), absorb a specific wavelength of light in the infrared region of the spectrum. Gases without dipole moments such as N_2 and O_2 do not absorb any power (infrared radiation). Carbon dioxide concentration can be measured by passing a particular wavelength of infrared light through a chamber containing a sample of carbon dioxide, noting the power of the infrared light transmitted to the sample, and the power of the infrared light transmitted through the sample. Using infrared light of a particular wavelength characteristic, Beer's Law yields CO_2 concentration in a mixture of gases

$$P_t = P_0 e^{-aLc} \tag{3.30}$$

where P_t is the power per unit area of transmitted infrared light received by the detector, P_0 is the power per unit area of infrared light entering the sample, a is the absorption coefficient, L is the path length of light through the gas, and c is the concentration of gas. Figure 3.13 shows the basic components of an infrared spectroscopy system (Primiano, 1998). The infrared source is modulated by rotating vanes similar to a windmill. The detector compares transmission through the sample cell with transmission through a reference cell.

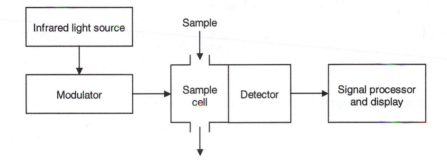

Figure 3.13 An infrared transmission spectroscope measures the absorption of infrared light by a sample drawn through the sample cell by a vacuum pump. From Webster, J. G. (ed.) *Medical Instrumentation: Application and Design.* 3rd Edition. Copyright © 1998 by John Wiley & Sons. Reprinted by permission of John Wiley & Sons.

3.13 Nitrogen by Emission Spectrometry

The concentration of nitrogen, N_2, in a sample mixture of gases can be measured by emission spectrometry. Nitrogen concentration in a mixture of gases is a measurement made in respiratory and pulmonary medicine. Emission spectrometry involves applying a voltage of 600 to 1500 V dc between two electrodes on opposite sides of an ionization chamber in which the pressure is low (1 to 4 mm Hg or 150 to 550 Pa). Under these conditions, the nitrogen in the chamber is ionized and emits ultraviolet light (wavelengths from 310 to 480 nm). Figure 3.14 shows the main components of a nitrogen analyzer (Primiano, 1998).

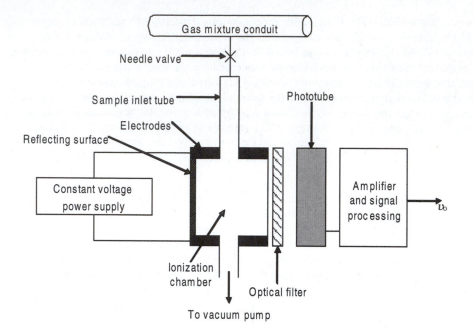

Figure 3.14 Emission spectroscopy measures ultraviolet light emitted by ionized nitrogen. From Webster, J. G. (ed.) *Medical Instrumentation: Application and Design.* 3rd Edition. Copyright © 1998 by John Wiley & Sons. Reprinted by permission of John Wiley & Sons.

3.14 Drugs by Fluorometry and Chromatography

3.14.1 Fluorometry

Fluorescence is the emission of energy in the form of light as a result of the electrons of a molecule returning to their ground state. When certain molecules absorb energy (in the form of electromagnetic radiation) their electrons are raised to higher energy levels. As the electrons of the molecule return to their ground states, they fluoresce. A fluorometer is an instrument that measures the intensity of light produced when the electrons of a molecule return to their ground state. Thus, fluorometry is defined as measuring the relationship between the concentration of a substance and the intensity of the fluorescence produced by that compound when it is excited by radiation. Fluorometry is used to measure therapeutic drugs such as phenobarbital (treatment of epilepsy), enzymes such as proteases (used to break down proteins), hormones such as cortisol, and analytes such as catecholamines (e.g. epinepherine and norepinepherine) and bilirubin. Figure 3.15 shows the components of a fluorometer, which include an excitation source, a primary (excitation) filter, a secondary (emission) filter, a cuvette sample, a detector, and a readout device.

Although the components of a fluorometer are basically the same as those of a spectrophotometer, fluorometry is up to four orders of magnitude more sensitive than spectrophotometry (Wheeler, 1998). In spectrophotometry, the concentration of the sub-stance is determined by the difference of the absorbance between a solution that has zero absorbance (the solution used to calibrate the spectrophotometer) and the absorbance of the unknown solution. When the absorbance of the unknown solution is relatively small, small errors in the measurement of the zero absorbance solution can cause relatively large errors in the final determination. Fluorometry avoids this problem by measuring fluores-cence directly from the sample without reference to another measurement. A disadvan-tage of fluorometry is that measurements are also susceptible to inner-filter effects, in-cluding excessive absorption of the excitation radiation (prefilter effect) and self-absorption of atomic resonance fluorescence (postfilter effect).

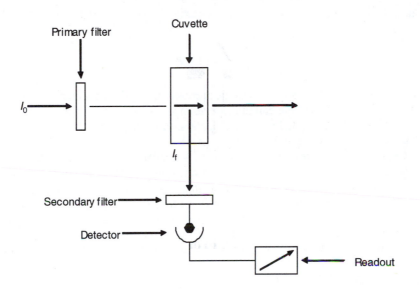

Figure 3.15 Block diagram of a fluorometer. The primary filter passes only wavelengths that excite the fluorescent molecule. The secondary filter blocks all scattered excitation wavelengths and passes only the scattered fluorescent wavelengths. The secondary filter and detector are at a right angle to the primary beam in order to avoid direct transmission of the light source through the sam-ple to the detector.

Theory of Fluorometry

The theory behind fluorometry can be derived from the Beer–Lambert law

$$I = I_0 10^{-aLc} \tag{3.31}$$

where

I = intensity of transmitted light
I_0 = intensity of the incident light
a = absorptivity of the sample

We rearrange to determine the amount of light absorbed:

$$I_0 - I = I_0(1 - 10^{-aLc})$$

The fluorescence intensity, I_F, is proportional to the amount of light absorbed, $I_0(1 - 10^{-aLc})$, and the fluorescence efficiency, ϕ, is the ratio of light emitted to light absorbed, giving

$$I_F = \phi I_0(1 - 10^{-aLc}) \tag{3.32}$$

We can expand through a Taylor's series, rearrange, and convert the logarithm base to yield

$$I_F = \phi I_0 (2.23^{aLc})$$

Thus the fluorescence intensity is directly proportional to the concentration of the fluorescent molecules in the sample and the excitation intensity.

Fluorometry is limited by several factors, including concentration of the sample. For more details, see (Burtis and Ashwood, 1994).

Spectrofluorometers replace the fixed secondary filter with an emission monochromator and sweep it at 5 to 50 nm increments to yield a spectrum of emission magnitude versus wavelength. Some spectrofluorometers also sweep an excitation monochromator instead of having a fixed primary filter. In time-resolved fluorometers, long-lived fluorescent molecules are excited by pulsed light and their decay is monitored over 0.6 to 100 µs. Laser-induced fluorometers measure emission from individual fluorescent or fluorescent-labeled molecules, cells, or particles passing through a flow cell cuvette.

3.14.2 Chromatography

Chromatography is an efficient technique used to separate components in a mixture, such as antibiotics, therapeutic drugs, or abused drugs.

The two main phases in a chromatographic system are a stationary phase and a mobile phase. Chromatographic separation relies on the components in a mixture being separated by the stationary phase. The basis for chromatography is the difference in affinity for the stationary phase versus the affinity for the mobile phase of the components in a mixture. The greater the affinity for the stationary phase, the less a component will move. Likewise, greater affinity for the mobile phase indicates that the component will migrate farther. In order for separation to occur, the mobile phase must move past the stationary phase. In gas chromatography, the mobile phase is the gas and the stationary phase is the solvent, which covers a solid support in the column. The components in a chemical mixture react differently with each of the phases in the column. For example, as the component mixture is injected into the column, some components interact with the stationary phase longer than others. As a result, the components in the mixture become separated in the column. The components that interacted for a shorter time with the stationary phase get washed out of the column by the solvent more quickly than the strongly interacting components.

A chromatogram is a plot of the relative amounts of the components in the mixture as a function of the amount of time it takes for them to get washed (eluted) off or pass through the column. The number of peaks plotted in a chromatogram indicates the complexity of the mixture. Comparing the position of the peaks relative to the position of standard peaks can provide qualitative information on the sample composition and also reveal the performance of the column. Measuring the area under each peak yields a quantitative evaluation of the relative concentrations of the components.

3.14.3 Gas Chromatography

Gas chromatography (GC) is a type of chromatography that uses an inert gas as the mobile phase. GC is used to separate compounds that are volatile at the temperature in the column of the chromatograph. These mixtures of volatile organic compounds, often small organic molecules such as therapeutic drugs, are measured from a sample of urine or blood. Figure 3.16 shows a gas chromatograph consisting of several components: a carrier gas, a column, a column oven, an injector, a detector, a recorder, and a flowmeter.

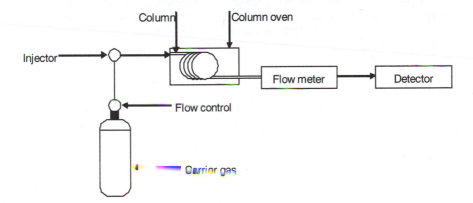

Figure 3.16 In a gas chromatography system, the sample is injected into a carrier gas and flows through a long column to the detector.

There are several advantages of GC: it is fast, extremely sensitive, and requires a very small sample.

Carrier Gas

The inert carrier gas, such as argon, helium, or nitrogen, is the mobile phase in the system because it moves the components in the evaporated solute through the column. The carrier gas has very little effect on the rate of component retention in the solute.

Column

The column in GC is usually enclosed in a computerized temperature regulated oven. Ovens are usually programmed to produce a linear increase in temperature over a span of time, depending on the solute. When the column is heated over a range of temperatures, components in the solute with lower boiling points elute off the column first, followed by components with higher boiling points. Thus, heating the column gradually allows the components to separate.

The column (stationary phase) separates the components in the sample because each has different retention times. The two most common types of columns used in GC are packed columns and capillary columns. Packed columns, typically between 1 and 4 mm in diameter and 1 to 4 m in length (Burtis and Ashwood, 1994), are packed with a solid support material such as diatomaceous earth (a type of algae with cell walls containing Ca^{++}) and then coated with a liquid (the stationary phase). Capillary columns are typically between 0.2 and 0.5 mm in diameter and 10 to 150 m in length (Burtis and Ashwood, 1994). In capillary columns, the inner wall of the tube is coated with the stationary phase.

Both types of columns are made with stainless steel or glass. The advantage of glass is that it is inert, while the advantage of stainless steel is that oxide films do not develop on the surface, thus preventing reactions with solutes. The advantage of capillary columns is that they can detect very small amounts of a component and are highly efficient. The advantage of packed columns is that they can handle a large sample size and require less maintenance.

Detector

The purpose of the detector is to quantify the components in a sample. It does this by providing an electric signal proportional to the amount of the component coming off the column. Some of the commonly used detectors in GC include the flame ionization detector, the thermal conductivity detector, the electron capture detector, the nitrogen phosphorus detector, and detectors based on mass spectroscopy. The combination of mass spectrometry and GC is ubiquitous in clinical chemistry laboratories in hospitals and is a very powerful tool.

Resolution

Figure 3.17 shows that the detector yields a chromatogram. For a flow rate F, the peak of a solute, A, appears after retention time t_r. The volume that has passed before the peak is the peak volume and is given by $V_r(A) = t_r F$. Band spreading broadens the peak to a Gaussian shape. An efficient column minimizes band spreading. The degree of efficiency is the number of theoretical plates, $N = [V_r(A)/s(A)]^2$, where $s(A)$ is the standard deviation of the peak volume. The width of a Gaussian peak at 1/2 its height, $w_{1/2}$, is 2.354 standard deviations, from which $N = 5.54[V_r(A)/w_{1/2}(A)]^2$.

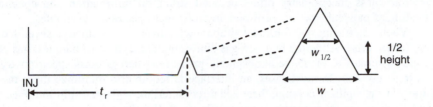

Figure 3.17 In chromatography, the peak appears after retention time t_r.

The resolution of two solute bands is a measure of their degree of separation in terms of relative migration rates and bandwidths and is

$$R_s = \frac{V_r(B) - V_r(A)}{[w(A) + w(B)]/2} \tag{3.33}$$

where $w(A)$ and $w(B)$ are the peak widths measured at the corresponding base. Inadequate separations occur for $R_s < 0.8$ and baseline separation occurs for $R_s > 1.25$ (Bowers et al., 1994).

Example 3.1 For a gas chromatograph with a flow rate of 1.5 L/min, solute A has a retention time of 1 min and solute B has a retention time of 2.5 min. If the peak width for solute A is 0.5 L and the peak width for solute B is 1.5 L, is the resolution sufficient? Since the flow rate of the solutes and peak widths are known, we only need to find the peak volume before we can use Eq. (3.33) to determine if the resolution is sufficient.

$$w(A) = 0.5 \, L$$

$$w(B) = 1.5 \, L$$

$$V_r(A) = t_{rA} F = 1 \, min \times 1.5 \, L/min = 1.5 \, L$$

$$V_r(B) = t_{rB} F = 2.5 \, min \times 1.5 \, L/min = 3.75 \, L$$

$$R_s = \frac{V_r(B) - V_r(A)}{[w(A) + w(B)]/2} = \frac{3.75 \, L - 1.5 \, L}{[0.5 \, L + 1.5 \, L]/2} = 2.25$$

Since $R_s > 1.25$, the base separation of the bands is sufficient.

Coupling GC to MS

GC separates mixtures of volatile organic compounds. Combining the gas chromatograph with the mass spectrometer (MS) is a powerful tool because GC has high separation efficiency while mass spectrometry offers detailed structural information and conclusive identification of molecules when corresponding reference spectra are available.

When interfacing gas chromatography with mass spectrometry, several compatibility issues arise. Although both are gas phase microanalytical techniques, gas chromatography operates at or above atmospheric pressure, whereas mass spectrometry requires a high vacuum. In most cases, an interface is required to minimize these incompatibilities. For a capillary column, there is a direct interface and open split interface. For a packed column, the jet separator is commonly used. For the capillary–direct interface, the column is directly coupled to the MS ion source. The advantage of the direct interface is total inertness from the head of column to the ion source. The disadvantages are that chromatography is affected by the mass spectrometer's vacuum (i.e. a shorter column is needed), venting the MS vacuum system to change the column may be required, and it is limited to > 12.5 m narrow bore (0.2 mm) and > 25 m wide bore (0.3 mm) columns.

The jet separator is the most efficient GC/MS interface. It can remove about 90% of a light carrier gas such as helium, yet pass all but a few percent of the heavier sample compounds. This is accomplished by capitalizing upon differences in gas diffusion rates in a supersonic jet stream. As the gas from the GC passes through a nozzle expanding into an evacuated region, the momentum of the heavier sample gas is increased relative to the carrier gas. The ratio of sample to carrier gas molecules in the region of the jet stream increases, enriching the sample passing through the nozzle leading to the mass spectrometer. The jet separator operates with maximum efficiency at column flows of 25 mL/min and is capable of transmitting a wide range of molecular weights from permanent gases to high molecular weight compounds. Efficiency improves as mass (m/e) or momentum (mv) of the sample gas increases relative to the carrier.

3.14.4 Liquid Chromatography

Liquid chromatography (LC) is a type of chromatography that uses a liquid of low viscosity as the mobile phase. The stationary phase through which this liquid flows is an immiscible liquid coated on either a porous support or absorbent support. One commonly used type of liquid chromatography is high-performance liquid chromatography (HPLC). HPLC is used to separate components dissolved in solution by conducting the separation process at a high velocity with a pressure drop. Figure 3.18 shows a schematic of an HPLC instrument, indicating the major components: the reservoir of mobile phase, a high-pressure pump, the sample injection port, a separation column, and a detector.

An HPLC instrument operates by injecting a liquid sample via a syringe into the stream of mobile phase that is being pumped into the column. A detector records the presence of a substance in the effluent, and can observe changes in refractive index, UV light absorption and fluorescence. Just as in gas chromatography, a mass spectrometer coordinates with the HPLC to help identify the components of the mixture.

3.14.5 Ion Exchange Chromatography

Ion exchange chromatography is another type of chromatography widely used in the separation and purification of biological material. Ion exchange chromatography operates on the principle that charged molecules in a liquid phase pass through a column where the stationary phase has a charge opposite that of the molecules in the liquid. Due to the high affinity of charged molecules for opposite charges, this method is highly selective. It requires a solution with a specific pH or ionic strength before molecules will elute from the column. Ion exchange chromatography is used in the separation of both proteins and amino acids.

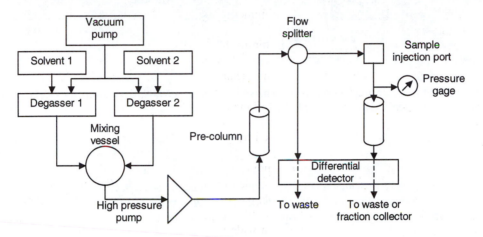

Figure 3.18 A schematic diagram of a high-performance liquid chromatography instrument.

3.15 Electrophoresis

Electrophoresis is a technique used to separate charged molecules in a liquid medium with an electric field. It is extensively used in separation of serum proteins, separation of proteins in urine, determination of molecular weight of proteins, DNA sequencing, genetic disease diagnosis, and comparison of DNA sequences in forensic testing.

3.15.1 Components of an Electrophoretic System

An electrophoretic system consists of a power supply, a chamber containing a support medium, and an anode and cathode made of carbon or a metal, such as platinum (Figure 3.19). Under the influence of an electric field, negatively charged anions in a liquid medium migrate to the anode (+). Positively charged cations migrate to the cathode (−). Velocity depends on their charge, and several factors influence the distance and rate of migration. Electrophoresis works when a charged molecule is placed in an electric field, and the molecule sustains a force, which is proportional to the charge of the molecule and the strength of the electric field.

native = size + charge
SDS = size
proteins

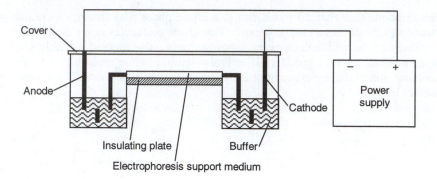

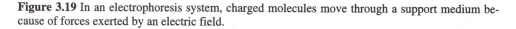

Figure 3.19 In an electrophoresis system, charged molecules move through a support medium because of forces exerted by an electric field.

Power Supply

The power supply provides a constant voltage, current, or power across the support medium. Most power supplies for electrophoresis systems provide between 50 and 200 V. The heat introduced into the electrophoretic system because of the resistance of the support medium is

$$\text{Heat (energy)} = Pt = VIt \tag{3.34}$$

where V is voltage, I is current, and t is time. There is a difference between an electrophoretic system that uses a constant voltage supply and one that uses a constant current supply. In a system with a constant voltage supply, the migration rate increases with time because the resultant current generates heat, resulting in thermal agitation of the dissolved ions. This causes a decrease in the resistance of the medium, further increasing the current, which generates more heat and evaporates water quickly. Water loss from the system causes an increase in the ion concentration of the medium, which also lowers the resistance and thus increases the migration rate of the different solutes. Migration rates that are too fast or that change with time are undesirable because of decreased resolution of the solutes in the gel. Therefore, it is more desirable to have a constant current through the gel than a constant voltage. Present-day systems regulate the voltage in an attempt to maintain constant current. Other systems regulate temperature in order to prevent the gel from melting.

Support Medium

One of the most commonly used support mediums is agar. Agar is so prevalent in clinical labs that electrophoresis is often referred to as agarose gel electrophoresis. Clinically, electrophoresis is performed on a plastic support covered with a thin (1 mm) layer of agarose gel. Agarose has the following advantages: After electrophoresis is performed and the plate is dried, the agarose is very clear, which enables a densitometer to easily examine it. Agarose absorbs very little water because it contains very few ionizable groups.

Another commonly used support medium is cellulose acetate. Cellulose acetate membranes are produced by reacting acetic anhydride with cellulose. These membranes consist of about 80% air in pockets in-between interconnected cellulose acetate fibers. These air pockets fill with liquid when the membranes are placed in a buffer. Cellulose acetate electrophoresis (CAE) has the advantage of being relatively quick (20 to 60 min). CAE membranes can be made clear for densitometry using a solution that dissolves the cellulose acetate fibers.

Electrodes and Chamber

Electrodes may be made of metal such as platinum, or of carbon. Gel trays may be made from UV transparent acrylic to enable direct observation of the migration of different molecules.

3.15.2 Electrophoretic Mobility

Several factors cause different migration rates among the molecules in a mixture. They are: the net charge of the molecule, q; the size, shape and mass of the molecule; the strength of the electric field, E; the viscosity of the medium; and the temperature of the medium. A molecule in the medium experiences a force $F_E = qE$ in the direction of the electric field that is proportional to the strength of the electric field, E, and the net charge of the molecule, q. The molecule also experiences a drag force $F_D = fv = 6\pi\eta rv$ in the opposite direction of movement which is proportional to the molecule's resistance f, velocity, v, the viscosity, η, of the buffer solution in which it is migrating, and the ionic radius, r, of the solute molecule, Q.

$$F_E = F_D \tag{3.35}$$

$$qE = fv$$

$$v_{ss} = \frac{qE}{f} \tag{3.36}$$

The mobility of a charged particle is defined as its steady state velocity divided by the strength of the electric field.

$$\text{electrophoretic mobility } \mu = \frac{v_{ss}}{E} = \frac{q}{f} \tag{3.37}$$

One mobility unit is defined as 10^{-5} cm^2/(V s).

Analysis and Quantification

Electrophoresis plates are usually dried before analysis to prevent diffusion of the migrated solutes. Most proteins, nucleic acids, and other molecules require staining in order to be seen. Cameras are available that can photograph the electrophoresis gel, using filters specific for different stains. A commonly used stain for visualizing DNA is ethidium bromide, which is fluorescent under a light known as a Woods lamp. The bands appearing in the gel, or on the plate, can be identified by inspection under the Woods lamp, but a densitometer is required for accurate quantification.

Densitometry is commonly used in clinical laboratories to measure the transmittance of light through a solid electrophoretic sample, similar to how a spectrophotometer operates. An electrophoresis plate is placed over the exit slit of the monochromator of the densitometer. The plate or sample is then slowly moved past the exit slit while measuring the amount of transmitted light. As the sample specimen moves past the exit slit, the location and absorbance of each band of the electrophoretic sample are recorded on a chart.

Serum Protein Electrophoresis

Electrophoresis is used to separate the proteins in blood serum. Figures 3.20 and 3.21 show that the first peak is albumin. It is produced in the liver and performs many tasks. For example, it thickens the blood, binds steroids, and binds carbon dioxide. The other three peaks are referred to as alpha, beta, and gamma. The proteins that make up gamma are referred to as gamma globulins. The most prevalent gamma globulins are antibodies (the term antibodies and immunoglobulins are synonymous since antibodies are part of the immune system). Various diseases can be diagnosed with the information contained in these peaks. For example, low albumin can be a sign of liver disease. The gamma globulin peak in Figure 3.21 is a spread of many types of antibodies as a result of a B-lymphocyte tumor (bone tumor).

3.16 DNA Sequencing

There are 20 standard amino acids that make up all proteins. Each has a carboxyl group and an amino group bonded to an α carbon. They are distinguished by their different side chains (R groups), and can be separated, identified, and quantified by ion-exchange chromatography. In high-performance liquid chromatography (HPLC), 0.4 µm spherical cation–exchange resin is packed into a cylindrical column about 5 mm in diameter and 200 mm long. Amino acids in a buffer solution are pumped through the column at about 13 MPa for about 40 min. Amino acids with the least positive charge bind weakly to the resin, move most rapidly through the column and elute first to be spectrophotometrically detected at wavelengths less than 300 nm. Amino acids with a more positive charge bind more tightly to the resin and elute later. The area under the peak of each amino acid on the chromatogram is proportional to the amount of that amino acid (Lehninger et al., 1993).

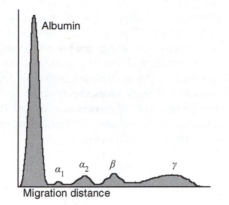
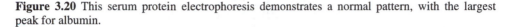

Figure 3.20 This serum protein electrophoresis demonstrates a normal pattern, with the largest peak for albumin.

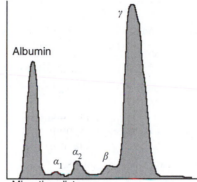

Figure 3.21 This serum protein electrophoresis demonstrates a decrease in the albumin and an increase in gamma globulins.

Amino acids can be covalently joined by peptide bonds to form polypeptides, which generally have molecular weights less than 10,000. Peptides are obtained by purification from tissue, by genetic engineering, or by direct chemical synthesis. Within the cell, amino acids are synthesized into a longer polypeptide sequence (a protein) through the translation of information encoded in messenger RNA by an RNA–protein complex called a ribosome.

To separate and purify proteins, cells are broken open to yield a crude extract. Differential centrifugation may yield subcellular fractions. Ion-exchange chromatography can separate proteins by charge in the same way it separates amino acids. Size-exclusion chromatography separates by size. Affinity chromatography separates by binding specificity to a ligand specific for the protein of interest. The purified protein is characterized by ion-exchange chromatography to measure the amount of the protein of interest and the contaminants.

A nucleotide consists of a nitrogenous base, a pentose sugar, and one or more phosphate groups. The nucleic acids RNA and DNA are polymers of nucleotides. The genetic code can be determined by sequencing the four nucleotides that form DNA: A = Adenine, C = Cytosine, G = Guanine, and T = Thymine. A dideoxynucleoside phosphate (ddNTP) analog specific for one of the nucleotides interrupts DNA synthesis to prematurely terminate the fragment at that nucleotide, for example A. Different analogs terminate C, G, and T fragments. When each of these radiolabeled (with a radioactive compound) fragments is separated electrophoretically, it yields the autoradiogram (by darkening photographic film) pattern in Figure 3.22.

	A	C	G	T
G			•	
T				•
G			•	
T				•
C		•		
A	•			

Figure 3.22 Nucleotide fragments ending in A, C, G, and T are injected into lanes at the top of electrophoresis columns. The sequence is read from the rows of bands from the bottom up as ACTGTG.

DNA sequencing is automated by labeling each of the four fragments with a different colored fluorescent tag. Then all four fragments are combined and analyzed by a single lane during electrophoresis. The eluting peaks from successively longer fragments are scanned by a 40 mW UV laser beam, which causes them to fluoresce with one of the four colors. A computer displays the colored peaks and sequence. Electrophoresis is carried out in 64 lanes of 200 μm thick acrylamide denatured gel between 300×200 mm optical grade glass for 7 h at 1700 V and 3 mA.

3.17 Problems

3.1   Give the equation for Beer's law, define each term, and give units.
3.2   Explain the operation of a spectrophotometer and its purpose. List the components of a spectrophotometer.
3.3   A sample concentration of 10 mg/dL yields a spectrophotometer transmission of 35%. Assume Beer's law holds and calculate the unknown concentration for a transmission of 70%.
3.4   A sample of concentration 20 mg/dL has an absorbance of 0.4 in a spectrophotometer. The sample is then diluted and yields an absorbance of 0.25. Calculate the new concentration.
3.5   Define oxygen saturation and state the physiological meaning of SaO_2.
3.6   Search the literature for a plot of SaO_2 versus PO_2 and sketch it.
3.7   Describe how NADH is used to measure lactate concentration and why lactate concentration isn't ascertained by measuring lactate directly.

3.8   Describe why creatinine is measured and the technique used to measure it.

3.9   Describe why and how to measure urea and the body fluids it can be measured in.

3.10   Describe why and how to measure glucose from a drop of blood.

3.11   Describe how to measure glucose in automated equipment.

3.12   Describe amperometry as used in the PO_2 electrode.

3.13   Describe the most common enzymatic electrode method for measuring glucose.

3.14   Calculate the pH for a hydrogen ion concentration of 10^{-7} mol/L.

3.15   Draw a pH electrode and explain its principle of operation. Explain why its amplifier input impedance is important. Explain the relation of the CO_2 electrode to the pH electrode.

3.16   Explain the principle of operation and give an example of use for flame photometry.

3.17   Explain the principle of operation and give an example of use for mass spectrometry.

3.18   Explain why and how CO_2 is measured by infrared transmission spectroscopy.

3.19   Explain why and how N_2 is measured by emission spectroscopy.

3.20   Explain why and how fluorometry is used. Describe one of the advantages of fluorometry.

3.21   Explain why and how chromatography is used. Explain the two principles that are the primary factors affecting interactions in chromatography.

3.22   Explain how the glucose sensor minimzes sensitivity to PO_2 variations.

3.23   Explain why and how electrophoresis is used. List the factors that cause differential migration rates among the component molecules of a mixture.

3.24   Explain how protein purity is measured.

3.25   Explain how the DNA code is determined.

3.18 References

Antolasic, F. 1996. *What is Mass Spectrometry?*
      [Online] minyos.its.rmit.edu.au/~rcmfa/mstheory.html,

Berger, S. A. 1996. *Introduction to Bioengineering.* New York: Oxford University Press.

Blum, L. J. 1991. *Biosensor Principles and Applications.* New York: Marcel Dekker.

Bowers, L. D., Ullman, M. D. and Burtis, C. A. 1994. Chromatography. In C. A. Burtis and E. R. Ashwood (eds.), *Tietz Textbook of Clinical Chemistry.* 2nd ed. Philadephia: W. B. Saunders.

Burtis, C. A. and Ashwood, E. R. (eds.) 1994. *Tietz Textbook of Clinical Chemistry.* 2nd ed. Philadelphia: W. B. Saunders.

Canh, T. M. 1993. *Biosensors.* London: Chapman & Hall.

Carrier, R. and Bordonaro, J. 1994. Chromatography. [Online] www.rpi.edu/dept/chem-eng/Biotech-Environ/CHROMO/chromequip.html.

Considine, D. M., and Considine, G. D. (eds.) 1995. *Van Nostrand's Scientific Encyclopedia.* New York: Van Nostrand Reinhold.

de la Peña, A., Liu, P., and Derendorf, H. 2000. Microdialysis in peripheral tissues. *Advanced Drug Delivery Reviews* **45**: 189–216.

DM Scientific 2000. DS-34 Camera for Electrophoresis Gel Recording by Polaroid. [Online] www.dmscientific.com/ds34.html.

Hall, E. A. H. 1991. *Biosensors.* Englewood Cliffs, NJ: Prentice-Hall.

Hicks, M. R., Haven, M. C., Schenken, J. R., and McWhorter, C. A. 1987. *Laboratory Instrumentation.* 3rd ed. Philadelphia: J. B. Lippincott.

Haven, M. C., Schenken, J. R., Tetrayktm J. R. 1995. *Laboratory Instrumentation.* 4th ed. New York: Van Nostrand Reinhold.

IQ Scientific Instruments. 2001. The ultimate source for non-glass ISFET technology pH systems [Online] www.phmeters.com.

Kurnik, R. T., Berner, B., Tamada, J., and Potts, R. O. 1998. Design and simulation of a reverse iontophoretic glucose monitoring device. *J. Electrochem. Soc.* **145**: 4119–4125. [Online] www.cygn.com/monitor/journals.html.

Lehninger, A. L., Nelson, D. L., and Cox, M. M. 1993. *Principles of Biochemistry.* 2nd ed. New York: Worth Publishers.

Mallinckrodt. 2003. OxiFirst™ Fetal Oxygen Saturation Monitoring System [Online] www.mallinckrodt.com/respiratory/resp/.

Melcher, U. 2003. Gel Electrophoresis. *Molecular Genetics.* [Online] opbs.okstate.edu/~melcher/MG/MGW4/MG422.html.

Narayanan, S. 1989. *Principles and Applications of Laboratory Instrumentation.* Chicago: American Society of Clinical Pathologists.

Pagana, K. D., and Pagana, T. J. 1995. *Mosby's Diagnostic and Laboratory Test Reference.* 2nd ed. St. Louis: Mosby.

Peura, R. A. 1998. Chemical biosensors. In J. G. Webster (ed.) *Medical Instrumentation: Application and Design.* 3rd ed. New York: John Wiley & Sons.

Primiano, F. P. Jr. 1998. Measurements of the respiratory system. In J. G. Webster (ed.) *Medical Instrumentation: Application and Design.* 3rd ed. New York: John Wiley & Sons.

Taylor, R. F., and Schultz, J. S. 1996. *Handbook of Chemical and Biological Sensors.* Bristol UK: IOP Publishing.

Tissue, B. M. 2000. High Performance Liquid Chromatography. [Online] www.chem.vt.edu/chem-ed/scidex.html.

Tissue, B. M. 2000. Fluorimetry. [Online] www.chem.vt.edu/chem-ed/scidex.html.

Tissue, B. M. 2000. Infrared Absorption Spectroscopy (IR). [Online] www.chem.vt.edu/chem-ed/scidex.html.

Tissue, B. M. 2000. Mass Spectroscopy (MS). [Online] www.chem.vt.edu/chem-ed/scidex.html.

Webster, J. G. (ed.) 1997. *Design of Pulse Oximeters.* Bristol, UK: IOP Publishing.

Webster, J. G. (ed.) 1988. *Encyclopedia of Medical Devices and Instrumentation.* New York: John Wiley & Sons.

Wheeler, L. A. 1998. Clinical laboratory instrumentation. In J. G. Webster (ed.) *Medical Instrumentation: Application and Design.* 3rd ed. New York: John Wiley & Sons.

Chapter 4

Surface Characterization in Biomaterials and Tissue Engineering

Jorge E. Monzon

This chapter discusses measurements of molecular variables needed for research and development in the rapidly growing areas of biomaterials and tissue engineering.

4.1 Molecules and Biomaterials

The following sections present some biomaterials fundamentals needed to understand the importance of measuring certain molecular properties.

4.1.1 Integrated Approach

Over the years, a number of different types of materials have been used in medical practice for treating wounds and diseases or cosmetic purposes.

A *biomaterial* is a synthetic material used to replace part of a living system or to function in intimate contact with living tissue (Park and Lakes, 1992). This definition suggests the development of materials for biological use is accomplished only through an integrated approach of several disciplines: science and engineering, especially material sciences for detailed study and testing of the structural properties and phenomena of materials; biology and physiology, to provide the necessary tools for experimentation based on immunology, anatomy, cellular and molecular biology; clinical sciences, as many medical specialties are directly related to the actual use of biomaterials (Park, 1995).

A device fabricated with a biomaterial that is placed inside the living system is called an *implant*.

4.1.2 Types of Biomaterials

Materials for biological use are classified according to their base structure as ceramics, composites, metals, and polymers. Table 4.1 summarizes different types of biomaterials and their most common applications (Silver, 1994; Park and Lakes, 1992). This listing is not comprehensive, as the search for new materials and their clinical applications is a dynamic process. Synthetic polymers constitute the vast majority of biomaterials used in humans (Marchant and Wang, 1994).

Table 4.1 Classification of biomaterials in terms of their base structure and some of their most common applications.

Biomaterials	Applications

Ceramics
  Aluminum oxide
  Carbon
  Hydroxyapatite
	Dental and orthopedic

Composites
  Carbon-carbon fibers
  and matrices
	Heart valves and joint implants

Metals
  Atoms
    Aluminum
    Chrome
    Cobalt
    Gold
    Iridium
    Iron
    Manganese
    Molybdenum
    Nickel
    Niobium
    Palladium
    Platinum
    Tantalum
    Titanium
    Tungsten
    Vanadium
    Zirconium
  Metallic alloys
    wide variety using metallic atoms
	Joint replacement components,
	  fracture fixation, dental implants,
	  pacemakers, suture wires,
	  implantable electrodes

Polymers
  Nylon
  Synthetic rubber
  Crystalline polymers
	Replacement of soft tissues: skin,
	  blood vessels, cartilage, ocular lens,
	  sutures
	Orthopedic

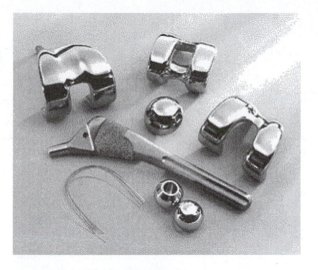

Figure 4.1 These titanium–alloy joint replacements are an example of the many applications in metal biomaterials for implantations (from www.spirebiomedical.com/).

4.1.3 Polymers

Structure

A polymer is characterized by a repeating subunit (monomer) covalently connected to form a macromolecule.

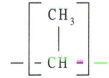

Figure 4.2 Polymers are made up of many monomers. This is the monomer for poly(ethylene), a common biomaterial used for medical tubing and many other applications.

The average number of repeating units is called the degree of polymerization (Lee et al., 1995). A larger number of these subunits (i.e. bigger molecules) leads to a polymer of higher molecular weight, and thus to a more rigid material (Park and Lakes, 1992).

Synthetic polymers are obtained through two basic chemical processes: addition polymerization and condensation polymerization. Some addition polymers are poly(ethylene), poly(methyl methacrylate), poly(vinyl chloride), and poly(ethylene-terephtalate). Condensation polymers include poly(esters), poly(amides), and poly(urethanes) (Marchant and Wang, 1994; Park and Lakes, 1992; Silver, 1994).

Complex chemical processes beyond the scope of this book allowed the development of polymers of very different physical and chemical properties to find assorted applications in health care. Table 4.2 illustrates the most common polymers and their clinical uses.

Table 4.2 The clinical uses of some of the most common biomedical polymers relate to their chemical structure and physical properties.

Biomedical polymer	Application
Poly(ethylene) (PE)	
Low density (LDPE)	Bags, tubing
High density (HDPE)	Nonwoven fabric, catheter
Ultra high molecular weight (UHMWPE)	Orthopedic and facial implants
Poly(methyl methacrylate) (PMMA)	Intraocular lens, dentures, bone cement
Poly(vinyl chloride) (PVC)	Blood bags, catheters, cannulae
Poly(ethylene terephthalate) (PET)	Artificial vascular graft, sutures, heart valves
Poly(esters)	Bioresorbable sutures, surgical products, controlled drug release
Poly(amides) (Nylons)	Catheters, sutures
Poly(urethanes) (PU)	Coat implants, film, tubing

Properties

Biomaterials for implants should be nontoxic, noncarcinogenic, nonallergenic, functional for its lifetime, and biocompatible (Billotte, 2000).

Polymeric biomaterials exhibit reasonable cost of processing and can be manufactured with various shapes having desired mechanical and physical properties. In addition, they can be sterilized through different methods (Lee et al., 1995).

However, polymers may lose some of their properties with time, a process called degradation, which is the consequence of chemical, mechanical, photo, or thermal factors.

Biocompatibility

The most important requirement for an implant is its *biocompatibility*; that is, the acceptance of an artificial device by the surrounding tissues and the body as a whole (Park,

1995). Greisler et al. (1996) suggest that biocompatibility does not mean lack of biological response—as might be expected—but rather the presence of a desirable set of biological responses.

Bulk and surface properties of polymer biomaterials, which are functions of their molecular structure and organization, determine their interaction with living organisms (Marchant and Wang, 1994).

Figure 4.3 This artificial heart valve is coated with a biocompatible material that allows the body to accept the implant (from www.sjm.com/devices/).

4.2 Molecules and Tissue Engineering

As is the case with biomaterials, molecular phenomena also play an important role in tissue engineering research. We next present some basic aspects of this engineering field.

4.2.1 Tissue Engineering

Tissue engineering can be defined as the application of engineering and life sciences toward the design, growth, and maintenance of living tissues (Berthiaume et al., 1994).

This discipline uses living cells and their extracellular products to develop more biological replacements instead of using only inert implants (Skalak et al., 1988). Tissue engineering activities can be performed either *in vivo* (in the body) or *in vitro* (in solution), the latter being of more interest to bioengineers.

Bioartificial tissues constructed *in vitro*, are composed of biologic and synthetic substances, and represent a valid alternative to organ transplantation in cases of impaired natural organ function, vital organ failure, or organ donor unavailability. Although many bioartificial tissues are still in a developmental stage, it is expected that in the future bioartificial organs and tissues will be able to regenerate and perform complex biochemical functions, a behavior that pure artificial implants cannot exhibit that is similar to the natural organs and tissue they will replace (Berthiaume et al., 1994).

4.2.2 Cellular Composites

Tissues are multiphase systems of cellular composites. We can identify three main structural components: cells organized into functional units, the extracellular matrix, and scaffolding architecture (Wintermantel et al., 1996).

The three-dimensional structures that characterize bioartificial tissues represent a cell mass that is greater, by several orders of magnitude, than the two-dimensional cell cultures traditionally developed by biologists (Berthiaume et al., 1994). Tissue engineering also concentrates on the development of *in vitro* models that overcome the absence of an adequate extracellular matrix in conventional cell cultures (Sittinger et al., 1996).

Table 4.3 shows some current examples of applications and *in vitro* research in tissue engineering (Berthiaume et al., 1994; Palsson and Hubbell, 1995; Skalak and Fox, 1988).

Table 4.3 Some examples of current applications in tissue engineering. Not all of the listed applications are at the same developmental stage.

Biological system	Example of application
Blood	Hematopoietic (production of red blood cells by) stem cells culture
Cardiovascular	Endothelialized synthetic vascular grafts (angiogenesis) Regeneration of the arterial wall Compliant vascular prostheses
Liver and pancreas	Bioartificial pancreatic islets Bioartificial liver
Musculoskeletal	Cartilage reconstruction Bone reconstruction
Neural	Neurotransmitter-secreting cells (polymer-encapsulated) Neural circuits and biosensors Peripheral nerve regeneration
Skin	Bioartificial skin substitutes

Of the structural components of tissues mentioned earlier, the scaffolding architecture is, for tissue engineers, a synthetic biomaterial. The interaction of cells with the biomaterial surface is a main topic in tissue engineering.

The characteristics of the synthetic material directly affect the recruitment, attachment, proliferation and differentiation of living cells, defining the quality of the new tissue. The behavior of molecules from the surrounding tissue and on the surface of the biomaterial determines the conditions of the interface to which cells respond (Boyan et al., 1996).

After our review of the main aspects of biomaterials and tissue engineering, it should be clear that measurements at the molecular scale constitute a key activity that must be routinely performed by biomaterials and tissue engineers.

In the next sections we discuss some of the most important molecular variables and present the necessary tools for quantitative study.

4.3 Surface Analysis

The successful use of a synthetic polymer—either as a biomaterial implant or as a scaffolding architecture for tissue growth—relies on the qualitative and quantitative knowledge of several properties of such a polymer.

Numerous physical and chemical variables arising at a microscopic scale eventually determine the overall quality of the interaction of synthetic polymers and living organisms. It is not possible to determine which of these variables prevails over the others, so researchers are continuously designing instruments, procedures and techniques to better understand these variables, and for their relation to the overall behavior of material implants.

Surface analysis of biomaterials is a set of procedures to evaluate the characteristics and properties of biomaterials at the microscopic level.

A complete description of the chemistry and morphology of a surface cannot be accomplished using one single technique. Surface characterization is based upon partial descriptions achieved from different methods (Castner and Ratner, 1988).

Basic characterization using microscopy, the study of the chemical composition by means of spectroscopy, and the evaluation of the polymer–water interface using contact angle methods, are the main variables and procedures currently used in polymer surface analysis (Andrade, 1985a).

This section reviews the most common variables of interest and some widely used instruments to measure them. References describe the principles governing the procedures and instruments that are presented.

Due to its importance for biomaterials and tissue engineering, we discuss protein adsorption—a mechanism related to surface behavior—in a separate section.

4.3.1 Topography

Andrade (1985c) points out that the very first characterization to be performed on any surface is visual inspection. Some properties such as color or reflectivity can be observed with the naked eye. Microscopes are needed for observing other qualities such as roughness, surface texture, or topography (or microtopography on the molecular scale).

Major developments in topographic studies of synthetic and biologic materials were achieved after the invention of the electron microscope. Next we discuss the different types of electron microscopes that are used for observation of surface topographies.

Transmission Electron Microscope (TEM)

This is the classical *electron microscope*. Developed several decades ago, TEM made possible a relevant increase in magnification compared to visible light optical microscopes. Instead of using a beam of light as the image-forming radiation, TEM relies on an electron beam that passes through the sample in a vacuum medium to develop a transmitted image of the specimen.

Resolution and Magnification: The resolving power, or resolution, of a microscope is its ability to discriminate between two closely placed structures. Magnification measures the increase in the diameter of a final image compared to the original. However, higher magnification does not necessarily mean better resolution. Useful magnification is calculated as the ratio of the resolution of the human eye to the resolution of the lens systems (Bozzola and Russell, 1991).

The resolving power of a microscope is a function of the wavelength, λ, of the source of illumination. For an electron, the wavelength is given by de Broglie's equation:

$$\lambda = \frac{h}{mv} \tag{4.1}$$

where h is Planck's constant, m is the mass of the electron, and v is the velocity. Better resolution is achieved by using lower wavelengths. It is therefore possible to increase the resolving power of the microscope by increasing the electron velocity, thus lowering λ. Electrons accelerated in an electric field will reach a velocity

$$v = \sqrt{\frac{2eV}{m}} \tag{4.2}$$

where V is the accelerating voltage, e is the charge of the electron, and m is its mass. By substitution in Eq. (4.1), and replacing the constants by their values (see Appendix), we obtain

$$\lambda = \frac{1.22}{\sqrt{V}} \, \text{nm} \tag{4.3}$$

which shows that short wavelengths can be obtained by using high voltages (on the order of kilovolts). Equation 4.3 is not useful for calculating the resolution because lens aberrations, refractive index of the medium, and aperture angles limit the resolution. However, it is possible to demonstrate that the highest practical resolution of TEM is 0.4 to 1 nm (4 to 10 Å), with a practical magnification of 100,000 to 200,000.

Operation: Figure 4.4(a) shows the basic components of a TEM. Not shown is the overall vacuum system for proper operation. The simplest electron source is a heated, pointed tungsten wire, although higher resolution microscopes are equipped with a lanthanum hexaboride (LaB_6) cathode to which typically voltages of 40 to 120 kV are applied (Dykstra, 1992).

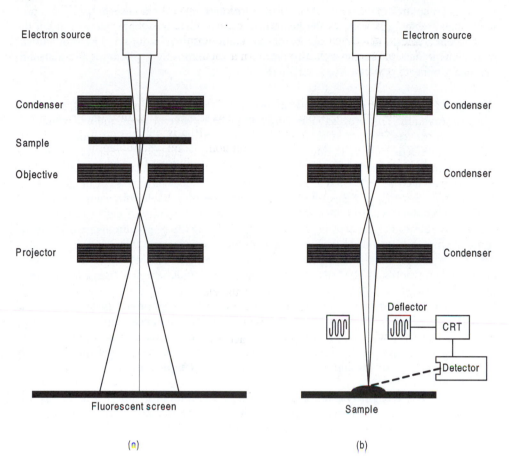

(a)   (b)

Figure 4.4 (a) TEM microscope. The electron beam passes through the sample, generating on the fluorescent screen a projected image of the sample, which can be recorded by photographic means. (b) SEM microscope. Condenser lenses focus the electron beam on the specimen surface leading to secondary electron emission that is captured by the detector and visualized on the CRT screen. Both TEM and SEM operate in a particle free (vacuum) environment.

A set of condenser lenses focuses electrons onto the specimen on the area under examination. Lenses for an electron microscope are not made of glass as for optical microscopes, rather they are magnetic coils (solenoids) that can bend the electron path (Bozzola and Russell, 1991).

Below the sample plane, the objective lens assembly focuses and magnifies the specimen image. By changing the aperture of the objective lens it is possible to control the specimen contrast and correct astigmatism. Scattered elements of the electron beam emerging from the sample are also eliminated by the objective lens system (Dykstra, 1992).

To control the magnification of the projected image, it is necessary to use a projector lens, which also focuses the beam of electrons for appropriate intensity upon the fluorescent screen. This screen can be observed through a viewing window (not shown) or the image can be photographically reproduced by exposing a photographic plate immediately beneath the fluorescent screen (Packer, 1967).

Specimen Preparation: A meticulous preparation of the sample is required for suitable TEM observation. This process is lengthy and involves several steps, most of which are chemical processes: fixation, washing, dehydration, infiltration with transitional solvents and resins, embedding, and curing. Once the specimen is chemically prepared it must be cut into extremely thin slices or sections (from 30 to 60 nm) in a procedure called *ultra-microtomy*, which is performed to allow the beams of electrons to pass through the sample material (Bozzola and Russell, 1991). Preparation of suitable thin samples for TEM studies in biomaterials is hard to accomplish due to the difficulty of performing ultramicrotomy on elastomeric materials, such as some polymers (e.g. polyurethanes). There are some alternative methods of specimen preparation that overcome this difficulty. Another problem with polymers is their low contrast. However, this can be overcome by using a defocused conventional TEM; a defocused beam also avoids sample damage and destruction (Goodman et al., 1988).

One major disadvantage of TEM is its limitation for a proper three-dimensional view (Dykstra, 1992). To obtain information on the surface and near surface morphology using TEM, the viewing of the sample must be done at several angles. Multiple views permit reconstruction of the three-dimensional structure; however, angle tilt is severely limited by the loss of resolution, especially with thick samples needing large tilt angles (Goodman et al., 1988).

For tissue engineers, TEM represents a powerful tool for studying the intrastructural features of the soft tissue–biomaterials interface, particularly surface related phenomena such as cell adhesion and biomaterial degradation (Sheffield and Matlaga, 1986).

Scanning Electron Microscope (SEM)

SEMs and TEMs have many features in common, primarily they both use electron beams to visualize a sample. However, they differ in the way the beam interacts with the specimen. The principle of SEM operation makes it very useful for topographic analysis, as we will see in the next paragraphs.

Principle: SEM is based upon the interaction of an electron beam with a specimen. The incident beam (primary electrons) displaces orbital electrons from the sample atoms, giving rise to secondary electron emission (Figure 4.5).

The number of secondary electrons emitted depends on the specimen's electron density, which is related to its elemental composition, so suitable detection of this secondary emission yields information on the atoms of the surface sample. Some primary electrons may not collide with orbital electrons. Instead they pass by the nucleus to become backscattered electrons (Dykstra, 1992).

These are not the only emissions arising at the atomic level, as we explain in Section 4.3.2 (see Figure 4.10), but these two are the ones of interest for SEM operations.

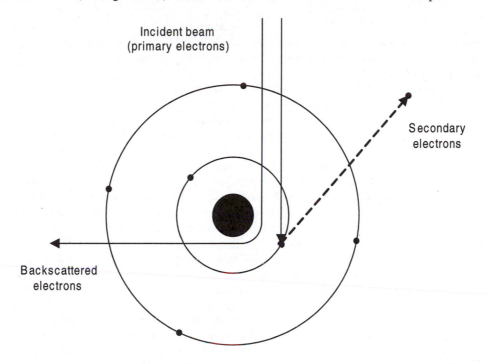

Figure 4.5 Principle of SEM operation. An incident beam of primary electrons displaces orbital electrons from the sample atoms, resulting in secondary electron emission, which is detected for image formation. Some primary electrons pass by the nucleus to become backscattered electrons.

Operations: Figure 4.4(b) shows a schematic diagram of a scanning electron microscope; note the similarities with the TEM in Figure 4.4(a). Some components of the SEM are identical to those of the TEM, such as a vacuum system for operation in a particle-free environment. Both microscopes use the same electron source, except SEM accelerating voltages are limited to around 40 kV. The TEM contains magnetic lenses of various types, while in the SEM they are all condensers. The purpose of the condenser lenses in the SEM is to demagnify the electron beam from approximately 50 μm at the electron gun down to about 2 nm or less near the specimen as small spot sizes are essential for high resolution (Bozzola and Russell, 1991).

Deflector coils bend the primary beam for scanning across the specimen. This high-energy beam produces secondary electrons near the surface of the sample. Some secondary electrons contact a positively charged detector, also called a collector, connected to a cathode ray tube (CRT). The contact generates a point of illumination on the

CRT screen with intensity proportional to the number of secondary electrons from the corresponding point on the specimen. Deflector coils in the SEM connect to the CRT deflector plates so that as the SEM beam scans over the specimen, the CRT gun synchronously sweeps over the screen. This imaging mode is known as SEI (secondary electron imaging) (Dykstra, 1992).

Resolution and Images: Image resolution in SEM is a more complex concept than in TEM because time is an integral part of the image formation process in scanning microscopes (Dykstra, 1992).

Synchronization of beam and CRT plate deflectors account for simultaneous emission of secondary electrons and their imaging on the CRT screen. However, some backscattered electrons, as they leave the atom (Figure 4.5) and just prior to their exit, may interact with electrons some distance from the initial point of entry in the specimen, which produces secondary electron emission. These extraneous secondary electrons produced by backscattered radiation are also detected by the collector and displayed on the CRT screen, giving a blurred or noisy signal. One way of reducing the noise is by increasing the true signal-to-noise ratio. Some backscattered electrons can also be collected by a special detector. By carefully adjusting the condenser lenses it is possible to obtain high resolution with suitable depth of field for working distances (distance from the final condenser to the specimen) of approximately 10 nm (Bozzola and Russell, 1991).

A very attractive feature of the SEM is its ability to generate images that appear to have three dimensions. The positioning of the primary scanning beam (incident angle) and the location of the detector affect the image contrast, therefore giving a sense of three dimensions. Secondary electrons are emitted from various parts of the specimen, but those facing the primary beam and in line of sight with the detector appear bright on the CRT screen, whereas a dark appearance corresponds to areas not reached by the beam and not emitting secondary electrons. Medium brightness represents partial specimen exposure to the beam, detecting fewer secondary electrons. This variation in contrast contributes to the generation of an image that appears to have depth, although that image is visualized on a two-dimensional CRT screen (Bozzola and Russell, 1991).

Sample Preparation: As for the TEM, specimen preparation for SEM viewing is also a lengthy process that must be carefully carried out (Dykstra, 1992).

The three dimensional effect on SEM-generated images is especially useful for topographical studies. Usually low voltage SEM (~5 kV) provides similar surface sensitivity to that obtained with XPS (see Section 4.3.2). The sample can be probed to depths of tenths of micrometers making it possible to convert a chemical structure into topographical relief. In this way, SEM may be used to view a topography that is a function of the structure with very high surface resolution (Goodman et al., 1988).

Mostly dry samples must be stable in the high vacuum environment. Additionally, specimens should be conductive (noncharging), therefore a thin metal coating on the sample is needed (Andrade, 1985c). The SEM's remarkable depth of focus allows the examination of uneven surfaces. The analysis of elemental composition of samples can be achieved by using additional detectors, especially for backscattered radiation (Dociu and Bilge, 1986).

Scanning Tunneling Microscope (STM)

The scanning tunneling microscope (STM), along with the scanning force microscope (SFM) that we discuss later, belongs to the more general group of scanning probe microscopes (SPM or SXM).

Very small sensors (probes) scan the surface of the sample in close proximity. In this way, it is possible to obtain information on surface topography, and on some mechanical and electronic properties at the atomic level (Marti, 1993).

The STM was initially designed to give a clearer view of the atomic nature of the surface of solids. Surface properties derive from atoms not being completely surrounded by other atoms, as they are in the interior of a solid (Dykstra, 1992).

Principle of Operation: Governed by quantum mechanics, STM operation is based upon the electron tunneling effect. This rather complicated function of kinetic and potential energies—which is related to the probability of electrons crossing a gap between two conducting surfaces—can be synthesized in our discussion of the STM by saying that if a voltage is applied between the probe tip shown in Figure 4.6 and the sample surface, a current will develop across the junction (tunneling current).

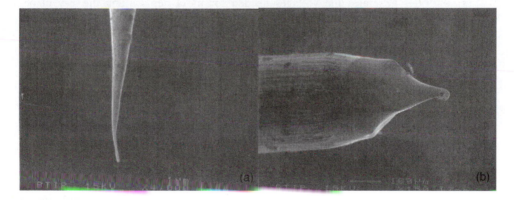

Figure 4.6 (a) An STM probe tip made of tungsten magnified 4,000 times. The tip is very small, and can be ruined on a sample, which is seen in Figure 4.6(b). (from www.orc.soton.ac.uk/~wsb/stm/photos.htm)

The variation in current is due to the fact that a cloud of electrons exists above the specimen surface and the number of electrons decreases exponentially with distance from the surface. Distances of an atomic diameter cause a relevant decrease in the tunneling current, thus allowing precise measurements of the vertical position of atoms at the specimen surface (Dykstra, 1992). Images are formed by monitoring voltages, tunneling currents, and/or probe positions.

STM resolution of is a function of the geometry of the tip, the topography of the sample, and their electronic structures. For large objects, in the micrometer range, the geometry of the tip plays a leading role. However, on an atomic scale, resolution is governed by the stability of the tunnel barrier width (i.e. the distance between tip and sample). Lateral resolution of the STM (as the tip scans the specimen surface) is related to the diameter of the tunneling current filament (Marti, 1993).

Calculation of STM resolution is not a straightforward process. The gap between probe and specimen is typically 1 nm. Stable and precise STMs can measure distance variations of 1 pm, provided the instrument is in a vibration-free environment (Dykstra, 1992).

STM Instrument: Figure 4.7 shows piezotranslators used to move the sample relative to the tip rather than vice versa. The piezoelectric effect is the mechanical expansion and contraction of a material in response to an electric field. In this way electric voltages are used to generate a three-dimensional (3-D) movement of the piezoscanner holding the probe. A widely used piezoscanner made of ceramic material is the piezotube, which has inner and outer electrodes for movements along the three Cartesian axes x, y, and z. The distance between the tip and the sample is kept at 1 nm. The sample is connected to a tunnel voltage source. The tunneling current from the sample to the tip is fed into a current-to-voltage converter. The output of the I/V converter is further processed at the voltage processor block for display. This voltage is also fed back to the system via the scanner voltage block, which provides the necessary voltages for movements along the z axis. A scan controller generates the voltages needed for movements along the x–y plane. Visualization and interpretation of the data provided by the STM must be accomplished using image processing methods and algorithms (Marti, 1993).

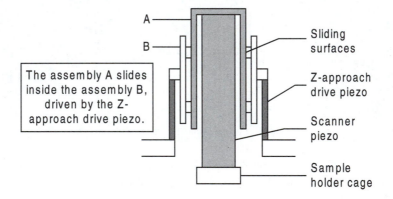

Figure 4.7 This is a sample of a piezotube. There are different approaches, but all use the same method of two opposing piezoelectric materials to move the sample in each axis. (from http://www.topac.com/).

Modes of Operation: Constant current and *constant height* are the two modes of STM operation. In the constant current mode, very high-gain voltage amplifiers are used to keep the tunneling current constant and force the tip to follow the contours. By recording the input of the scanner voltage block (in fact, the z voltage amplifier) as a function of x and y, it is possible to determine the topography of the specimen surface. Any suitable recording device can be used for such a purpose: chart recorder, oscilloscope, or computer. The constant height mode of operation is appropriate only for flat surfaces. In this mode the tip is held at a constant height and no longer follows the sample topography. The tunneling current is then recorded as a function of x and y (Marti, 1993).

Three disadvantages of the STM are the need for an environment free of vibrations, very stable temperatures, and the requirement for highly conductive samples to develop tunneling currents.

The low cost and simplicity of the STM are attractive, however, careful data acquisition and interpretation are highly recommended (Ratner, 1988). Figure 4.8 shows a block diagram of the STM.

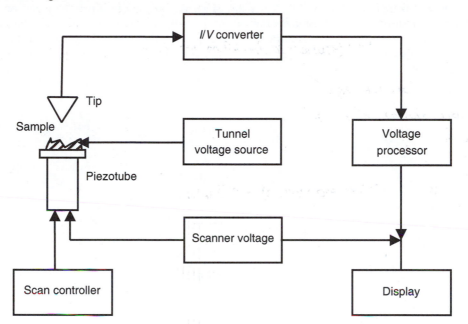

Figure 4.8 STM schematics. The tip of a probe scans the surface of the sample. Three-dimensional movements of the sample under the tip are accomplished using a voltage-controlled piezoscanner. The tunneling current crossing from the sample to the tip is further processed leading to a topographical image.

Scanning Force Microscope (SFM)

The Scanning Force Microscope (SFM), also known as an Atomic Force Microscope (AFM) was developed after the STM and shares many of the same features. Other types of force microscopes are the magnetic force microscope, the attractive mode force microscope, the friction force microscope, and the electronic force microscope. All of these microscopes measure forces generated between a probe and a sample surface (Dykstra, 1992).

Theory of the SFM: As with the STM, operation of the SFM is governed by principles of quantum mechanics. The force between a tip and sample is used to obtain an image of the surface topography. In the absence of other magnetic or electrostatic potential, this small force—on the order of 100 nN—depends upon the interaction potentials between the atoms of the tip and sample (Marti, 1993).

Contact SFMs require the tip to actually touch the sample; these are the most common microscopes, although noncontact SFMs are also available (Bustamante et al., 1996).

Instrument Operation: Figure 4.9 shows the SFM. A sharp probe tip is mounted on a flexible cantilever. As the tip moves toward the sample, two types of forces develop: at large distances the interaction is attractive due to the van der Waals forces; at short distances repulsive forces develop due to the exclusion principle of quantum mechanics (Marti, 1993).

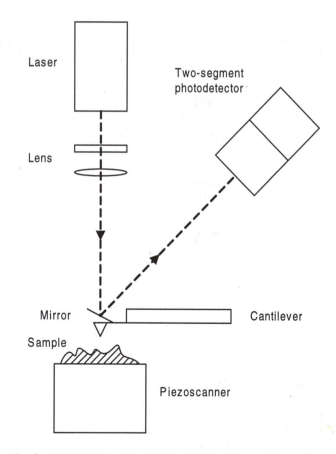

Figure 4.9 Sketch of an SFM. A laser beam is focused on the cantilever, and reflected back to a two-segment photodetector. The difference in output from each segment is proportional to the deflection amplitude of the cantilever scanning the sample.

The attractive and repulsive forces between tip and sample lead to downward and upward deflection of the cantilever, respectively. Although the tip radius is very small (~10 nm), the forces acting on it can deflect the entire cantilever. As the sample scans under the probe, the cantilever is deflected in correspondence with the surface topography (Bustamante et al., 1996).

Several methods are used to detect cantilever deflections less than 1 nm. The most common method is the optical lever, as shown in Figure 4.9. A laser beam focuses on the cantilever and reflects back to the middle of a two-segment photodiode. Any deflection of the cantilever results in an unbalanced number of photons hitting each of the segments, which in turn unbalances the electric currents in the photodiodes. This differential current signal at the output of the photodetector is proportional to the cantilever deflection and thus to the surface topography. The differential current signal is amplified for further processing and also used in a feedback loop (not shown in Figure 4.9) to control the movements of the sample with respect to the tip according to the modes of operation discussed next. The optical lever acts as a motion amplifier, as a 100 pm deflection of the cantilever can easily be detected by the differential photodiode (Bustamante et al., 1996; Marti, 1993).

Modes of Operation: Similar to the STM, piezoelectric elements are used to control the motion of the sample or the tip.

In the *constant force* mode, the contact force between probe and sample is kept constant by an electronic feedback loop that controls the piezoscanner. We said previously that these forces are in the nanonewton range, but due to the action on tiny areas of contact (about 0.1 nm^2), very high pressures develop. As the tip scans the surface, an upward deflection of the cantilever indicates a topographic obstacle and the sample retracts, by means of the feedback loop, to keep the force constant. The movement of the sample correlates to the surface topography. In the *constant height* mode, the cantilever deflects in both directions, and the sample is kept at a constant height. In this mode, direct recording of the cantilever deflection is performed. A combined mode, the *error signal* mode, requires fast feedback response to follow the surface as in the constant force mode, but recordings correspond to cantilever deflection, not to the sample motion. This mode allows enhancement of sharp features (Bustamante et al., 1996).

Thermal drift is not usually a problem using optical detectors, although careful calibration is required to focus and position the laser beam on the cantilever, an operation requiring the help of optical microscopes. In contact SFM (repulsive force operation), some lateral forces between the tip and the sample may develop, giving distorted images and thus reduced resolution. These lateral tensions deform the sample. The magnitude of the lateral forces is a function of the tip dimension and shape, so careful design of the tip is necessary. One possible solution to the unwanted shearing forces is offered by noncontact SFMs. Although very useful for imaging soft samples, they exhibit some mechanical instability due to their attractive force operation (Bustamante et al., 1996; Marti, 1993).

The AFM is widely used in surface analyses of biomaterials. It has great advantages over the STM in that it does not need electrically conducting samples. In addition, the AFM works very easily in liquid (Descouts, 1995).

4.3.2 Chemical Composition

Several properties of a biomaterial are understood by analyzing the chemical composition in order to identify elements and their quantity in a sample. Accurate quantitative information on chemical composition is achieved by means of spectroscopic methods. We present three of these methods.

X-ray Photoelectron Spectroscopy (XPS)

X-ray Photoelectron Spectroscopy (XPS), also known as Electron Spectroscopy for Chemical Analysis (ESCA), is a widely used technique for surface analysis. Some major advantages accounting for its broad acceptance as a key tool include: it is nondestructive, it provides chemical bonding information, it has ease of data interpretation, and—except for hydrogen and helium—it is sensitive to all chemical elements (Andrade, 1985b).

Researchers rely on XPS's high surface specificity, which takes its information from within 10 nm (100 Å) of the surface, because it allows the study of biomaterial interfaces with living tissue. Concentration and chemical state of every element in the sample can be obtained using XPS (Paynter, 1988).

XPS is based on the photoelectric effect, as shown in Figure 4.10. When an X-ray photon interacts with an atomic orbital electron, if the photon carries enough energy to overcome the binding energy of the electron, the electron is liberated from the atom as a photoelectron. The basic equation for this photoemission process and thus for XPS is

$$E_k = h\nu - E_b - \varphi_s \qquad (4.4)$$

where E_k is the electron kinetic energy determined by an analyzer. As we will explain shortly, E_b represents the electron binding energy, $h\nu$ is the incident photon energy (ν is the X-ray frequency and h is Planck's constant) (Andrade, 1985b), and ϕ_s is a work function required for photoemissions from solids—it is the extra energy required to transfer the electron from the surface to the vacuum. The value of ϕ_s is predetermined for each spectrometer.

The interaction of a monochromatic X-ray photon beam with the surface of the sample results in photoelectrons produced with different kinetic energies according to their orbital position in the atom (i.e. different bonding energies). An electron analyzer, also known as an electron spectrometer, measures the number of these electrons. Processing the electrons' kinetic energies yields a spectrum of photoelectron intensity as a function of binding energy.

Figure 4.11 shows a typical XPS spectrum. Distinct peaks correspond to specific atomic orbitals so that binding energy position of each peak allows elemental identification. Tabulations of orbital binding energies are readily available and usually preprogrammed into the data system of the instrument. All of the elements in the periodic table exhibit binding energies in the approximate range of 0 to 1000 eV (electron volts), making positive identification possible. Further quantization is achieved by measuring the area under the peaks (Andrade, 1985b; Paynter, 1988). Many detected electrons lose energy in interactions with the sample, leading to lower kinetic energies and higher binding energies recordings (see Eq. (4.4)) adjacent to every intense peak, giving the spectrum a stair-step shape (Andrade, 1985b).

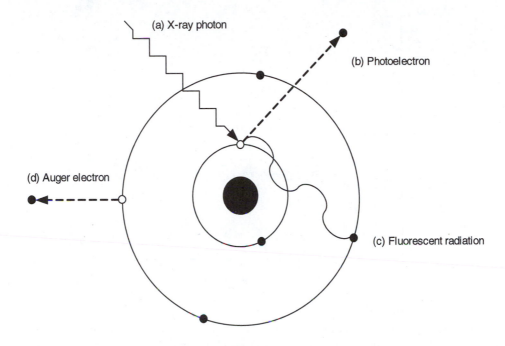

Figure 4.10 When an X-ray photon (a) interacts with an atomic orbital electron of the sample, a photoelectron (b) is emitted verifying Eq. (4.4). The now unstable atom must relax to the ground state. The relaxation process can be accomplished by either of two mechanisms: (1) an outer orbital electron releases energy as fluorescent radiation (c) while occupying the place of the emitted photoelectron, or (2) the excess energy is used to unbind and emit another outer orbital electron called an Auger electron (d). These mechanisms operate for different sample depths, yielding the Auger electron emission characteristic of the outermost surface of the sample.

Figure 4.12 shows the basic schematic of an XPS instrument. The X-ray source provides a monochromatic photon beam that hits the surface of the sample, which is placed on a moving holder for appropriate scanning. The resulting photoelectron radiation is captured by an analyzer. The excitation and emission processes take place in an ultrahigh vacuum (UHV) chamber (not shown in Figure 4.12) for minimal sample contamination and easy passage of the photoelectrons to the analyzer (Paynter, 1988).

The analyzer consists of two concentric and electrostatically charged hemispheres. An electrically biased grid at the entrance of the analyzer retards the kinetic energy of the photoelectrons, which are then spatially dispersed as they initiate their semicircular trajectory between the hemispheres. The voltage across the hemispheres is kept constant (fixed analyzer transmission or FAT mode), therefore the energy resolution is also constant. Ramping voltages in the preretardation stage allows actual kinetic energy scanning with a constant resolution mode, typically ~0.1 to 1.0 eV.

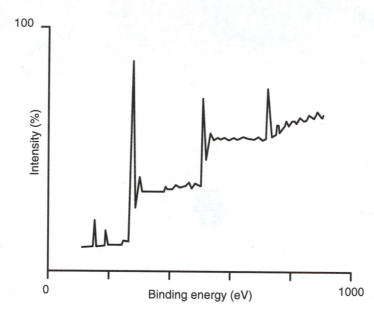

Figure 4.11 A typical XPS spectrum, showing photoelectron intensity as a function of binding energy. Each peak may correspond to a distinct element of the periodic table or to different orbital electrons of the same element. Some peaks may also represent Auger radiation.

At the other end of the analyzer an electron multiplier detector increases by several orders of magnitude the number of electrons reaching the data processing and display sections, which usually consist of a phosphorescent screen and a detector device that counts the number of light pulses arising at the screen (Andrade, 1985b; Paynter, 1988).

The use of an X-ray monochromator in XPS eliminates undesirable radiation coming from the X-ray source (see Figure 4.12). Although monochromated beams minimize the radiation damage to the sample, they also decrease the intensity of the flux to the sample (Andrade, 1985b). The detection system must compensate for this reduced intensity in order to obtain suitable signal levels. This is done by incorporating multichannel photoelectron detectors (Paynter, 1988).

High cost of the instrument, large area of analysis, high vacuum required and low speed (several minutes per peak) are major disadvantages of the XPS technique (Andrade, 1985b). Also the behavior of a biomaterial in a vacuum may differ considerably from its behavior when in contact with living tissues (Paynter, 1988).

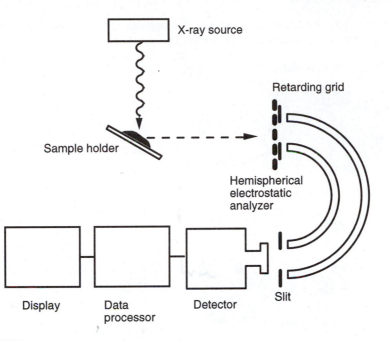

Figure 4.12 Basic schematics of an XPS instrument. An X-ray beam strikes the sample surface, giving photoelectron radiation. These electrons enter the hemispherical analyzer where they are spatially dispersed due to the effects of the retarding grid and of the electrostatic field of the concentric hemispheres. Ramping voltages at the retarding grid allow kinetic energy scanning. At the other end of the analyzer electrons are detected, counted, and displays a spectrum of photoelectron intensity versus binding energy.

Imaging X-ray Photoelectron Spectroscopy (iXPS)

The lack of spatial resolution is an inherent weakness of XPS. A system like ES-CALAB™, shown in Figure 4.13, uses the capability of parallel imaging to produce photoelectron images with spatial resolution better than 5 mm.

The parallel imaging method collects photoelectrons simultaneously from the whole field of view. A combination of lenses before and after the energy analyzer focus the photoelectrons on a channel plate detector. The hemispherical analyzer permits electrons with only a narrow band of energy to reach the detector. Neither the X-ray nor the photoelectrons are scanned to produce an image. Spectroscopy and imaging use the same analyzer to avoid the risk of misalignment. In ESCALAB, images are formed on a channel plate detector, while spectroscopy uses an array of channel electron multipliers.

The advantage of iXPS is that it produces images much more quickly than the conventional methods (for comparable spatial resolution) because the entire field of view is imaged simultaneously. Using an electron flood source, it is possible to get a real time (i.e. close to TV rate) physical image of exactly the same position as the images. This makes setting up experiments much quicker. Moreover, the introduction of a magnetic objective lens has improved the sensitivity for a given spatial resolution, enabling the development of a new technique of XPS imaging.

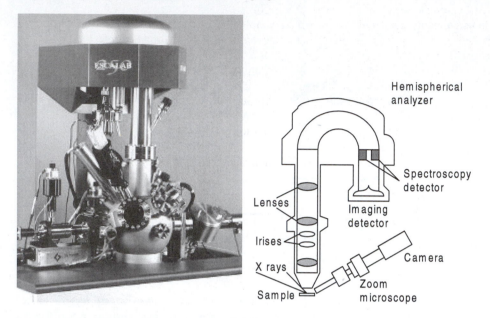

Figure 4.13 Photograph (from www.thermo.com/eThermo/CDA/Products/Product_Detail/1,1075,15885-158-X-1-1,00.html) and schematics of an ESCALAB. This iXPS instrument offers the capability of parallel imaging, which obtains positional information from dispersion characteristics of the hemispherical analyzer and produces photoelectron images with spatial resolution better than 5 μm.

Auger Electron Spectroscopy (AES)

AES, based on an electron beam and not an X-ray photon beam, and Auger electron emission (refer to Figure 4.10) are complementary to XPS for surface analysis, being especially useful for quantitative analysis of certain elements. The Auger electron can only be characterized by its absolute kinetic energy and not by a binding energy. AES found practical applications earlier than XPS, but due to the need of electron bombardment to the sample (instead of photons), it is a highly destructive technique not recommended for modern polymer surface analysis (Andrade, 1985b). AES is also known as SAM (Scanning Auger Microprobe) (Ratner, 1988).

Secondary-Ion Mass Spectroscopy (SIMS)

Secondary-Ion Mass Spectroscopy (SIMS) is another technique for studying the nature of biomaterial surfaces. Due to its destructive nature, it has been applied to biomedical polymers, however, it provides useful information for understanding the composition and structure of an implant material surface (Andrade, 1985c; Castner and Ratner, 1988).

SIMS is based on the principle of mass spectrometry, a proven sensitive and powerful analytical technique in which sample molecules are ionized. If enough ionizing energy is present, sample molecules are even dissociated, forming ionic fragments. The mass-to-charge ratios of the molecular and fragment ions are determined and plotted against abundance (mass-spectrum). The pattern of product ions formed is dependent on the structure of the intact compound (Wait, 1993).

Figure 4.14 illustrates the SIMS procedure and instrument schematics. A SIMS instrument, and in general all mass spectrometers, consists of a means of ion generation, a mass analyzer for their separation, and an ion detector (Wait, 1993). As with XPS, ultrahigh vacuum (UHV) chambers are used to avoid not only external contamination but also collisions with particles present in a nonvacuum environment.

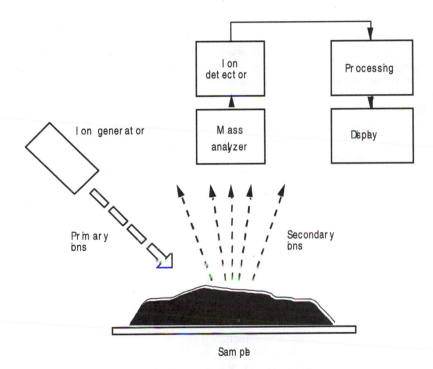

Figure 4.14 Schematic diagram of a SIMS instrument. Bombardment of primary ions on the sample surface leads to secondary ion emission. A mass analyzer separates these ions in terms of their mass-to-charge ratio. The ion detector converts this ionic current into an electric signal for further processing. The display presents the SIMS spectra, consisting of the count of ions versus their mass-to-charge ratio.

Ion Generator: The generator or gun provides a beam of charged particles that once accelerated into the surface sample, lead to the ejection of secondary ions. The size, energy, mass, and flux of the incident beam are directly proportional to secondary ion emission and sample damage. A good compromise between suitable secondary emission and damage is a 2 keV Xe^+ ion beam, which gives a sampling depth of approximately 1 nm (10 Å). The use of such low-energy inert gas ion beams, referred to as static SIMS, is the most commonly used technique. It allows the analysis of the topmost surface of insulating materials such as polymers. A high-energy and high-flux beam technique, called dynamic SIMS, is used to profile the depth of minerals and semiconductors (Andrade, 1985c).

The use of a charged incident beam as mentioned previously can significantly alter the electrical characteristics of the sample surface, particularly insulators, as surface charge buildup affects the secondary ion detection. One way of avoiding this charge buildup is by using a neutral atom incident beam, a technique referred to as fast atom bombardment (FAB) SIMS (Castner and Ratner, 1988).

Charge Neutralization: For suitable reading of secondary ion emission, the charge that builds up on the surface of an insulating sample must be neutralized. This can be accomplished by a variety of devices that generate electron beams for the case of positively charged incident ion beams. Even for the case of neutral FAB SIMS, the use of neutralizing electrons has shown to increase the quality of detected data (Castner and Ratner, 1988).

Mass Analyzer: The counts of secondary emitted ions versus their mass-to-charge ratio (measured in Daltons, where 1 D = 1/12 the mass of C^{12}) constitute the data obtained through a SIMS instrument.

The well known expression for kinetic energy in terms of mass and velocity can be applied to an ion, for which its number of charges z (in units of the charge of an electron) and its accelerating voltage V are known, leading to Eq. (4.5)

$$zV = \frac{1}{2}mv^2 \tag{4.5}$$

which can be rearranged to give the fundamental mass spectrometer equation

$$\frac{m}{z} = \frac{2V}{v^2} \tag{4.6}$$

where v is the ion velocity and m its mass (Wait, 1993).

The quadrupole mass spectrometer, which determines mass divided by charge m/z by scanning voltages, analyzes fragments up to 500 D. Most polymers yield fragments in this range. Quadrupole spectrometers are commonly used because they are inexpensive and most polymers yield fragment ions in their detection range (Castner and Ratner, 1988).

For sensitivities higher than 500 D, the time-of-flight (TOF) mass analyzer is used. The TOF mass analyzer measures velocity instead of voltages. It measures the time through a path of fixed length (see Eq. 4.6) to calculate the ratio m/z. For precise time measuring, pulsed ionization techniques must be employed for primary beam generation, which accounts for the higher cost of TOF-based instruments.

Data Processing and Interpretation: Once the ions are separated according to their mass-to-charge ratio *m/z*, a detector converts the ion current into an electric signal by a type of electron multiplier. The electric signal is amplified, computer processed, and visualized.

In modern instruments, computers are used to select parts of spectra for analysis and also to control some instrument features such as beam generation and intensity, neutralization energies, and sample scanning (Castner and Ratner, 1988; Wait, 1993).

Static SIMS determines the types, structures, and quantity of the chemical compounds on a surface. Note that relating the numerous peaks of the SIMS spectra to the surface chemical group from which they originated is not an easy task. This is mainly due to the different ionic particles—some of them fragments of molecules—that are detected.

Reference materials, with standard or at least well known spectra, must be used for comparative purposes. Nevertheless, careful SIMS experiments allow identification of polymeric chains and some composition to depths of 1.5 nm (15 Å). SIMS complements XPS for polymer composition studies (Castner and Ratner, 1988).

ISS: Another ion beam technique called Ion Scattering Spectroscopy (ISS) consists of a low-energy inert gas beam that interacts with the sample surface atoms. Incident ions lose energy in these elastic collisions, a loss that depends on the mass of the atom in the surface. There is no secondary ion emission as in SIMS, and therefore primary ions are the ones being detected in ISS (Andrade, 1985c).

Infrared Spectroscopy (IR)

Next we present some infrared spectroscopy techniques, which are usually combined to conduct surface analysis. Infrared spectroscopy is a versatile, fast, and inexpensive method that leads to conclusive molecular identification of samples in any of the three phases: gas, liquid, or solid (Grim and Fateley, 1984).

Dispersive Infrared Spectroscopy: The infrared region of electromagnetic radiation includes wavelengths in the range of 700 nm to 1.0 mm. Usually only the midinfrared region is used for measuring and includes wavelengths from 2.5 μm to 50 μm (Knutson and Lyman, 1985). For convenience, spectroscopists use the wavenumber, expressed in cm^{-1} and defined as $1/\lambda$, rather than wavelengths. The near infrared region (NIR) is conventionally defined as the region from 20,000 to 4,000 cm^{-1}; the far infrared (FIR) region is defined to be from 200 to 10 cm^{-1} (Grim and Fateley, 1984).

The underlying principle of infrared spectroscopy states that the absorption of electromagnetic radiation of the appropriate energy excites the atom. This absorbed energy is converted to vibrational and rotational motion governed by quantum mechanics rules. The absorption process depends on the relative masses and geometry of the atoms and on the forces between bonds within the sample molecules. The wavelengths of the radiation that can excite a molecule from one vibrational level to another fall in the infrared region. The vibrational and rotational motions take place at particular frequencies (or wavelengths) according to the molecular structure. The absorbed energy associated with the vibration is described by quantum mechanics in terms of discrete vibrational energy levels (Knutson and Lyman, 1985).

A spectrum (energy vs. wavelength) allows identification of the structural features of the unknown sample compound, and unlimited spectral libraries are available to verify tentative identifications (Grim and Fateley, 1984).

The dispersive infrared spectroscope is similar to the spectrophotometer (see Chapter 3) in that they both measure absorbance or transmittance as a function of wavelengths (or wavenumbers), and they use a prism or grating to disperse radiation from the source into spectral elements. Dispersive spectrometry has some limitations, which are overcome by the use of interferometric techniques, as we discuss next.

Interferometric Infrared Spectroscopy: The basics of interferometry are related to the interference of monochromatic radiation: two waves will interfere with each other if their optical paths intersect. This interference can be constructive (sum of amplitudes of radiations in phase) or destructive (amplitudes of out-of-phase radiations cancel each other). For intermediate phase differences, the resulting amplitude of the interacting radiation waves will be a function of the phase angle. An interferogram is the plot of amplitude vs. phase (Grim and Flateley, 1984).

A major component in interferometry instruments is the interferometer, a device for developing a difference in phase between two waves (Grim and Flateley, 1984). Figure 4.15 shows schematically the optical configuration of a Michelson interferometer, the first designed and most common interferometer in use today. It consists of a beamsplitter set at 45° from a set of two mirrors, one fixed, and the other able to slide perpendicularly to the fixed mirror. The beamsplitter transmits half of the source radiation to the fixed mirror and the other half to the sliding mirror (Fink and Larson, 1979). A phase difference between the beams can be induced by sliding the mirror. The changes in the mirror position change the path length for each beam. The beams are recombined before passing through a sample to reach the detector. If the path lengths to the fixed and sliding mirrors are the same, constructive interference occurs, yielding a maximum at the detector. If the path lengths differ by $\lambda/2$, destructive interference occurs, yielding a minimum at the detector. For a monochromatic source, as the sliding mirror moves distance x, the detector yields an output $I = I_0 \cos(2\pi x/\lambda)$. Thus by counting maxima as the sliding mirror moves, we measure the number of wavelengths of movement.

Fourier Transform Infrared Spectroscopy (FT-IR): For a polychromatic source, the interferogram will show a center burst (maximum) only when the two path lengths are equal. We use equal path lengths with no movement. Then if absorption from one path is present at one wavelength, an inverted cosine wave will appear across the interferogram. We use a mathematical procedure: the Fourier Transform (FT) or the Fast Fourier Transform (FFT) to measure all wavelengths simultaneously and convert from the spatial domain of the interferogram to the absorption spectrum of spectroscopy (Grim and Fateley, 1984).

This operator allows the display of the interferogram (and any time signal) in the frequency domain. The interferogram represents all wavelengths in one graph, an advantage over dispersive spectroscopy in which energy levels are measured for each wavelength at a time. FT calculations for displaying interferograms in the frequency domain were a lengthy and tedious process, one that has been all but eliminated by the use of digital computers. This powerful mathematical tool has also allowed the use of interferometry in the far infrared region, which required almost endless calculations. It has improved the benefits of interferometry over dispersive infrared spectroscopy, and most current interferometric instruments have a built-in computer for FT calculations (Buijs, 1984).

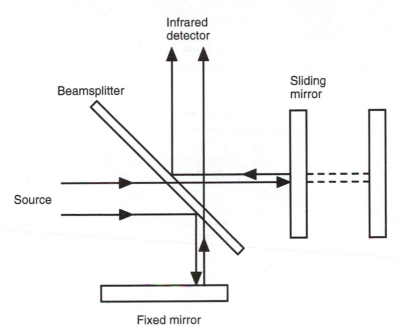

Figure 4.15 Michelson interferometer. A beamsplitter transmits half of the source radiation to the fixed mirror and the other half to the sliding mirror. A phase difference between the beams can be induced by sliding the mirror, which causes detection of the two beams at different times. The detector provides the interferogram, a plot of energy as a function of differences in optical paths, Beams have been slightly shifted in the drawing to allow easy following of their path.

Attenuated Internal Total Reflection (ATR): Total internal reflection is a special case of the more general phenomenon of reflection and refraction of incident electromagnetic radiation at an interface of two media having different indexes of refraction (Knutson and Lyman, 1985).

Figure 4.16 shows that when an incident beam with an angle θ traveling in a medium of refractive index η_c encounters the interface with a second medium of different refractive index η_s, the beam will be reflected back into the first medium and transmitted into the second medium with an angle other than the incident (refraction) according to Snell's Law of Refraction

$$\eta_c \sin\theta = \eta_s \sin\phi \qquad (4.7)$$

where the indexes of refraction represent relationships of the speed of light in the medium to the speed of light in a vacuum. From Eq. (4.7) and from Figure 4.10 it becomes evident that if the first medium is denser that the second (i.e. $\eta_c > \eta_s$), then ϕ is greater than θ. Also, there is a value for θ that will make $\phi = 90°$ and is called the *critical angle* θ_c. Eq. (4.7) becomes

$$\theta_c = \sin^{-1}\left(\frac{\eta_s}{\eta_c}\right) \tag{4.8}$$

Values of θ larger than θ_c will cause *total reflection* of the incident beam (i.e. no refractive beam develops).

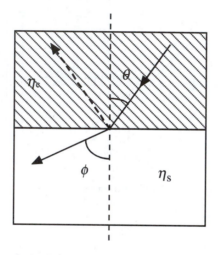

Figure 4.16 When an incident beam traveling at an angle θ in a medium of refractive index η_c encounters another medium of refractive index η_s, it will reflect in a direction given by θ and refract in the direction given by ϕ, verifying Snell's Law of Refraction (Eq. (4.7).

We have highly simplified the presentation of these phenomena. A more rigorous treatment would have required the use of tools provided by electromagnetic wave theory.

Surface Infrared Spectroscopy, FTIR-ATR: The combination of the techniques that we have just presented represent a very useful method for studying the surface properties of materials.

The electromagnetic radiation is totally internally reflected (ATR) through an optically transparent material in contact with the sample (see Figure 4.16). The reflectivity exhibited by the surface depends upon the interactions of the electromagnetic field established within the sample. Using infrared spectroscopy (IR), it is possible to obtain a spectrum representing the characteristics of the sample surface within a finite distance of the interface, providing information on its chemical composition (Knutson and Lyman, 1985).

Finally, the addition of the Fourier Transform led the method known as *Fourier Transform Infrared Spectroscopy-Attenuated Total Internal Reflection,* FTIR-ATR, which ultimately combines the versatility of the infrared spectrum and the convenience of computers for fast digital processing, represents a key technique for surface analysis.

4.3.3 Polymer–water Interface

Biomedical polymers are usually implanted within a liquid environment. The polymer-liquid interface in terms of surface free energy and surface tension can be accurately measured with the contact angle method.

Contact Angle Method

The contact angle method requires relatively simple and inexpensive equipment, although it is one of the most sensitive methods used to obtain information about the outermost layer of solid surfaces (Andrade et al., 1985).

This method allows the determination of the excess free energy per unit area, the surface tension, whose SI unit is N/m. The surface is actually a discontinuous boundary between three different phases: liquid, gas, and solid. A surface tension develops in these phase boundaries when the equilibrium is altered giving rise to an excess energy, which minimizes the surface area. Adsorption (attraction of foreign materials) or chemisorption (bonding with the adsorbent) are other means of minimizing the surface energy (Park and Lakes, 1992).

Figure 4.17 illustrates the three surface tension components that interact to limit the spread of a drop on top of a rigid surface. At equilibrium, Young's equation is

$$\gamma_{SV} - \gamma_{SL} = \gamma_{LV} \cos \theta \tag{4.9}$$

where γ is the interfacial free energy for the solid–vapor, solid–liquid, and liquid–vapor boundaries, and θ is the contact angle (Andrade et al., 1985). Rearranging Eq. (4.9) leads to

$$\cos \theta = \frac{\gamma_{SV} - \gamma_{SL}}{\gamma_{LV}} \tag{4.10}$$

The values of θ describe the wetting characteristic of a surface; a complete wetting is represented by a value of zero for θ. Nonwetting corresponds to $\theta > 90°$ and to a more spherical drop (see Figure 4.17). Equation 4.10 shows that the value of θ represents ratios and not absolute values of surface tension. For a contact angle greater than zero, the lowest surface tension of a liquid in contact with a solid surface is called the *critical surface tension,* γ_c, which is obtained by measuring contact angles of homologous liquids (Park and Lakes, 1992).

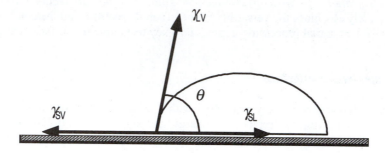

Figure 4.17 Surface tension components of a three-phase system to limit the spread of a drop on top of a surface. γ is the interfacial free energy for each of the phases. θ is the contact angle.

Several conditions must be satisfied for a straightforward interpretation of the contact angle: the solid surface is rigid, very smooth, uniform, homogenous, and does not interact with the liquid other than the three-phase equilibrium; the liquid surface tension is well known, constant, and the liquid vapor does not adsorb on the solid surface to change its surface free energy (Andrade et al., 1985).

The contact angle can be measured, among another methods by: (1) direct microscopy with a goniometer or protractor for direct measurement of the angle, (2) calculation of the angle from known dimensional measures of the drop, such as diameter, volume and trigonometric relationships and (3) capillarity of a liquid of known surface tension (Andrade et al., 1985).

4.4 Protein Adsorption

Adsorption is the increased concentration of a substance at the surface of a substrate. The interaction of proteins with solid surfaces, which is protein adsorption, represents an important process for the fields of biomaterials due to its role in blood coagulation and in the reaction of tissues to artificial implants. For tissue engineers, cell-surface adhesion—which is mediated by an interlayer of adsorbed protein—is usually required for cell culture *in vitro* (Andrade, 1985d).

Diverse processes are affected by protein adsorption: thrombosis and hemostasis, healing of hard and soft tissues, infection, and inflammation. Some types of artificial heart valves require permanent anticoagulation treatment, which in turn requires knowledge of the cell–protein interface. In orthopedic applications, surface properties affect bone healing (Chinn and Slack, 2000).

It is generally accepted that the first interaction of blood with a foreign surface is the adsorption of proteins and that the nature of the adsorbed proteins will affect further interactions of the blood (Choi and Callow, 1994). This blood compatibility or incompatibility is considered to be related to adsorption processes involving gamma globulins, fibrinogen, fibronectin, albumin, and several other proteins (Andrade, 1985d).

Tissue engineering promotes new tissue formation. Whether or not the biomaterial used for such purpose is biodegradable, its surface properties (affected by adsorbed proteins) will influence the initial cellular events (cell attachment) at the cell-material interface (Boyan et al., 1996). Implantation of biomaterials or engineered tissues, for example, generally leads to an inflammatory response aimed at destroying the foreign material. A central aspect of inflammatory responses is the adhesion of leukocytes to blood vessel walls before emigrating into tissues. This cell adhesion to the vessel walls is mediated by a family of protein receptors (Jones et al., 1996).

Adsorbed proteins may exhibit a complex behavior not only related to the direct interaction of surface and proteins, but also to the structural rearrangements (i.e. conformational changes, see Section below) that proteins may undergo during the adsorption process (Boyan et al., 1996).

4.4.1 Protein Molecule

Proteins are polymers of biologic origin, and among organic molecules they constitute the largest group.

Carbon, nitrogen, oxygen, and hydrogen atoms are the basic elements in a protein molecule—in many cases, sulfur atoms are also a part of the protein molecule. Proteins, as we mentioned, are polymeric chains with subunits called *amino acids*. There are 20 different amino acids. Figure 4.18 shows the structure of an amino acid molecule, which has a central carbon atom. An amino group (—NH$_2$), a hydrogen atom, and an acid group (—COOH or carboxyl group) are bonded to the central carbon atom. Also, in Figure 4.18, R represents the rest of the amino acid molecule, a part that is different for each of the 20 amino acids.

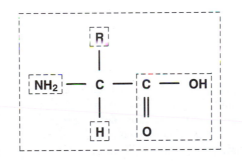

Figure 4.18 The amino acid molecule. To a central carbon atom, an amino group, a carboxyl group and a hydrogen atom are bonded. R represents the rest of the molecule, which is different for each of the 20 amino acids.

Different amino acids combine to form polypeptides and these in turn give rise to a very high number of complex structures (proteins). The combination of polypeptides gives each protein a distinct three-dimensional shape. This spatial relationship between the amino acid chains giving rise to the structure of a protein is called *conformation*. On the surface of a protein molecule there are hydrophobic, charged and polar domains. The particular regions of a protein to which antibodies or cells can bind are called *epitopes*. The existence, arrangement, and availability of different protein epitopes is called *organization* (Chinn and Slack, 2000).

4.4.2 Protein Adsorption Fundamentals

For a single protein solution, the rate of adsorption to the substrate depends upon transport of the protein to the substrate. Four primary transport mechanisms have been identified: diffusion, thermal convection, flow convection, and combined convection-diffusion. For constant temperature and static systems, transport is exclusively by diffusion, and the net rate of adsorption can be described by Langmuir's theory of gas adsorption (Chinn and Slack, 2000).

$$r_A = k_A\, C_b\, (1-\theta) - k_D \theta \tag{4.11}$$

where r_A is the net rate of adsorption, k_A is the adsorption rate constant, C_b is the bulk concentration of the protein in solution, θ is the fraction of surface occupied by the adsorbed molecules and k_D is the desorption rate constant. At equilibrium, the rate of adsorption equals the rate of desorption, so the net rate r_A is zero and from Eq. (4.11), the fractional surface coverage θ is:

$$\theta = \frac{\dfrac{k_A}{k_D} C_b}{1 + \dfrac{k_A}{k_D} C_b} \tag{4.12}$$

Equation (4.12) is called the Langmuir Adsorption Isotherm, and is most applicable to dilute solutions (Chinn and Slack, 2000).

The initial adsorption and the eventual deposition of various proteins are different. The *Vroman effect* designates a set of transient phenomena related to protein adsorption: the layer composition on the surface appears to be evolutionary and varies from surface to surface, and proteins may replace one another in some ordered sequence during adsorption, giving rise to the concept of protein multilayers. For the special case of blood compatibility of a material, the sequence of different protein interactions is not well understood (Choi and Callow, 1994).

4.4.3 Protein Adsorption Measurements

In order to understand protein adsorption, quantitative evaluation of the adsorption process must be performed, along with information on gathering the properties and characteristics of the protein molecule itself that will affect the adsorption mechanisms.

Although a set of useful information is obtained from a protein adsorption experiment, the available techniques are limited and in general do not provide all the information desired. The amount of adsorbed protein (adsorbed isotherms) is measured using XPS and ATR-IR (see Section 4.3.2). Adsorption as a function of time is also measured using ATR-IR and fluorescence spectroscopy. The conformation and conformational changes (denaturation) of the protein molecule can be determined using IR techniques (Section 4.3.2), circular dichroism (CD), and calorimetry. The thickness of the adsorbed layer is measured by ellipsometry. For studying the heterogeneity of adsorption, electron microscopy (Section 4.3.1) and microautoradiography are used. The use of radiolabeled proteins allows direct measurement of protein adsorbed to a substrate.

Differential Scanning Calorimetry (DSC)

Protein adsorption involves redistribution of charge, changes in the state of hydration and rearrangements in the protein structure (Norde, 1985). The result of these processes is a generation of heat, which is measured in a calorimeter. Most commercial calorimeters are of the conduction type, and heat developed in the reaction vessel is completely transferred through a thermopile to a surrounding heat sink of large capacity. The voltage signal generated by the thermopile is proportional to the heat flux. In a twin calorimeter, two vessels are within a single heat sink. The reaction takes place in one vessel, while the other vessel serves as the reference. The detectors are in opposition so the differential signal indicates the heat from the reactor vessel and rejects disturbances from the surroundings that affect both vessels (Feng and Andrade, 1994).

Circular Dichroism (CD)

Many biologically important molecules are not functionally identical to their mirror images. They are called optically active or chiral. These asymmetric molecules exhibit optical rotation. That is, a vertical band of polarized light will rotate to the left or right if passed through an optically active substance. Left- and right-circularly polarized light (LCP and RCP) exhibit circular dichroism (CD), and this is the difference in absorption of LCP and RCP light (Hatano, 1986). Measurements can also be carried out in the infrared (Ferraro and Basile 1985).

Total Internal Reflection Fluorescence (TIRF)

Most proteins fluoresce naturally due to the intrinsic fluorescence of one or more of the aromatic amino acids comprising them: tyrosine (Tyr), phenylaline (Phe), or tryptophan (Trp) (Hlady et al., 1985). Light from xenon, or another light source, passes through an excitation monochromator, emitting a narrow band of usually ultraviolet wavelengths. The protein in a flow cell fluoresces at a longer wavelength. The emission passes through an emission monochromator and is measured by a photomultiplier.

Ellipsometry

Ellipsometry operates on the principle that light changes its polarization upon reflection from a solid–solution interface. It requires optically flat, reflective surfaces and provides information on the thickness and optical contents of adsorbed film as a function of time. Morrissey (1977) used ellipsometry to measure differences in adsorbed amount and thickness of gamma globulin onto flat silica plates.

Autoradiography

The distribution of a radiolabeled compound within a specimen can be imaged using autoradiography, which is more sensitive than pulse counters (Dykstra, 1992). Radioisotopes are usually administered at 10 to 40 µCi/g (Ci = curie) body weight for in vivo studies. Silver halide photographic emulsions are placed over the radiolabeled specimen, then developed. This method is slow and exposure times can be up to 3 weeks.

Radiolabeling

Radiolabeling permits measurement of the affinity for a given protein onto materials (Baquey et al., 1991). During testing, peristaltic pumps cause biological fluids to flow through tubing segments of material samples such as PVC, silicone elastomer, polyurethane or polyethylene. Human albumin may be labeled with 99mTc or radio-iodinated. Detectors based on junctions of semiconductors are preferred because of their good energy resolution. A scintillation camera (used for viewing emissions of radioactive substances) can be used to image the amount of adsorbed protein on a given surface and to study its distribution on this surface.

4.5 Molecular Size

For molecular size below a few micrometers, or 10,000 D (daltons), particle size is measured by light scattering. A laser emits polarized light focused on the solution in a scattering cell. Light intensity from a particle in solution that is small compared with the wavelength, λ, of the incident light is proportional to the concentration, c, multiplied by the molecular weight Mw. This is known as static or Rayleigh scattering.

In dynamic light scattering, the diffusion constant of the molecules moving in the solution is calculated from the autocorrelation function of the scattered light. This is also known as quasi-elastic or photon correlation spectroscopy (PCS).

The thermal agitation of particles causes the random movements of Brownian motion. Coherent light from two particles scatters and can add or interfere at the photosensor. Photomultiplier tubes or low-noise, high gain avalanche photodiodes (APDs) measure the scattered light at various angles. The photosensor output fluctuates as the particles move relative to each other. Small particles move rapidly, whereas large particles move slowly, thus the fluctuations are slow for large particles because they do not move very far between measurements. Photon counter signals are amplified then discriminated to produce pulses, which are fed to digital signal processor (DSP) correlator. Figure 4.19 shows that the autocorrelation function (ACF) versus time yields an exponential decay from which particle size is calculated (Brown, 1993).

$$G(\tau) = <I(t) \times I(t + \tau)> \tag{4.13}$$

where:

$G(\tau)$ = ACF
$I(t)$ = intensity at time t
$I(t + \tau)$ = intensity at $(t + \tau)$
τ = delay time
$< >$ = time average

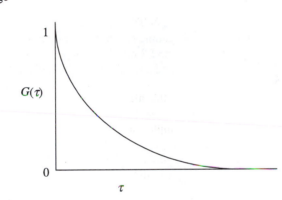

Figure 4.19 The autocorrelation function $G(\tau)$ is 1 when two signals have delay time $\tau = 0$, then decays to 0 for long delay time.

A PCS measurement can be thought of as a measure of the time it takes for a particle to diffuse a distance $1/K$, where K is the scattering vector

$$K = 4\pi n \frac{\sin(\theta / 2)}{\lambda} \tag{4.14}$$

where:
n = refractive index of the diluent (n = 1.333 for water)
θ = scattering angle
λ = wavelength (λ = 633 nm for He–Ne laser)

The ACF function for identical particles is a single decaying exponential

$$G(\tau) = \exp(-2\Gamma\tau) \tag{4.15}$$

where:

$G(\tau) = \text{ACF}$

$\Gamma = $ decay constant of particle

Note that when $G(\tau) = 0.37$, $1/(2\Gamma) = \tau$. We use Γ to calculate the diffusion co-efficient $D_T = \Gamma/K^2$. From the diffusion constant, we use the Stokes–Einstein equation to calculate the diameter of a spherical particle.

$$D_T = \frac{kT}{3\pi\eta d} \tag{4.16}$$

where D_T is the diffusion coefficient, the Boltzmann constant is $k = 1.38 \times 10^{-23}$ J/K, T is the absolute temperature, η is the viscosity of the medium ($\eta = 0.001$ Pa· s for water at 20 °C), and d is the diameter of the particle.

Example 4.1 Given autocorrelation function data obtained from experiment, calculate the particle size of polystyrene latex in the sample solution. The data were obtained at a temperature of 20 °C with $\theta = 90°$ and $\lambda = 632.8$ nm. Table 4.4 gives the data for τ and the corresponding intensity. The solution is considered water ($n = 1.33$, $\eta = 0.001$ Pa· s).

Table 4.4 Experimental data for Example 4.1.

τ	$G(\tau)$
0	1
0.001	0.35
0.002	0.15
0.003	0.051
0.004	0.012

The data from Table 4.4 yield an exponential plot, but they don't provide a useful mathematical model for calculation. Therefore, take a natural log of the data and then linearize the data using mathematical software or, for example, LINEST in Microsoft Excel to obtain a linear equation. The given data yield the linear approximation

$$\ln(G(\tau)) = -1077.2(\tau) + 0.0852$$

To calculate the diffusion coefficient, Γ needs to be calculated first. Since $G(\tau) = 0.37$ when $1/(2\Gamma) = \tau$, the linear approximation can be used to find τ when $G(\tau) = 0.37$.

$$\tau = \frac{\ln(0.37) - 0.0852}{-1077.2} = 1.00209 \times 10^{-3}$$

Now using the relationship $1/(2\Gamma) = \tau$, $\Gamma = 498.956$.
Solve for K using Eq. (4.14) and the known constants

$$K = 4\pi n \frac{\sin(\theta/2)}{\lambda} = 4\pi(1.33)\frac{\sin(45)}{632.8\times10^{-9}} = 1.87\times10^7$$

Finally obtain the diffusion coefficient $D_T = \Gamma/K^2 = 1.427\times10^{-12}$ and solve for the particle diameter using the rearranged Stokes–Einstein equation, Eq. (4.16)

$$d = \frac{kT}{3\pi\eta D_T} = \frac{1.38\times10^{-23}(293)}{3\pi(0.001)\left(1.427\times10^{-12}\right)} = 301\,\text{nm}$$

4.6 Problems

4.1   Search the literature to find a biomaterial listed in Table 4.1 and describe its application.

4.2   Search the literature to find a polymer listed in Table 4.2 and describe its application.

4.3   Describe how polymers are made more rigid.

4.4   Describe the three main structural components of cellular composites.

4.5   Search the literature to find an application listed in Table 4.3 and describe it.

4.6   State the highest practical resolution of TEM and calculate the accelerating voltage required to achieve it. Compare with typical accelerating voltage.

4.7   Describe the principle and sample preparation for a TEM.

4.8   Describe the principle and sample preparation for a SEM.

4.9   Describe the principle and sample preparation for 2 modes of STM.

4.10   Describe the principle and sample preparation for 2 modes of SFM.

4.11   Describe the principle and sample preparation for XPS.

4.12   Describe the principle and sample preparation for AES.

4.13   Describe the principle and sample preparation for SIMS.

4.14   Describe the principle and sample preparation for ISS.

4.15   Describe the principle and sample preparation for dispersive infrared spectroscopy.

4.16   Describe the principle and sample preparation for interferometric infrared spectroscopy.

4.17   Describe the principle and sample preparation for FT-IR.

4.18   Describe the principle and sample preparation for ATR.

4.19   Describe the principle and sample preparation for FTIR-ATR.

4.20   Describe the principle of the contact angle method.

4.21   Describe the principle of DSC.

4.22   Describe the principle of CD.

4.23   Describe the principle of TIRF.

4.24   Describe the principle of ellipsometry.

4.25   Describe the principle of autoradiography.

4.26   Describe the principle of radiolabeling.

4.27   Given the data in Section 4.5 and when $G(\tau) = 0.37, \tau = 100$ ns, calculate particle diameter d.

4.28   We wish to view the topography of a biomaterial surface and the molecules that adsorb to it in a liquid. Select an instrument for imaging. Name other instruments you rejected and why you rejected them. Describe the principle and operation of your selected imaging instrument.

4.7 References

Andrade, J. D. 1985a. Introduction to surface chemistry and physics of polymers. In J. D. Andrade (ed.) *Surface and Interfacial Aspects of Biomedical Polymers*, Vol. 1 Surface Chemistry and Physics. New York: Plenum.

Andrade, J. D. 1985b. X-ray photoelectron spectroscopy (XPS). In J. D. Andrade (ed.) *Surface and Interfacial Aspects of Biomedical Polymers*, Vol. 1 Surface Chemistry and Physics. New York: Plenum.

Andrade, J. D. 1985c. Polymer surface analysis: conclusions and expectations. In J. D. Andrade (ed.) *Surface and Interfacial Aspects of Biomedical Polymers*, Vol. 1 Surface Chemistry and Physics. New York: Plenum.

Andrade, J. D. 1985d. Principles of protein adsorption. In J. D. Andrade (ed.) *Surface and Interfacial Aspects of Biomedical Polymers*, Vol. 2 Protein adsorption. New York: Plenum.

Andrade, J. D., Smith, L. M. and Gregonis, D. E. 1985. The contact angle and interface energetics. In J. D. Andrade (ed.) *Surface and Interfacial Aspects of Biomedical Polymers*, Vol. 1 Surface Chemistry and Physics. New York: Plenum.

Billotte, W. G. 2000. Ceramic biomaterials. In J. D. Bronzino (ed.) *The Biomedical Engineering Handbook*. 2nd ed. Boca Raton, FL: CRC Press.

Baquey, C., Lespinasse, F., Caix, J., Baquey, A. and Basse-Chathalinat, B. 1991. In Y. F. Missirlis and W. Lemm (eds.) *Modern Aspects of Protein Adsorption on Biomaterials*. Dordrecht, Holland: Kluwer Academic Publishers.

Berthiaume, Fl, Toner, M., Tompkins, R. G. and Yarmush, M. L. 1994. Tissue engineering. In R. S. Greco (ed.) *Implantation Biology: The Host Response and Biomedical Devices*. Boca Raton, FL: CRC Press.

Boyan, B. D., Hummert, T. W., Dean, D. D. and Schwartz, Z. 1996. Role of material surfaces in regulating bone and cartilage cell response. *Biomaterials*, **17** (2): 137–46.

Bozzola, J. J. and Russell, L. D. 1991. *Electron Microscopy: Principles and Techniques for Biologist*. Boston: Jones and Bartlett Publishers.

Brown, W. (ed.) 1993. *Dynamic Light Scattering: The Method and Some Applications*. Oxford: Clarendon Press.

Buijs, H. 1984. Advances in instrumentation. In T. Theophanides (ed.) *Fourier Transform Infrared Spectroscopy*. Dordrecht, Holland: D. Reidel Publishing.

Bustamance, C., Eire, D. A. and Yang, G. 1996. Scanning force microscopy of biological macromolecules: present and future. In H. C. Hoch, L. W. Jelinski and H. G. Craighead (eds.) *Nanofabrication and Biosystems: Integrating Materials Science, Engineering and Biology*. New York: Cambridge University Press.

Castner, D. G. and Ratner, B. D. 1988. Static secondary ion mass spectroscopy: a new technique for the characterization of biomedical polymer surfaces. In B. D. Ratner (ed.) *Surface Characterization of Biomaterials*. Amsterdam: Elsevier Science Publishers.

Chinn, J. A. and Slack, S. M. 2000. Biomaterials: protein-surface interactions. In J. D. Bronzino (ed.) *The Biomedical Engineering Handbook*. 2nd ed. Boca Raton, FL: CRC Press.

Choi, E. T. and Callow, A. D. 1994. The effect of biomaterials on the host. In R. S. Greco (ed.) *Implantation Biology: The Host Response and Biomedical Devices*. Boca Raton, FL: CRC Press.

Descouts, P. A. R. 1995. Nanoscale probing of biocompatibility of materials. In A. A. Gewirth and H. Siegenthaler (eds.) *Nanoscale Probes of the Solid/Liquid Interface*. Dordrecht, The Netherlands: Kluwer Academic Publishers.

Dociu, N. and Bilge, F. H. 1986. Evaluation by scanning electron microscopy. In A. F. von Recum (ed.) *Handbook of Biomaterial Evaluation: Scientific, Technical and Clinical Testing of Implant Materials*. New York: Macmillan.

Dykstra, M. J. 1992. *Biological Electron Microscopy: Theory, Techniques and Troubleshooting*. New York: Plenum.

Feng, L. and Andrade, J. D. 1994. Protein adsorption on low-temperature isotropic carbon: I. Protein conformational change probed by differential scanning calorimetry. *J. Biomed. Materials Res.*, **28**: 735-43.

Ferraro J. R. and Basile, L. J. (eds.) 1985. *Fourier Transform Infrared Spectroscopy: Applications to Chemical Systems*, Vol. 4. Orlando, FL: Academic Press. p.61.

Fink, U. and Larson, H. P. 1979. Astronomy: planetary atmospheres. In J. R. Ferraro and L. J. Basile (eds.) *Fourier Transform Infrared Spectroscopy: Applications to Chemical Systems*, Vol. 2. New York: Academic Press.

Fischbach, F. T. 1992. *A Manual of Laboratory & Diagnostic Tests*. 4th ed. Philadelphia: J. B. Lippincott.

Goodman, S. L., Li, C., Pawley, J. B., Cooper, S. L. and Albrecht, R. M. 1988. Surface and bulk morphology of polyurethanes by electron microscopies. In B. D. Ratner (ed.) *Surface Characterization of Biomaterials*. Amsterdam: Elsevier Science Publishers.

Greisler, H. P., Gosselin, C., Ren, D., Kang, S. S. and Kim, D. U. 1996. Biointeractive polymers and tissue engineered blood vessels. *Biomaterials*, **17** (3): 329-36.

Grim, W. M. and Fateley, W. G. 1984. Introduction to dispersive and interferometric infrared spectroscopy. In T. Theophanides (ed.) *Fourier Transform Infrared Spectroscopy*. Dordrecht, Holland: D. Reidel Publishing.

Hatano, M. 1986. Induced circular dichroism in biopolymer-dye systems. In S. Okamura (ed) *Advances in Polymer Science*, Vol. 77. Berlin: Springer Verlag.

Hlady, V., Van Wagenen, R. A. and Andrade, J. D. 1985. Total internal reflection intrinsic fluorescence (TIRIF) spectroscopy applied to protein adsorption. In J. D. Andrade (ed.) *Surface and Interfacial Aspects of Biomedical Polymers*, Vol. 2 Protein adsorption. New York: Plenum.

Hutton B. 1999. Introduction to XPS. [Online]
calcium.chem.ucl.ac.uk/webstuff/people/williams/xps/xpsintro.html

Jones, D. A., Smith, C. W. and Mcintire, L. V. 1996. Leucocyte adhesion under flow conditions: principles important in tissue engineering. *Biomaterials*, **17** (3): 337-47.

Knutson, K. and Lyman, D. J. 1985. Surface infrared spectroscopy. In J. D. Andrade (ed.) *Surface and Interfacial Aspects of Biomedical Polymers*, Vol. 1 Surface Chemistry and Physics. New York: Plenum.

Lee, H. B., Khang, G., Lee J. H. 2000. Polymeric biomaterials. In J. D. Bronzino (ed.) *The Biomedical Engineering Handbook*. 2nd ed. Boca Raton, FL: CRC Press.

Marchant, R. E. and Wang, I. 1994. Physical and chemical aspects of biomaterials used in humans. In R. S. Greco (ed.) *Implantation Biology: The Host Response and Biomedical Devices*. Boca Raton, FL: CRC Press.

Marti, O. 1993. SXM: an introduction. In O. Marti and M. Amrein (eds.) *STM and SFM in Biology*. San Diego, CA: Academic Press.

Morrissey, B. W. 1977. Adsorption of proteins at solid surfaces. *Ann. N. Y. Acad. Sci.*, **288**: 50-64.

Norde, W. 1985. Probing protein adsorption via microcalorimetry. In J. D. Andrade (ed.) *Surface and Interfacial Aspects of Biomedical Polymers*, Vol. 2 Protein adsorption. New York: Plenum.

Palsson, B. O. and Hubbell, J. A. 2000. Tissue engineering. In J. D. Bronzino (ed.) *The Biomedical Engineering Handbook*. 2nd ed. Boca Raton, FL: CRC Press.

Packer, L. 1967. *Experiments in Cell Physiology*. New York: Academic Press.

Pagana, K. D. and Pagana, T. J. 1997. *Mosby's Diagnostic and Laboratory Test Reference*. 3rd ed. St. Louis: Mosby.

Park, J. B. 2000. Biomaterials. In J. D. Bronzino (ed.) *The Biomedical Engineering Handbook*. 2nd ed. Boca Raton, FL: CRC Press.

Park, J. B. and Lakes, R. S. 1992. *Biomaterials: An Introduction*. 2nd ed. New York: Plenum.

Paynter, R. W. 1988. Introduction to x-ray photoelectron spectroscopy. In B. D. Ratner (ed.) *Surface Characterization of Biomaterials*. Amsterdam: Elsevier Science Publishers B.V.

Ratner, B. D. 1988. The surface characterization of biomedical materials: how finely can we resolve surface structures? In B. D. Ratner (ed.) *Surface Characterization of Biomaterials*. Amsterdam: Elsevier Science Publishers B.V.

Ratner, B. D., Hoffman, A. S., Schoen, F. L. and Lemons, J. E. (eds.) 1996. *Biomaterials Science*. San Diego: Academic Press, 1996.

Sheffield, W. D. and Matlaga, B. F. 1986. Evaluation by transmission electron microscope. In A. F. von Recum (ed.) *Handbook of Biomaterial Evaluation: Scientific, Technical and Clinical Testing of Implant Materials*. New York: Macmillan.

Silver, S. H. 1994. *Biomaterials, Medical Devices and Tissue Engineering: An Integrated Approach*. London: Chapman & Hall.

Sittinger, M., Bujia, J., Rotter, N., Reitzel, D., Minuth, W. W. and Burmester, G. R. 1996. Tissue engineering and autologous transplant formation: practical approaches with resorbable biomaterials and new cell culture techniques. *Biomaterials*, **17** (3): 237–42.

Skalak, R. and Fox, C. F. (eds.) 1988. *Tissue Engineering*. New York: Alan R. Liss.

Skalak, R., Fox, C. F. and Fung, B. 1988. Preface. In Skalak, R. and Fox, C. F. (eds.). *Tissue Engineering*. New York: Alan R. Liss.

Spire Biomedical. [Online] www.spirebiomedical.com/

Thermo VG Scientific. 2003. [Online]
www.thermo.com/eThermo/CDA/BU_Home/BU_Homepage/0,1285,158,00.html

University of Alabama in Huntsville. Materials Science Program. Electron Spectro-
scopies for Surface Characterization [Online]
matsci.uah.edu/CourseWare/mts723/lectures/

Wait, R. 1993. Introduction to mass spectrometry. In C. Jones, B. Mulloy and A. H.
Thomas (eds.) *Spectroscopic Methods and Analyses: NMR, Mass Spectrometry and
Metalloprotein Techniques.* Totowa, NJ: Humana Press.

Wintermantel, E., Mayer, J., Blum, J., Eckert, K. L., Lüscher, P. and Mathey, M. 1996.
Tissue engineering scaffolds using superstructures. *Biomaterials*, **17** (2): 83–91.

Chapter 5

Hematology

Susanne Clark Cazzanti

This chapter discusses various measurements and bioinstrumentation related to blood and its function in the human body. Blood is composed of cells as well as other unformed elements. A cell is the smallest unit of a living organism capable of functioning independently as it contains at least one nucleus in which there are genetic codes for maintaining, growing, and reproducing the organism.

5.1 Blood Components and Processing

A typical adult has approximately 5 to 6 L of blood at any given time, or about 7% to 8% of a person's body weight. Blood is composed of formed elements, cells, and unformed elements. Formed elements make up approximately 45% while unformed elements account for the remaining 55%.

5.1.1 Components of Blood

Red blood cells (RBCs), white blood cells (WBCs), and platelets are formed elements. Red blood cells, also known as erythrocytes, contain hemoglobin, which makes oxygen delivery possible by RBCs throughout the body (see Chapter 3). Erythrocytes are relatively large compared to other types of blood cells. In fact, they are the limiting factor in capillary size and must deform slightly to squeeze through. White blood cells, also known as leukocytes, help defend the body against foreign bodies. There are many different types of leukocytes, each much smaller than red blood cells. Platelets are essential for clotting.

Water is the primary unformed element, but others include proteins, carbohydrates, vitamins, hormones, enzymes, lipids, and salts. These are distributed throughout the body via the circulatory system.

When anticoagulants are used, the protein fibrinogen is present, and separation yields the straw-colored liquid plasma. When blood is allowed to clot, the protein fibrinogen is not present, and separation yields the straw-colored liquid serum.

5.1.2 Basic Techniques

Described in Chapter 3, spectrophotometers are frequently used for hematology. Electron microscopy, explained in Chapter 4, is another frequently used technique in hematology. Phase microscopy, explained in Chapter 6, is also frequently used.

A centrifuge is also frequently used for separating the components of blood. A centrifuge sediments particles that are suspended in a liquid by spinning the solution at high speeds. The solution may be held at an angle or horizontally during spinning.

5.1.3 Blood Collection

Figure 5.1 shows the evacuated tube system used for blood collection. Ethylenediaminetetraacetic acid (EDTA), heparin, or sodium citrate is used to prevent coagulation of the samples, depending on the test.

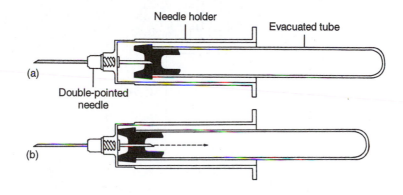

Figure 5.1 (a) A double-ended needle partially pierces the rubber stopper in an evacuated tube. (b) After the left needle pierces the vein, the evacuated tube is pushed left so the right needle pushes though the rubber stopper and blood fills the evacuated tube. (From Turgeon, M. L., *Clinical Hematology: Theory and Procedures*. 2nd Edition. Copyright © 1993 by Little, Brown and Company. Reprinted by permission of Little, Brown and Company.)

A blood smear can be made at the bedside or in the laboratory. A small drop of blood is placed near one end of a slide. A second slide is placed at 45° and wipes the blood into a thin film, which dries and is then stained with methylene blue or eosin.

Blood-borne pathogens, such as hepatitis and HIV, are a major concern to the health care industry. Precautions should be taken in every health care environment to avoid contact with blood. Devices such as the Unopette system (Figure 5.2) may be used in clinical settings to prevent contact.

The reservoir contains a premeasured volume of diluting fluid. Blood is frequently diluted to facilitate counting because it allows individual cells to be seen via microscopy. The diluting fluid may also be an anticoagulation agent, which prevents clotting of the blood, thus creating plasma. Without an anticoagulation agent the blood clots and creates serum. A hemolyzing (lysing) agent may be added to the blood to break up red blood cells, which allows for measurement of hemoglobin, platelets, and white blood cells.

The pipette has an overflow chamber and pipette shield. The overflow chamber prevents capillary action from overfilling the pipette. The pipette shield prevents technician injury while handling the pipette. When the pipette is filled with blood, the blood mixes with the diluting fluid in the reservoir. Different sizes and types of pipettes are available.

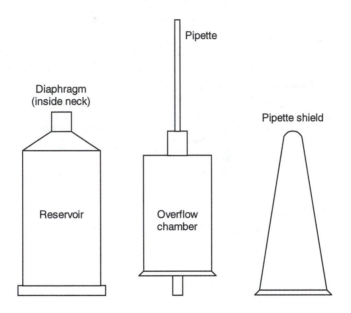

Figure 5.2 The Unopette system pipette fills by capillary action from the blood collection tube or free-flowing capillary blood from the lanced fingertip or from blood collection tube. The pipette covered with the shield punches through the diaphragm, then is removed and the shield discarded. The reservoir contains a premeasured amount of diluting solution. The reservoir is squeezed slightly, the pipette inserted on the top and pressure released. This draws the blood into the reservoir, where it is mixed. Slight squeezing then expels the diluted blood.

5.1.4 Centrifugal Method

Figure 5.3 shows the centrifugal analyzer tube, which offers accurate and rapid screening in the physician's office. The centrifugal analyzer examines the buffy coat layer in centrifuged whole blood samples. The float expands the layers by a factor of 10.71 and is only two to three cells thick. An ultraviolet light source, color filters, and a micrometer measure the thickness of each layer.

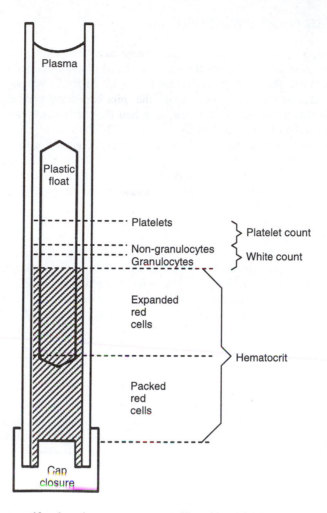

Figure 5.3 In the centrifugal analyzer, venous or capillary blood fills a capillary tube coated with acridine orange and oxalate stains and a float is introduced. After centrifugation, specific gravity variations separate the blood and an ultraviolet viewer shows different fluorescent colors for each layer. (From Stiene-Martin, E. A., Lotspeich-Steininger, C. A., and Koepke, J. A. (eds.) *Clinical Hematology: Principles, Procedures, Correlations.* 2nd Edition. Copyright © 1998 by Lippencott-Raven Publishers. Reprinted by permission of Lippencott-Raven Publishers.)

5.2 Red Blood Cells

Red blood cells are essential for oxygen delivery throughout the body. Red blood cells contain hemoglobin, a molecule that bonds with oxygen in the lungs and carbon dioxide in other tissues of the body.

5.2.1 Factors Affecting Red Blood Cell Count

The concentration of red blood cells is a basic indicator of how well the body is able to deliver oxygen. Average values for the number of red blood cells per liter vary by sex and age. For females, there are typically 3.6 to 5.6×10^{12} RBC/L. Males generally have a higher RBC count, between 4.2 to 6.0×10^{12} RBC/L. Newborns typically have an even higher RBC count, between 5.0 to 6.5×10^{12} RBC/L. After about one year the RBC count stabilizes at 3.5 to 5.1×10^{12} RBC/L.

There are many factors that can affect RBC count. Some of these include: children and adolescents, who have slightly lower counts than adults; people over 50 years of age, who have slightly lower counts than younger adults; strenuous physical activity, which increases RBC count; daily fluctuations between morning (highest count) and evening (lowest count).

RBC count may be indicative of some diseases. One example is polycythemia vera, a disease that results in increased red blood cells, white blood cells, and platelets. It is treated with bloodletting, radiotherapy, or cytotoxic drugs. Secondary polycythemia results in increased hemoglobin concentration caused by either a primary condition of respiratory or circulatory disorders that cause an oxygen deficiency in the tissues or with certain tumors, such as carcinoma of the kidneys. High altitude effects and dehydration are examples of respiratory or circulatory disorders. Decreased RBC count may be due to anemia, a reduction of hemoglobin in the blood, or secondary effects of several other disorders.

5.2.2 Hematocrit

Hematocrit is the volume of red blood cells. Red blood cells are heavier than other components of blood. Thus, using a centrifuge causes the red blood cells to separate and fall to the bottom. The hematocrit (HCT) is the ratio of the volume of all the formed elements to the total volume, normally 40% to 54% in men and 35% to 47% in women. It is called packed red cell volume (PCV). This measurement can be made manually or through automated instrumentation.

5.2.3 Hemoglobin

Hemoglobin (Hb) is a protein within the RBC that allows it to transport oxygen. Its concentration is normally 13.5 to 18 g/dL in men and 12 to 16 g/dL in women. If the RBCs are lysed (split open), the Hb is released and the concentration can be measured in a spectrophotometer.

5.2.4 Red Blood Cell Indices

Red blood cell indices characterize the RBC volume and Hb concentration. The mean corpuscular volume (MCV) is the hematocrit divided by the RBC count

$$MCV = \frac{HCT(L/L)}{RBC\,count(1/L)} = \frac{0.45(L/L)}{5\times10^{12}\,(1/L))} = 90\,fL = 90\,\mu m^3$$

The mean corpuscular hemoglobin (MCH) within the cell is the hemoglobin concentration divided by the RBC count

$$MCH = \frac{Hb(g/dL)}{RBC\,count(1/L)} = \frac{15(g/dL)\times10(dL/L)}{5\times10^{12}\,(1/L)} = 30\,pg$$

The mean corpuscular hemoglobin concentration (MCHC) within the cell is the hemoglobin concentration divided by the hematocrit

$$MCHC = \frac{Hb(g/dL)}{HCT(L/L)} = \frac{15(g/dL)}{0.45(L/L)} = 33.3\,g/dL = 33.3\%$$

The red cell volume distribution width (RDW) measures the spread of the RBC volume distribution, as shown in Figure 5.7(c).

$$RDW = \frac{s}{MCV}\times100 = \frac{11\,fL}{90\,fL}\times100 = 12.2\%$$

It is somewhat similar to the standard deviation, s, of a Gaussian distribution. A portion of the MCV distribution at the extreme ends is trimmed to exclude clumps of platelets and RBCs. An increased RDW beyond the normal range of 11.5 to 14.5% using the impedance method indicates a RBC production disorder.

5.2.5 Red Blood Cell Count

Neubauer hemocytometer

Figures 5.4 and 5.5 show the Neubauer hemocytometer, which is used to define the number of RBC/L of blood. A whole blood specimen with EDTA or heparin as the anticoagulant is mixed with an isotonic diluting fluid to assist counting and prevent lysis (splitting) of the red blood cells. Two sets of dilutions are made, and are inserted into the V slash at the edge of the central platforms and flow over the ruled areas.

Using a microscope, the RBC/L for each counting chamber is found by

RBC/L = (total cell count in all five squares)(volume counted)(0.02 μL)(dilution ratio)(10^6)

The results from each chamber are averaged. Error is generally 10% to 20%. This manual method of counting serves as a backup for automated counting.

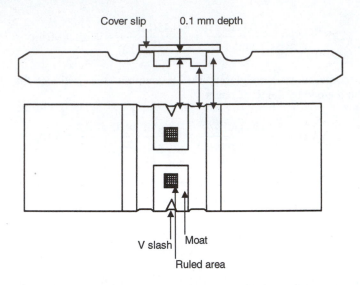

Figure 5.4 Neubauer hemocytometer, side and top view. The central platforms contain the ruled counting areas and are 0.1 mm under the cover slip, which is suspended on the raised ridges. From McNeely J. C. and Brown D. 1992. Laboratory evaluation of leukocytes. (From Stiene-Martin, E. A., Lotspeich-Steininger, C. A., and Koepke, J. A. (eds.) *Clinical Hematology: Principles, Procedures, Correlations.* 2nd Edition. Copyright © 1998 by Lippencott-Raven Publishers. Reprinted by permission of Lippencott-Raven Publishers.)

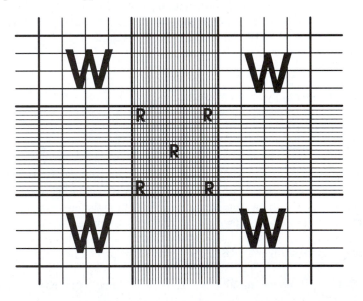

Figure 5.5 The Neubauer cytometer ruled counting area is 3 mm × 3 mm. The red blood cell counting area is marked by R and is 200 μm × 200 μm. The white blood cell counting area is marked by W.

Automatic Analyzers

Figure 5.6 shows the fluid diagram of a typical automatic analyzer. The whole blood is aspirated, leaving only RBCs, which are then diluted and mixed. Some are lysed to release the hemoglobin, which is analyzed spectrophotometrically in a colorimeter cuvette. Some RBCs are diluted a second time and analyzed by light scattering or impedance. The rest are lysed so the smaller WBCs and platelets can be analyzed without the error caused by the larger RBCs.

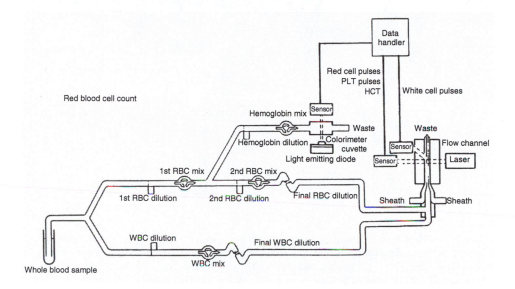

Figure 5.6 An automatic analyzer aspirates whole blood, divides it, dilutes it, mixes it and then analyzes it for hemoglobin and cell characteristics. (From Lotspeich-Steininger, C. A., Stiene-Martin, E. A. and Koepke, J. A. (eds.) 1992. *Clinical Hematology: Principles, Procedures, Correlations.* Copyright © 1992 by Lippencott Publishers. Reprinted by permission of Lippencott Publishers.)

Coulter Counter

Coulter counters determine the count and volume of RBCs. The Coulter counter uses blood diluted in an isomolar conducting solution. A vacuum draws a specific amount of the diluted sample through an aperture of 100 μm in diameter. Figure 5.7 shows an electric current passing between the internal and external electrode. As a nonconducting cell passes through the aperture, the resistance of the aperture increases and the voltage between the electrodes increases. The magnitude of the voltage change indicates cell volume.

Figure 5.7(a) shows a noncentral cell passing through a region of high current density, which causes an incorrectly higher voltage change. Figure 5.7(b) shows hydrodynamic focusing using a surrounding fluid stream sheath to force the cells through a central path. Figure 5.7(c) shows the resulting narrower cell volume distribution.

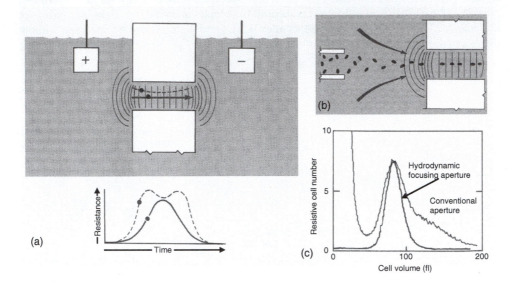

Figure 5.7 (a) Nonconducting cells passing through the aperture of the Coulter counter increase the resistance between the electrodes. (b) A surrounding fluid sheath forces cells to flow centrally in hydrodynamic focusing. (c) Hydrodynamic focusing yields a narrower, more accurate cell volume distribution. (From Handin, R. I., Lux, S. E. and Stossel, T. P. (eds.) *Blood: Principles & Practice of Hematology.* Copyright © 1995 by J. B. Lippincott Company. Reprinted by permission of J. B. Lippincott Company.)

Figure 5.8 shows the oscilloscope plot of aperture voltage versus time. Figure 5.9 shows that the Coulter counter calculates each cell volume and produces histograms of number of cells versus volume of cells.

Light Scattering

Figure 5.10 shows a flow cytometer. A fluid containing cells is pumped through a viewing window and the number of cells is counted. Whole blood is diluted with a fluid that will be absorbed by the red blood cells to make them round. Their total volume is unchanged (isovolumetric sphering). With a laser located on one side of the flow cell and a light scatter detector is on the opposite side, a diaphragm pump pushes the sample toward the flow cell. The volume of red blood cells is determined as they pass in front of the light beam by comparing high angle and low angle light scatter. Cells are labeled with dyes that fluoresce to aid in counting the different cell types.

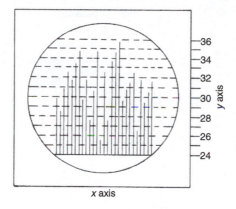

Figure 5.8 Each cell passing through the Coulter counter aperture causes a resistance change proportional to the cell volume. Thus measuring the height of each voltage spike yields the cell volume. (From Turgeon, M. L. 1993. *Clinical Hematology: Theory and Procedures.* Copyright © 1993 by Little, Brown and Company. Reprinted by permission of Little, Brown and Company.)

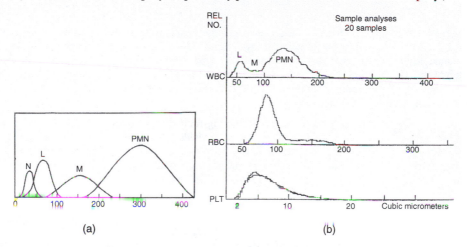

(a)   (b)

Figure 5.9 Number of cells versus cell volume from a Coulter counter. (a) Nucleated RBCs (N), lymphocytes (L), mononuclear cells (M), and polymorphonuclear leukocytes (PMN). (b) Leukocyte differential distribution (WBC), RBC distribution (RBC), and platelet distribution (PLT). (From Handin, R. I., Lux, S. E. and Stossel, T. P. (eds.) *Blood: Principles & Practice of Hematology.* Copyright © 1995 by J. B. Lippincott Company. Reprinted by permission of J. B. Lippincott Company.)

Figure 5.11 shows the Technicon H*1 instrument nomogram. It measures RBC volume [Mean Corpuscular Volume (MCV)] by low-angle (2° to 3°) forward light scatter after isovolumetric sphering and intracellular hemoglobin concentration (HC) by high-angle (5° to 15°) light scatter. This yields corpuscular hemoglobin concentration (CHC). For example, if the low and high angle pulse heights are both 30, then the corpuscular volume is 120 mL and the corpuscular hemoglobin concentration is 35 g/dL.

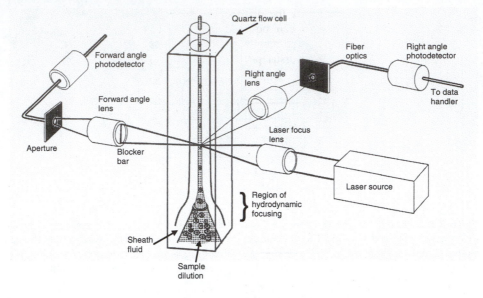

Figure 5.10 In flow cytometry, a sheath surrounds the sample to hydrodynamically focus the cells to the center, where they are illuminated by a laser. Forward (low) angle scatter measures cell volume. Right (high) angle scatter measures cell type. (From Stiene-Martin, E. A., Lotspeich-Steininger, C. A., and Koepke, J. A. (eds.) *Clinical Hematology: Principles, Procedures, Correlations.* 2nd Edition. Copyright © 1998 by Lippencott-Raven Publishers. Reprinted by permission of Lippencott-Raven Publishers.)

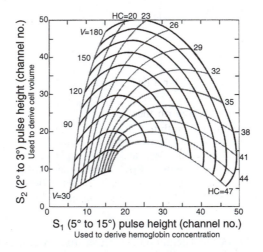

Figure 5.11 Mia analysis of RBCs measures low-angle and high-angle scattering to yield corpuscular volume *V* in fL and corpuscular hemoglobin concentration CHC (HC) in g/dL. (From Handin, R. I., Lux, S. E. and Stossel, T. P. (eds.) *Blood: Principles & Practice of Hematology.* Copyright © 1995 by J. B. Lippincott Company. Reprinted by permission of J. B. Lippincott Company.)

The leukocyte differential presents the distribution of leukocyte types. The RBCs are lysed so the smaller leukocytes can be analyzed. Peroxidase cytochemical staining permits detection of granulocytes and monocytes (types of white blood cells). Figure 5.12 shows the forward light scatter versus peroxidase absorption and the location of the different cell classes.

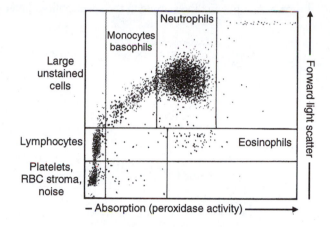

Figure 5.12 Leukocyte differential classifies the five basic leukocyte classes by forward light scatter versus peroxidase absorption. (From Handin, R. I., Lux, S. E. and Stossel, T. P. (eds.) *Blood: Principles & Practice of Hematology.* Copyright © 1995 by J. B. Lippincott Company. Reprinted by permission of J. B. Lippincott Company.)

Cell Sorting

Figure 5.13 shows a cell sorter. After a cell is identified, the computer waits a specified time until the cell is within a formed droplet. A surrounding charging collar is pulsed to apply a positive or negative charge on the droplet. Electrostatic deflection plates attract the charged droplet and deflect its path to the selected collection plate. Sorted cells can be further analyzed or used for therapy.

5.2.6 Hemoglobin

In order to measure hemoglobin, it must first be converted to cyanmethemoglobin. Red blood cells are lysed with a nonionic detergent. Potassium cyanide and potassium ferricyanide are added to convert all forms of hemoglobin, except sulphemoglobin, to cyanmethemoglobin. Potassium cyanide and potassium ferricyanide are added to lower the pH and speed the reaction. The solution is then examined by light at 540 nm using a spectrophotometer (see Chapter 3).

If high levels of sulphemoglobin are present, the hemoglobin levels are slightly underestimated. For example, a sulphemoglobin level of 5% at a concentration of 15 g/dL of hemoglobin yields a hemoglobin measurement of 14.8 g/dL (Bain, 1995).

Carboxyhemoglobin is present in high levels in heavy smokers so errors occur if the test is read before it is all converted into cyanmethemoglobin. A maximum error of 20% occurs if the level of carboxyhemoglobin is 6%. This error is avoided by waiting for a longer period of time before placing the sample in the spectrophotometer (Bain, 1995).

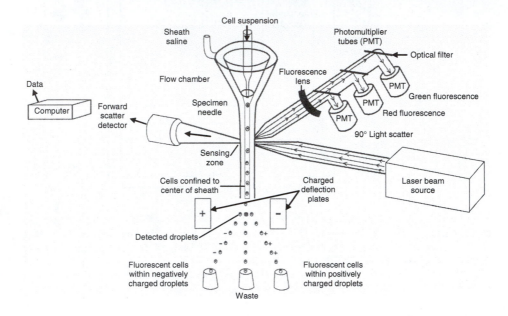

Figure 5.13 After cell identification by fluorescent light scattering, the cell sorter charges each cell droplet, and electrostatically deflects it to separate cells by type. (From Stiene-Martin, E. A., Lotspeich-Steininger, C. A., and Koepke, J. A. (eds.) *Clinical Hematology: Principles, Procedures, Correlations.* 2nd Edition. Copyright © 1998 by Lippencott-Raven Publishers. Reprinted by permission of Lippencott-Raven Publishers.)

5.2.7 Reticulocyte Count

In the six stages of red blood cell development, the first four stages are contained in the bone marrow, but the fifth stage, reticulocyte, is present in the bone marrow and the peripheral blood. Red blood cells are replaced at about 1% per day, thus the count of reticulocytes is a good measure of RBC production in the bone marrow.

Small amounts of RNA are present in the reticulocytes. Methyl blue is used to stain the reticulocytes and then by using a microscope, the reticulocytes and other red blood cells can be counted. The reticulocyte count is represented as a percentage of the total red blood cells counted. Error within the normal range is approximately 25%. For a result of 5%, the error is approximately 10%.

Reticulocyte count can also be done using flow cytometry. The cells are first diluted with an acidic solution to clear the cells of hemoglobin. Cells in their earlier stages have more RNA and absorb more dye (New Methylene Blue), thus scattering more light than later stage cells. Because of this, cells of different stages can be differentiated and counted.

5.2.8 Sickle Cell Test

Sickle cell anemia is caused by an abnormal type of hemoglobin in red blood cells. The sickling occurs when the cell is deprived of oxygen. To test for anemia, whole blood is mixed with sodium metabisulfite, a strong reducing agent that deoxygenates the hemoglobin. The sample is examined under the microscope after one hour. If sickling is present, this indicates sickle cell anemia. If sickling occurs after 24 hours the sickle cell trait is present.

5.3 White Blood Cells

White blood cells are an essential part of the body's immune system. The normal values of white blood cell counts are independent of sex, as are RBCs. Average values for adults are between 4.0 to 11.0×10^9/L. A newborn's white blood cell count ranges from 10.0 to 30.0×10^9/L. By one year of age, average values are between 6.0 and 17.0×10^9/L.

Many factors can affect the white blood cell count. One cause of elevated levels may be infection. Other causes are bacterial infections, appendicitis, leukemia, pregnancy, hemolytic disease of newborn, uremia, and ulcers.

Decreased white blood cell counts may indicate viral infection, brucellosis, typhoid fever, infectious hepatitis, rheumatoid arthritis, cirrhosis of the liver, lupus, or erythema.

White blood cell counts can identify infections or be used to follow the progress of a disease or treatment. The Neubauer hemocytometer described in Section 5.2.5 is used for the white blood cell count. The blood is diluted with a solution that hemolyzes red blood cells and stains the nucleus of white blood cells. The counting volume is 0.4 µL, and for normal values, the error is approximately 15%.

5.3.1 Differential White Blood Cell Counts

There are several different types of white blood cells. Each type of cell serves a different function. Thus, it is of interest to know the differential white blood cell count. Table 5.1 lists each type of white blood cell.

A granulocyte is a leukocyte that characteristically contains many granules in its cytoplasm. Granulocytes include neutrophils, eosinophils, basophils, and their precursors.

A neutrophil is stainable by neutral dyes. Neutrophilia is an increased level of neutrophils, and may indicate appendicitis, myelogenous leukemia, and bacterial infections. A band neutrophil is a neutrophil in which the nucleus is elongated but not segmented.

An eosinophil is a cell that stains intensely with the acid dye eosin. Eosinophilia is an increased level of eosinophils, which may indicate allergies, allergic reaction, scarlet fever, parasitic infections, or eosinophilic leukemia.

A basophil is a cell that stains with basic dyes. An increase in the number of basophils in blood is characteristic of chronic granulocytic leukemia, but may also be observed following infections such as pertussis, measles, tuberculosis, and hepatitis.

A lymphocyte is a leukocyte that has a round nucleus with well-condensed chromatin, no identifiable nucleolus, and usually cytoplasm with no granules. Lymphocytosis is an increased level of lymphocytes, which may indicate viral infections, whooping cough, infectious mononucleosis or lymphocytic leukemia.

A monocyte differs from the granular leukocytes by its larger size (12 to 20 µm in diameter), and by a round or indented nucleus. Its cytoplasm has no granules. Monocytosis, an increased level of monocytes, may indicate brucellosis, tuberculosis, monocytic leukemia, subacute bacterial endocarditis, typhoid, rickettsial infections, collagen disease, Hodgkin's disease, or Gaucher's disease.

Table 5.1 Normal values of white blood cell types. (From Brown, B.A. *Hematology: Principles and Procedures.* 6th ed. Copyright © 1993 by Lea & Febiger. Reprinted by permission of Lea & Febiger.)

Cell type	Percentage of WBC	Absolute number ($\times 10^9$/L)
Neutrophil	35–71	1.5–7.4
Band	0–6	0.0–0.7
Lymphocyte	24–44	1.0–4.4
Monocyte	1–10	0.1–1.0
Eosinophil	0–4	0.0–0.4
Basophil	0–2	0.0–0.2

Wedge Smear and Staining

The blood sample is diluted with a solution that lyses the red blood cells. A sample of blood is placed on a microscope slide using the wedge smear technique. Each type of white blood cell is stained, and the sample is examined under the microscope where each cell type is counted. The accuracy of such a technique is usually fairly good, but precision is poor (Bain, 1995).

Coulter Counter

The Coulter counter can be used to compute white blood cell count and differential white blood cell count. Whole blood is mixed with a lysing solution that splits red blood cells, but not the white blood cells. The same process is used as described in Section 5.2. With the size information found with WBC, three different cell types can be identified in Table 5.2.

Table 5.2 Sizes of white blood cell types. (From Brown, B.A. *Hematology: Principles and Procedures.* 6th ed. Copyright © 1993 by Lea & Febiger. Reprinted by permission of Lea & Febiger.)

Cell type	Size (fL)
Lymphocytes	35–90
Monocytes	90–160
Granulocytes	160–450

Flow Cytometry

To perform a five-part differential counting, flow cytometry is used. Three simultaneous measurements are made as the sample passes through the flow chamber. The volume measurements are made from aperture impedance. Conductivity measurements use a high-frequency electromagnetic probe to examine the cell's internal structure. Light scatter is a function of the cell surface features and internal structure. Light scattering is measured using a helium–neon laser and a photodetector. These measurements are integrated to determine cell type. All six types of white blood cells can be identified by this technique (Brown, 1993).

5.4 Platelets

Platelets are essential for clots to form, otherwise excessive bleeding occurs. However, too many platelets may lead to clotting and blockage within the blood stream. Normal values range from 150 to 400×10^9/L. Abnormal factors causing increased platelet count include thrombocytosis, polycythemia vera, idiopathic thrombocythemia, chronic myelogenous leukemia, and splenectomy. Factors causing a decreased platelet count include thrombocytopenia, thrombocytopenia purpura, aplastic anemia, acute leukemia, Gaucher's disease, pernicious anemia, chemotherapy, or radiation treatments.

5.4.1 Phase Hemocytometer

Platelets are counted using a phase microscope and a phase hemocytometer, a thin device with a flat bottom. Before viewing the sample, a thin disposable cover slip should be used. Then a phase microscope (see Section 6.2.2) uses objectives and condensers to retard transmitted light with respect to light through the sample by 1/2 to 1/4 of a wavelength in order to look at the sample (Bancroft & Stevens, 1990). Whole blood is diluted with 1% ammonium oxalate, which hemolyzes the red blood cells, thus reducing the hemoglobin. Now the platelets can be counted as in Figure 5.14 using the phase hemocytometer and the phase microscope. Results are checked by examining the platelets on a Wright stained smear. The same formula can be used to calculate the platelet count as used in Section 5.2.5.

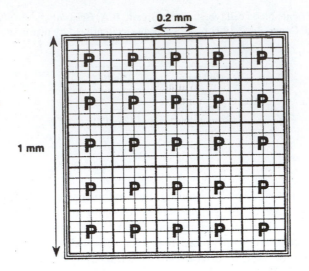

Figure 5.14 Suggested squares to use for platelet count.

Coulter Counter

In the red blood cell count with the Coulter counter, particles between 2 to 20 fL in size are defined as platelets.

Flow Cytometry

The technique described in Section 5.2.5 is used. Using the size information derived in this measurement, platelets are defined.

Blood Clotting

In the electromechanical method, an electric current flows from a fixed electrode to a moving electrode through plasma samples and a thromboplastin substrate. When a fibrin strand forms between the two electrodes, this completes the circuit and indicates clotting time.

In the photooptical method, red light from a light-emitting diode passes through a cuvette. When the rate of change of absorbance exceeds a predetermined level, this indicates clot formation (Turgeon, 1993).

5.5 Complete Blood Count

The complete blood count (CBC) is a common panel of tests ordered by clinicians. This panel includes hematocrit, red blood cell count, hemoglobin, platelet count, and white blood cell count. A typical option available is the differential white blood cell count. Many devices such as the Coulter counter incorporate all of these measurements into one device. Flow cytometry can also be used to make all of these measurements. This is much more efficient than manual techniques.

5.6 Problems

5.1   Explain how the Unopette system achieves accurate dilution.

5.2   Explain how the centrifugal analyzer tube expands the cell layers.

5.3   Calculate the hematocrit, given the mean corpuscular volume is 100 fL and the red blood cell count is 4×10^{12}/L.

5.4   Explain why we measure red blood cell counts and how we do it, both with automated and manual techniques.

5.5   Diluted blood with a resistivity of 160 $\Omega \cdot$cm passes through a Coulter counter cylindrical aperture 100 μm in diameter and 1 mm long. Calculate the resistance of the liquid filled aperture.

5.6   Blood with a HCT of 0.40 L/L and MCV of 90 fL is diluted 1:10,000. The volume of the RBC carrying central stream in a Coulter counter aperture is 100 pL. Use Poisson statistics to calculate the probability of 2 RBCs being counted as 1.

5.7   Explain how hydrodynamic focusing improves cell volume measurement.

5.8   Sketch the circuit used for measuring red blood cell (RBC) volume using impedance. Explain the reason for each circuit component. Sketch the resulting RBC histogram, label axes and units and explain how the circuit yields the histogram.

5.9   In the Technicon H*1 flow cytometer, low-angle pulse height = 20 and high-angle pulse height = 40. Determine the corpuscular volume and corpuscular hemoglobin concentration.

5.10   Explain how cells are automatically sorted by type.

5.11   Explain why we measure hematocrit and how do it, both directly and indirectly.

5.12   Explain why we measure hemoglobin concentration and how we do it.

5.13   Explain why we measure white blood cell counts, and how we do it, both with automated and manual techniques.

5.14   A white blood cell count determined manually is 9.0×10^9 cells/L. If the volume counted was 0.4 μL and the cells were diluted 1:10, determine the average number of cells counted in each square of the Neubauer cytometer. Explain how this number is used to yield the total number of cells per unit volume, that is, what each of the factors represents in the calculation.

5.15   Explain why we would measure the differential white blood cell count, and how we do it, both with manual and automated techniques.

5.16 Explain why we measure platelet counts, and how we do it, both automatically and manually.
5.17 Describe the complete blood count.
5.18 Describe automated devices used to measure the complete blood count.

5.7 References

Bain, B. J. 1995. *Blood Cells: A Practical Guide.* London: Blackwell Science.

Bancroft, J. D. and Stevens, A. (eds.) 1990. *Theory and Practice of Histological Techniques.* 3rd ed. London: Churchill Livingstone.

Beck, W. S. (ed.) 1991. *Hematology.* 5th ed. London: The MIT Press.

Brown, B.A. 1993. *Hematology: Principles and Procedures.* 6th ed. Philadelphia: Lea & Febiger.

Handin, R. I., Lux, S. E. and Stossel, T. P. (eds.) 1995. *Blood: Principles & Practice of Hematology.* Philadephia: Lippincott.

McKenzie, S. B. 1996. *Textbook of Hematology.* 2nd ed. Baltimore: Williams & Wilkins.

Stiene-Martin, E. A., Lotspeich-Steininger, C. A., and Koepke, J. A. (eds.). 1998. *Clinical Hematology: Principles, Procedures, Correlations.* 2nd Edition. Philadelphia: Lippencott-Raven Publishers.

Turgeon, M. L. 1993. *Clinical Hematology: Theory and Procedures.* 2nd ed. Boston: Little, Brown.

Wheeler, L. A. 1998. Clinical laboratory instrumentation. In Webster, J. G. (ed.) *Medical Instrumentation: Application and Design.* 3rd ed. New York: John Wiley & Sons.

Chapter 6

Cellular Measurements in Biomaterials and Tissue Engineering

Jeffrey S. Schowalter

This chapter discusses the different types of instrumentation used in the investigation of biological processes on the cellular level. Major advances have been made over the past several years allowing for the observation of cells and their structure. Many of the mysteries of how various inter- and intracellular processes work have been solved in recent years due to these advances in bioinstrumentation. One area that has benefited the most is tissue engineering and the development of biomaterials. Tissue engineering draws from a number of different disciplines including cell biology. Researchers use live cells, chemicals, and other synthetic materials to develop biomaterials for implantation in the human body (Lewis, 1995). A key reason researchers have been able to make this progress is due to the significant advances in the area of cellular bioinstrumentation.

6.1 Cell Measurement Overview

Scientists have been observing cells since the invention of the microscope by Antoni von Leewenhoek in the early seventeenth century. *Microscopy*, or using a microscope to study an object, has made significant advances since then. With the advent of newer manufacturing processes and new technology, the microscope is a major tool for today's biological scientist. First let's briefly look at a basic cell, different types of cellular measurements, and some of the parameters of interest to the biomedical researcher.

6.1.1 What Is a Cell?

A cell is the most basic structural unit that makes up plants, animals, and single cell organisms. It is the basic unit of life, and all organisms are composed of one or more cells. There are two types of cells: those with a nucleus (eukaryotic cells) and those without (prokaryotic cells). Eukaryotic cells are composed of three main features: a nucleus, surrounded by cytoplasm, which in turn is surrounded by a sack-like plasma membrane.

This membrane is made from a phospholipid bilayer and is semipermeable, which means it allows only certain things to pass in or out of the cell. Cells are 90% fluid (cytoplasm), which consists of amino acids, proteins, glucose, and numerous other molecules. On an elemental level, the cell contains 59% hydrogen, 24% oxygen, 11% carbon, 4% nitrogen, and 2% others including phosphorus and sulfur. On a molecular level, the cell contains 50% protein, 15% nucleic acid, 15% carbohydrates, 10% lipids, and 10% other. Table 6.1 shows the contents of the typical cell.

Table 6.1 Typical cell content

Structure	Description
Cytoplasm	The inside of the cell not including the organelles.
Organelles	Membranous sacs within the cytoplasm.
Cytoskeleton	Structural support made of microtubules, actin and intermediate filaments.
Endoplasmic Reticulum (ER) (two types)	Site of protein and lipid synthesis and a transport network for molecules.
Rough ER	Has ribosomes and tends to be flat.
Smooth ER	Does not have ribosomes and is tubular.
Golgi Apparatus	Modifies molecules and packages them into small membrane bound sacs called vesicles.
Lysosomes	Main point of digestion.
Microtubules	Made from tubulin and make up centrioles, cilia, cytoskeleton, etc.
Mitochondria	Site of aerobic respiration and the major energy production center.
Nucleus	Location of DNA; RNA transcription.
Peroxisomes	Use oxygen to carry out catabolic reactions.
Ribosomes	Located on the Endoplasmic Reticulum in the cytoplasm. RNA goes here for translation into proteins.

Table 6.2 Typical sizes of cellular features

0.1 nm	Diameter of hydrogen atom
0.8 nm	Amino acid
2 nm	Thickness of DNA membrane
4 nm	Protein
6 nm	Microfilament
7 to 10 nm	Cell membranes
17 to 20 nm	Ribosome
25 nm	Microtubule
50 to 70 nm	Nuclear pore
100 nm	AIDS virus
200 nm	Centriole
200 to 500 nm	Lysosomes
200 to 500 nm	Peroxisomes
1 μm	Diameter of human nerve cell
2 μm	Bacteria
3 μm	Mitochondrion
3 to 10 μm	Nucleus
9 μm	Human red blood cell
90 μm	Amoeba
100 μm	Human egg

To provide a rough idea of the scale that cell biologists and other related researchers are working on, Table 6.2 shows the typical sizes for some of the cell features.

Many processes that occur within cells are not well understood. Individual cells are complex systems by themselves but when added to their multicellular environment, the system becomes quite intricate. The human body contains over 200 types of cells of various shapes, sizes, and functions, yet each contains the same genetic material.

6.1.2 Fixed Versus Live Cells

The study of cells by a microscope is divided into two broad categories. The traditional method of fixed-cell evaluation is through viewing cells that have been permanently affixed to a microscope slide. The other way to examine cells is while they are living (e.g. as part of a tissue or in a controlled environment like a petri dish).

Fixed Cells

In some cases, with relatively loose or transparent tissue, the object under examination can be placed on a slide under the microscope. If the item of interest is part of a larger object, such as an organ or bone, then the tissue must be sliced. If the object is generally opaque, then the slice either must be very thin or some other method must be used as discussed in Chapter 4. Structures inside most cells are almost colorless and transparent to light. Therefore, dyes are used which react with some of the cell's structures. Staining the cells may alter the structure of the cell or kill it. Because of this, cells are normally preserved or fixed for observation. This is normally done with some fixing agent such as formaldehyde. The general structure of the cell and organelles may be preserved, but enzymes and antigens within the cell may be destroyed. Usually, the fixing process takes at least 24 h.

Another method involves the freezing or *cryofixing* of cells for observation. Typically, a section of tissue is quick–frozen to –15 to –20 °C using liquid nitrogen and sectioned in a device called a cryostat, which is a microtome mounted in a freezing chamber. Thin sections can be prepared in minutes, and the sections usually retain a frozen snapshot of their enzymes and antigens. However, this method has the disadvantage that ice crystals formed during the freezing process may distort the image of the cell, and the specimens also have a tendency to freeze-dry, so these sections must be used immediately. This is good when trying to make a diagnosis while a patient is on the operating table but not so good for long-term research studies.

To get some semblance of time responses using fixed cells, a stimulus is provided to the cells and they are quick–frozen moments later. The characteristics of individual cells at different stages of response to the stimulus are examined, and these individual responses are pieced together into one response that describes the entire process.

Sometimes the researcher may want to isolate a certain component of the cell for study. The process of breaking open cells is called *homogenization*, and the separation of organelles is called *fractionation*. Isolating organelles requires techniques such as gravity sedimentation and centrifugation.

Homogenization is normally accomplished by using force to break the cell membrane. The most common procedures use glass mortar and pestle arrangements with controlled bore sizes. When the pestle is inserted and turned rapidly in the tube containing small pieces of tissue, intracellular connections are broken, and membranes of whole cells are ruptured. The resulting mixture is a suspension of organelles and other material from the broken cells. Ultrasonic waves are also used to break the cell membrane, leaving the organelles intact.

After homogenization, the various components must be separated. For some materials, this is accomplished by gravity, which separates the organelles naturally due to differences in their density. The most widely used instrument for separation, however, is the centrifuge, in which a mixture of cell components is subjected to increasing centrifugal forces. At each level, the material that sediments is saved and the supernatant can be centrifuged at higher forces. Organelles of low density and small size sediment at high forces, while organelles of high density and large size sediment at low forces. A centrifuge that operates at speeds greater than 20,000 rpm is called an ultracentrifuge. Speeds can reach 110,000 rpm yielding 199,000 g.

Live Cells

Live cells have the advantage of time not typically being a factor when viewing. The cells are viewed in their natural environment and time studies are easily accomplished using video or other recording devices.

One of the three general categories of live cell research is the division of cell population studies. This is done while the cells are free to interact with one another. With this configuration, not only the activity within the cell can be studied, but also the interaction between cells and other substances or objects external to the cellular structure, such as man-made materials.

A second category of cell research involves examining the characteristics of an individual cell. This may be the physical features of the cell, its topology, or the relative chemical/molecular structure of the cell. Another area of study involves the response of the cell to experimental treatments.

Single cell imaging involves one of two different techniques. The first is the 3-D reconstruction of a scanned object. Using digital (computer) techniques, characterization of single cells can be done by the cell researcher. The second technique involves time sequencing of the movement, or morphology, of a cell. This involves the use of video cameras and video recording systems. For example, cell adhesion may be studied by video and fluorescence imaging of the spreading and detachment of adherent cells subjected to a range of fluid shear stresses in a radial flow chamber. Cell motility is assessed by time-lapse video microscopy of individual cells crawling on sheets of polydimethylsiloxane.

6.2 Light Microscopy

Since there is a theoretical limit (Rayleigh limit) on how much an object can be magnified and still seen clearly, scientists have resorted to a variety of new technologies and developed a number of techniques to enhance their understanding of the cellular world.

Light microscopes rely on the bending, or refraction, of light. When a curved lens is used, light rays pass straight through the center of the lens while rays passing through the outer edges of the lens are bent. The farther away from the center light passes, the more the bending, as shown in Figure 6.1(a). All light rays pass through one point, the focal point, F, placed at the focal length L_F.

Figure 6.1(b) shows that when an object is placed outside the focal length L_F, it is magnified by the magnification $mm = L_I/L_O$, where L_I is the image length, and L_O is the object length, and lengths are related by the lens makers' equation

$$\frac{1}{L_O} + \frac{1}{L_I} = \frac{1}{L_F}$$

(6.1)

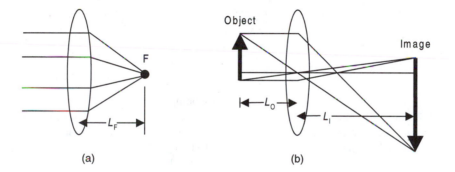

(a)   (b)

Figure 6.1 (a) Light rays through a lens and the corresponding focal point F. (b) The light rays are bent by the lens and refocus as an image.

Figure 6.2 shows a conventional (wide field and bright field) compound microscope composed of two lenses. The primary lens (objective) has a very short focal length placed close to the sample. It forms a much larger primary (intermediate) image just below the eyepiece. The secondary lens (eyepiece) alters the primary image by forming parallel rays that are focused on the retina by the lens of the eye. This is the enlarged virtual image seen when looking into a microscope. Light is intensified by the condenser lens so it more readily transmits through the sample.

If an electron beam replaces the beam of light, the microscope becomes a transmission electron microscope (Chapter 4). If light is reflected from the object instead of passing through, the light microscope becomes a dissecting microscope. If electrons are reflected from the object in a scanned pattern, the instrument becomes a scanning electron microscope (Chapter 4).

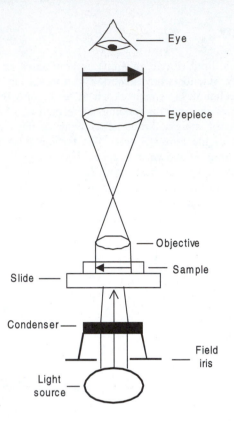

Figure 6.2 Compound microscope. The compound microscope contains a light source for sample illumination, a field iris to control the light field, a condenser to focus the illuminating light, an objective lens, and an eyepiece.

6.2.1 Resolution Versus Magnification

The microscope's function is to allow viewing of objects that cannot be seen with the human eye. Because we are trying to make an object look larger, the term resolution is often confused with magnification. As previously discussed in Chapter 4, magnification refers to the size of the image as compared to the original object and is usually expressed as × *mm*, where *mm* is the amount of magnification. However, resolution determines whether small objects that are close together can be distinguished as separate objects. As a general rule, the greater the magnification, the greater the resolution. However, there are several practical limitations of lens design that result in increased magnification without increased resolution. If an image of a cell is magnified by a factor of ten, the image will get larger, but not necessarily any clearer. Without resolution, no matter how much the image is magnified, the amount of observable detail is fixed. Regardless of how much image size increases, no more detail can be seen. At this point, the limit of resolution has been reached. This is also known as the resolving power of the lens and is only a function of the lens.

The human eye can only distinguish between two objects when they are about 0.1 mm apart at a distance of 100 mm. If two objects are 0.01 mm apart, they cannot be distinguished unless we magnify their image by times 10. In this case, the limit of resolution has changed from 0.1 mm to 0.01 mm, or inversely, our resolving power has increased by a factor of 10.

Unfortunately, a lens can magnify an image without increasing the resolution. Lenses are not perfect and cause objects to become blurry at the edges. Thus, even though they can be made to appear 0.1 mm apart, the edges are so blurry that we lose the ability to see them as two distinct objects. While microscope lenses are typically expressed in terms of magnification, the important value is the resolution.

The resolution of a lens is a function of how it is configured and the wavelength of light that is passed through the lens.

$$\text{Resolution} = \frac{0.61\lambda}{n \sin \alpha} \tag{6.2}$$

where λ is the wavelength of the illuminating light, n is the refractive index of the medium between the objective and the specimen, and the cone angle α is 1/2 the angle of the acceptance of light onto the lens (Figure 6.3). The value $n \sin\alpha$ is known as the numerical aperture (n.a.).

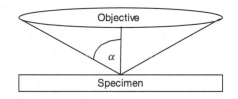

Figure 6.3 Cone angle α

There are three methods to improve resolution. One is to use shorter wavelengths of light, typically done using filters. Another is to increase the cone angle by placing the objective as close as possible to the specimen, practically the largest value of $\sin\alpha$ is 0.95 (GEK, 1997). Finally, the refractive index, n, can be increased. Immersion oil is sometimes used in place of the air between the objective and the slide to increase n.

With bright field microscopy, work is usually done in the visible light range so the image can be seen directly with the human eye. Since the shortest wavelength of visible light is blue, almost all microscopes incorporate a blue filter, often referred to as a daylight filter. However, the human eye has a maximum sensitivity to green and since specimens contain various colors, the resolution of an image typically varies between 0.2 to 0.35 µm depending on the color of the light source. Resolution can be enhanced by reducing the wavelength to the ultraviolet range. However, ultraviolet (UV) light cannot be seen directly by the human eye. Thus, this form of microscopy requires a detector or photographic film that is sensitive to UV, because it improves the resolution by a factor of two over visible light.

The best resolution is not attainable, however, unless the lenses are corrected for problems common to lens design. Modern microscopes normally use a set of lenses assembled to maximize the refractive index while minimizing chromatic and spherical distortions (aberrations), which occur in all lenses to some extent due to the fact that a 3-D object is being viewed as a 2-D image. However, aberrations are typically determined by how well the lenses were designed and manufactured. In general, the smaller the optical distortion, the greater the cost.

6.2.2 Light Microscope Modes

Bright Field

The light microscope can use several methods to view objects. Bright field microscopy is used to view stained or naturally pigmented organelles in the cell, and it involves the use of color detection. If an object contains colored features, the eye detects those colors. The stains or pigments absorb and alter the light that passes through the sample. Therefore, light passing through the sample is seen as a color that is distinctly different from the background or direct light not passing through the sample.

Most of the features in the cell are relatively colorless and difficult to see under the microscope. The easiest way to enhance the contrast of the relatively transparent cells is to stain them. Different colored dyes are used that stain different parts of the cell. If the cells are fixed, this is not a problem except for possible damage to the cellular integrity from the dye.

Further image enhancement is accomplished by using a specific monochromatic light source. One way to achieve this is by using a filter, the most common is one that allows only blue light to pass through. Blue light, about 450 nm, is usually the shortest wavelength used in light microscopy. It determines what cell features you can and cannot see under the microscope. In order to see a specific structure of the cell, it must be large enough to be able to perturb the wave motion of the light rays that strike it. Therefore, mitochondria, about 3 μm, are one of the smallest organelles to be seen through microscopy. Organelles smaller than mitochondria cannot be resolved because they are small compared to the wavelengths of visible light and are not able to perturb the wave motion of light traveling to our eyes.

Phase Contrast

Another method of viewing objects is called phase contrast microscopy. This type of microscopy is used to view living cellular structures because they are not pigmented, thus unable to be seen under bright field microscopy, and the structures within the cell tend to differ in refractive indices.

Phase contrast takes advantage of the fact that when rays of light pass through areas of differing refractive index or different lengths of optical path (thicker sections), the light rays get out of phase with one another. This causes interference, which in turn causes different areas to appear as differences in brightness when seen by the eye. Phase contrast microscopy makes use of these minute shifts in the phase as the light passes through an object. If the normal phase shift is increased by 1/4 or 1/2 wavelength, the differences will be magnified in intensity. Phase contrast microscopes that provide 1/4 wavelength shift are called medium phase while those that provide 1/2 wavelength shift are called dark phase.

A standard compound microscope is modified for phase contrast microscopy, as shown in Figure 6.4, by adding annular transparent rings, one below the eyepiece and another below the condenser. Some light transmitted from the lower annular ring passes through the specimen and is imaged on an upper semitransparent ring. Other light transmitted from the lower annular ring is refracted slightly by different parts of the specimen, passes outside the upper semitransparent ring, and is imaged on a surrounding transparent diaphragm, which is one-quarter wavelength thicker. Since the rays of light travel slightly different distances, when they meet up again, they will be slightly out of phase with one another. If the phase differences add to each other, referred to as constructive interference, the image will be brighter (Figure 6.5). If they cancel out, referred to as destructive interference, the image will be darker (Figure 6.6).

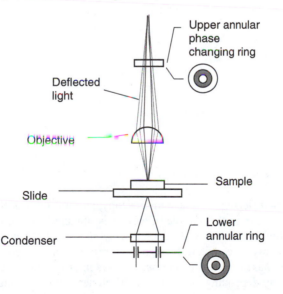

Figure 6.4 In phase contrast microscopy, light from the lower annular ring is imaged on a semitransparent upper annular ring. This microscope uses the small changes in phase through the sample to enhance the contrast of the image.

If the upper annular ring is opaque instead of semitransparent, it blocks all light except that which is refracted, thus yielding a dark field phase contrast microscope.

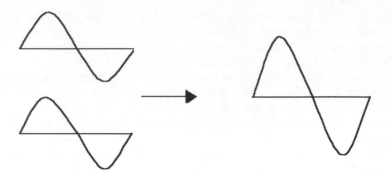

Figure 6.5 The interaction of light waves that meet in phase results in constructive interference. The amplitude of the wave is doubled by this interaction.

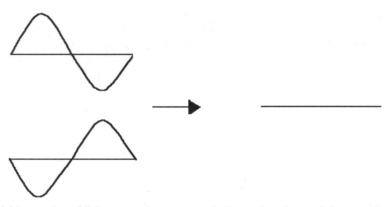

Figure 6.6 Interaction of light waves that meet one-half wavelength out of phase results in destructive interference. The waves cancel each other when they intersect.

Differential Interference Contrast (DIC)

DIC imaging is a modification of phase microscopy, and it produces contrast from differences in refractive index in the sample. Contrast is obtained by converting these phase changes into amplitude changes. DIC does this by splitting light with a prism into two parallel rays. These rays are routed through two different optical paths and then recombined. As with the phase contrast technique, the rays interfere with one another constructively or destructively depending on the phase difference. The advantage to this method is that the phase difference can be shifted in phase by a variable amount. The most commonly used configuration of differential interference microscopy is called the Normarski Interference Microscope, named for the optical theoretician George Normarski. In microscopy, the term Normarski is synonymous with the differential interference contrast technique. This technique is used for measuring thickness of cells and observing other items with little or no contrast. DIC imaging provides good rejection of out-of-focus interference. Essentially, it provides an image of the rate of change of refractive index in the sample. The technique essentially acts as a filter that emphasizes edges and lines. Out-of-focus refractive index changes will be blurred and have a small rate of change in the focal plane so they will not contribute much to the contrast of the image.

Dark Field

Figure 6.7 shows another method called *dark field* microscopy. With dark field illumination, the background field is dark because the condenser light comes from outside the collecting angle of the objective. This is due to an opaque stop below the condenser. Small objects, even those below the limits of resolution, can be detected easily since objects scatter light and appear to emit light on a dark field.

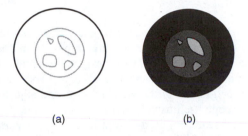

(a)   (b)

Figure 6.7 Example of darkfield illumination. (a) shows the typical brightfield view while (b) shows the dark-field view of the same object. Imagine the object in the darkfield view glowing like the moon at night.

Rheinberg illumination, also called optical staining, is especially useful for looking at unstained samples. This method is similar to conventional darkfield illumination except that the central opaque light stop at the base of the substage condenser is substituted for a colored central stop. A sample viewed by this method looks white on a background that is the color of the stop. If a colored transparent annular filter is added, the sample is colored.

Inverted Microscope

If a light microscope is functionally turned upside down to the object and viewed from under the sample, the microscope is called an *inverted* microscope. The light source and condenser are above the stage pointing downward, and the objective is below the stage pointing upward. As a result, one is looking up through the bottom of the specimen and rather than looking at the specimen from the top, typically through a cover glass, as on a conventional microscope. This is particularly useful with live cells since they can be seen through the bottom of a container, without opening the container, and without the air interface normally present between the objective and the surface of the culture. By adding phase contrast capabilities, cells can be observed without stains.

Common biological stereo–binocular (dissecting) microscopes are also used with reflected light to examine tissue that is too thick to allow light to pass through it.

Near Field Scanning Optical Microscopy (NSOM)

NSOM offers dramatic improvement in resolution from about 1 μm to about 10 nm (Paesler and Moyer, 1996). The lens, which has a far field diffraction limit, is replaced by a tapered optical fiber probe with an apex of curvature of about 10 nm. The sample is raster scanned in close proximity to the fixed probe, which detects light only from the surface in the immediate vicinity, and is placed on a vertical piezoelectric tube. Electrodes on the left and right sides are driven with 200 V to achieve 15 μm side movement. Electrodes on the front and back permit forward and backward movement, and an interior electrode permits 1.7 μm of vertical movement.

6.2.3 Electronic Microscopy

A photographic, video, or solid-state camera can be attached to a microscope to make a permanent record of an observation. There are three types of microscope viewing ports: monocular (one eyepiece), binocular (two eyepieces), and trinocular (two eyepieces and a photography port). A camera can be adapted to each of these viewing heads. Camera adapters usually are attached to one of the viewing ports and adjusted to be parallel with the eyepieces. The microscope itself usually acts as the camera lens.

For many years, researchers used only still cameras or video camera systems to acquire light microscope images (Shotton, 1993). The use of a still camera is called photomicroscopy and the still picture is called the *micrograph*. The use of a video camera is called video microscopy. The most recent instrumentation involves the use of specially configured microscopes, new types of high-resolution solid-state cameras, image digitization, and digital image processing. Some of these techniques and methods were the result of technological advances while others, most notably video enhanced contrast microscopy, were discovered by accident. However, as a result of these developments, the field is now usually called electronic microscopy or optodigital microscopy (Shotten, 1990).

Solid State Cameras

Solid-state cameras are ideally suited for many cellular microscope applications, and typically use a technology called a Charge-Coupled Device (CCD). These devices consist of a large square array of tiny detectors (called pixels) each of which produces an electric charge in response to incident photons (see Figure 6.8). Each pixel is about 20 μm^2. The charge for each pixel is collected and sent to an output amplifier so that the level of light incident to each pixel is recorded. Charge accumulates over time so the total amount of charge is proportional to the product of the light intensity and the exposure time. The intensity of each pixel is displayed as a video matrix on the computer, yielding a two-dimensional image of incident light. Tiny red, green, and blue filters placed in front of sequential pixels yield a color camera.

The CCD camera has very high signal-to-noise ratio and is available in high resolution arrays of 2032 × 2044 pixels or greater at 1.6 frames/s. Array cameras have a response in which the output is linearly related to the input light intensity over a very wide range of light levels. In addition, they are not susceptible to accidental light damage as are the high sensitivity video cameras. A lower resolution array of 1300×1030 pixels at 12 frames/s is still below the 60 Hz standard video scan rate. This limits these devices to studying either static or slow dynamic processes. Also, many times these devices are operated at low temperatures to maximize the signal-to-noise ratio and are therefore termed *cooled*.

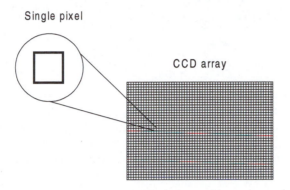

Figure 6.8 CCD Array. The CCD array is made up of many light-sensing elements called pixels. Each pixel can independently sense light level and provide this information to a computer.

Wide Field Versus Point Scan Imaging

It is important to distinguish between the two types of image acquisition (see Figure 6.9) that can be done with these microscopes. Image plane scanning, or wide field imaging, involves the acquisition of a two-dimensional image by the recording device. The device is focused on a plane of the object and a two-dimensional picture is directly recorded. This is the type of image that you would see if you looked into a traditional light microscope. In contrast, a point-scanned device is focused on a single point of the object. The instrumentation must have some way of scanning or focusing on different points. This is typically done in the two horizontal (x–y) planes so the resulting picture would look the same as the wide field image (Shotten, 1990). An example of this is the scanning electron microscope discussed in Chapter 4.

6.2.4 Cellular Markers

There are two general methods for highlighting or marking different structures within a cell or to mark different types of cells. These markers are used for a variety of purposes from locating a certain structure within a cell, identifying which cells contain a particular substance, how much of the substance they contain, where it is located within the cell, how it moves over the cell's cycle, if the substance moves between cells, and even how much of the substance is present within a cell.

These markers can be put into the cells in several different ways. Some are naturally taken up by the cells when introduced into their environment. Others are injected into cells using special techniques or devices such as the tiny syringe-like microinjector. Once put into cells, some are analyzed immediately, others are given some time to bond to specific cellular structures, while others are activated at a later time when the researcher wants to observe time specific cellular dynamics.

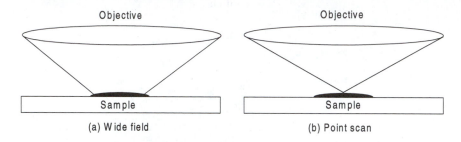

Figure 6.9 Comparison of the wide field versus the point scan techniques. (a) wide field collects an entire image while (b) point scan image must be assembled point by point.

Fluorescent Probes

The use of fluorescent markers or probes is revolutionizing the imaging of live cells and their associated contents, structures, and dynamics. These systems all involve the excitation of a fluorescent material, the subsequent fluorescence emission by the material, and the detection of this fluorescence by a detector (Molecular Probes, 2003).

The probes, also called fluorochromes and fluorophores, are fluorescent materials that are introduced into the cellular environment. There are a number of different ways that these probes are used with the cells:

- The probe binds directly to the cellular structure of interest.
- The probe binds to another molecule that has an affinity or reactivity towards some cellular structure.
- The probe binds to an antibody that in turn binds to some cellular structure.
- The probe does not bind, but its fluorescence depends on local concentration of a solute.
- Nonfluorescent molecules are converted to fluorescent molecules through an enzymatic reaction.

These probes are especially useful in microscopy for several reasons. First the excitation wavelength is different from the emission wavelength, which allows for easy detection because the excitation wavelength can easily be filtered out. Also, different types of probes emit fluorescence at different wavelengths. This is especially useful because several probes can be used to mark different structures within a cell and these can be detected at the same time. In addition, the entire excitation/fluorescence cycle can be repeated.

Two considerations are important regarding probe selection. First, probes must be chosen based on *sensitivity*, the amount of substance, labeled within the cell. Also the probe's *specificity*, the ability to only label the substance of interest, is important. Specificity is provided by fluorochromes, which are either synthetic chemicals or natural products that can have innate specificity, binding to certain classes of molecules, or can be attached to molecules that have specificity (for example, antibodies). Some types of fluorochromes are generated by the activity of enzymes or are sensitive to the presence of specific ions. A final class of fluorochrome is generated by gene expression.

The advantages of using probes include the high signal levels against a low background (noise) level, they are not prone to motion artifact, and they can be used to monitor very small changes.

Care must be taken regarding the intensity level of the light used to excite the probe. *Photobleaching* is the breakdown of the fluorescent probe within the sample as it is exposed to light. Typically, the more intense the light, the more photobleaching occurs. Photobleaching causes several problems. It reduces the fluorescent signal emitted, it generates oxygen free radicals, which can damage cells, and it may generate forms of dye that distort the relationship between the substance being marked and other areas of the cell.

In a flow cytometer, a laser excites fluorescence of individual cells in an interrogation volume. The fluorescence is then separated on the basis of color, through the use of optical filters, and the results displayed as a histogram of cell count vs. cell type. Cells may also be selectively isolated or sorted to high purity as they pass through this system.

Fluorescence video microscopy can capture multiple images at differing wavelengths of light simultaneously at video rates. The fluorescence event follows exponential kinetics with a lifetime of approximately 350 ms.

Radioactive Tracers

Radioactive tracers are used in a number of areas of cell research as an effective and safe means of monitoring cellular interactions. This was the primary technique used for years, but is being used less and less with the development of more and better fluorescent probes. These tracers have the advantages of high precision and specificity, but suffer the disadvantages of extensive radiation handling safety procedures, the inability to distinguish more than one probe at a time, and a relatively long time (hours to days) to obtain results.

Radioactivity is caused by the spontaneous release of either a particle and/or electromagnetic energy from the nucleus of an atom. Tritium (^{3}H) is an important tracer used for cellular research. Other radioactive elements used include ^{14}C (Carbon–14), ^{32}P (Phosphorus–32), ^{125}I (Iodine–125). Tritium is most prominent because replacing hydrogen with tritium in a molecule does not cause chemical or physiological changes in molecular activity (Heidecamp, 2003).

Radioactive elements emit radiation spontaneously, but over time a percentage of all radioactive elements in a solution decays. It is nearly impossible to predict which individual element will radioactively decay, but predictions are made about large numbers of the elements. The radioactivity of a material, the rate of how many decays occur each second, is measured in becquerels.

When an alpha or beta particle, or a gamma ray passes through a substance, ions are formed. To monitor this ionization effect, a device called an ionization chamber is used. The most common forms of ionization chambers are the Geiger counter and the pocket dosimeter. These chambers have two electric plates separated by an inert gas with a voltage applied across them by a battery or power supply. When a particle enters the chamber, ions form and are drawn to one of the electric plates as shown in Figure 6.10. The negative ions attract to the positive plate (anode) and the positive ions attract to the negative plate (cathode). This causes a small voltage change in the form of a pulse to temporarily occur across the plates. This pulse is passed by the capacitor, which blocks the dc voltage from the power supply. The output pulse is then read by some measuring device. The sensitivity of the system depends on the voltage applied between the electric plates. High-energy alpha particles are significantly easier to detect than beta particles, and they require lower voltage. In addition, alpha particles penetrate through the metal casing of the counter tube, whereas beta particles can only pass through a quartz window on the tube. Consequently, ionization chambers are most useful for measuring alpha emissions. High-energy beta emissions can be measured if the tube is equipped with a thin quartz window and if the distance between the source of emission and the tube is small.

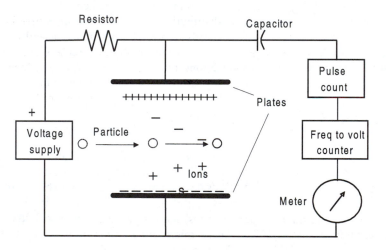

Figure 6.10 Basic radiation counting system. The power supply has a voltage source and series resistor. Ionization by a radiation particle causes a voltage pulse, which passes through the capacitor and is read by the meter.

A pocket dosimeter is a modification of the basic ionization chamber, using a miniaturized tube that can be charged to hold a voltage without constantly recharging the unit with a power source. Essentially, it is a capacitor charged by a base unit and carried as a portable unit. It is small enough to be easily carried in a pocket. When exposed to ionizing radiation, the capacitor discharges slightly. Over a period of time, the charge remaining on the dosimeter can be measured and radiation exposure can be calculated. Since dosimeters carry lower voltage potentials, they are used for the measurement of X-ray and high-energy gamma radiation, but do not detect beta emissions.

Liquid scintillation counters are also used to measure radiation. They use a substance in the counting chamber, called a scintillant, to produce a flash of light. This flash is produced for each radiation particle detected in the chamber. A circuit then converts each light flash to an electric current pulse. This signal is then sent to a counting system, which keeps track of total counts or counts per minute (cpm). If the efficiency of the system (or the percent of actual radioactive decays that are detected) is known, the disintegrations per minute (dpm) can be calculated using Eq. (6.3).

$$dpm = cpm/efficiency \qquad (6.3)$$

Autoradiography is the process of using radioactivity to trace cellular activity. Tritium is used in cell analysis because it is relatively weak and can be localized within cell organelles. Carbon and phosphorous are also used, but are more radioactive so they require more handling precautions and provide poorer resolution of intracellular details. They are more commonly used to analyze details at the tissue or organ level.

The picture of a cell on photographic film that is labeled with radioactive tracers is called an *autoradiogram*. When silver ions in a photographic emulsion are exposed to radiation they change to silver grains. The number and distribution of these grains are therefore proportional to the amount and distribution of the cellular radioactive label.

The quantity of silver grains depends on the type of emulsion and the kind of radiation emitted from the cell. Alpha particles produce straight, dense tracks a few micrometers in length. Gamma rays produce long random tracks of grains, which are useless for autoradiograms. Beta particles or electrons produce single grains or tracks of grains. High-energy beta particles (^{32}P) may travel more than a millimeter before producing a grain. Low-energy beta particles (^{3}H and ^{14}C) produce silver grains within a few micrometers of the radioactive area, thus providing excellent resolution.

6.2.5 Confocal Laser Scanning Microscopy

Confocal laser scanning microscopes (CLSMs) are a significant innovation for the light microscope industry (Wright, 1993). These microscopes remove the out-of-focus blur from the 2-D image directly through hardware rather than requiring a software deblurring technique (discussed in Section 6.4.1). Also, this technique yields images from relatively thick biological tissue, and successive 2-D optical sections (known as a Z-series) can be assembled in an imaging system to construct a 3-D image. A standard light microscope can be used for CLSM with some technically simple, yet expensive, modifications.

As opposed to a conventional wide field microscope which produces a 2-D image directly, the CLSM is a point scan device, which builds a 2-D image by scanning and assembling the image point by point with an integrated imaging system. A CLSM normally uses a laser beam as a light source because a laser provides high-intensity, well-focused, monochromatic light, which easily illuminates individual points within a specimen. Using a point light source avoids the interference of scattered light when the entire field is illuminated, known as *halation*.

The scanning is typically accomplished by a series of x–y plane rotating mirrors in the light path between the dichroic mirror and the objective that focus the laser beam on different points of the specimen in the x–y plane. To obtain an image of the specimen, the light from the specimen is routed to a detector. A pinhole allows only the light originating from the focal plane to reach the detector, as shown in Figure 6.11. The detector (usually a CCD camera or photomultiplier tube) measures the intensity of light and produces a signal proportional to the light at each point. This signal is then typically sent to an imaging system where the current is converted to a voltage, digitized, processed to reduce noise, and an image is reconstructed for display on a monitor.

CLSM has several advantages over conventional microscopy. CLSM provides high-contrast images of specimens without the image artifacts normally present with conventional contrasting techniques. It also has the ability to view relatively thick specimens and is well suited for fluorescence applications, as will be discussed later. The disadvantages of this technique are relatively high cost and possible damage to the specimen from the laser illumination. In an attempt to minimize this damage, each slice may be scanned several times at relatively low laser power and the results integrated.

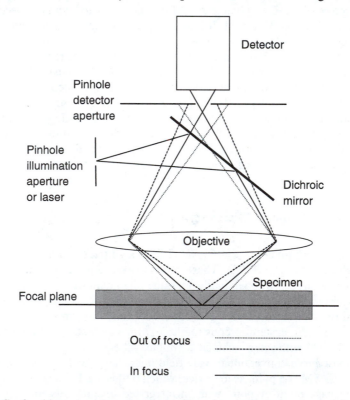

Figure 6.11 Confocal laser scanning microscope. The microscope removes out-of-focus (z-plane) blur by keeping out of focus light from reaching the detector using a pinhole.

6.2.6 Two-Photon Excitation Microscopy

Another technique developed to minimize the excitation laser's potential damaging effects is two-photon excitation microscopy (TPEM). This technique is based on the principle that two photons of longer wavelength light simultaneously excite a probe that would normally be excited by one photon of a shorter wavelength, as shown in Figure 6.12. This method has the advantage that less photobleaching occurs because only the focal point is excited as compared to the entire depth of the specimen, as in confocal and wide-field imaging. (LOCI, 2003). In addition, TPEM can obtain optical sections from deeper into a specimen. This is possible because the excitation light is not absorbed by the probe above the plane of focus and the longer wavelengths are scattered less within the sample. Two-photon excitation has several limitations. A given probe has lower resolution when compared to confocal imaging. This can be eliminated by the use of a confocal aperture at the expense of a loss in signal. Also, thermal damage may occur if the specimen contains substances that absorb the excitation wavelengths.

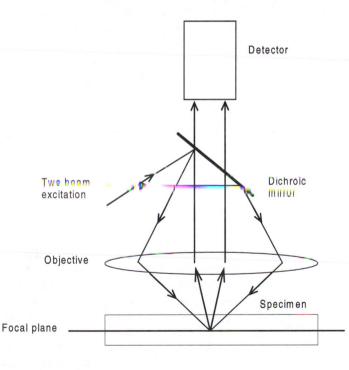

Figure 6.12 Two photon excitation microscope. This microscope doesn't require a pinhole but is able to excite single points within the sample by having two photons excite the sample only at the exact location of interest.

6.2.7 Image Processing

The advent of faster and cheaper personal computers has permitted a myriad of image processing techniques that are ideally suited for the microscope. Several of these techniques correct the image in software as opposed to hardware. Others take the large volumes of data generated by digital imaging and manipulate it into a form that is more readily usable by the researcher.

Computational Deblurring

A computational deblurring (digital deconvolution) light microscope system is a relatively new system that offers an alternative to confocal microscopy. These types of systems use a computer interfaced with a conventional light microscope to remove the blur (haze) from the 2-D image by removing the effects of out-of-focus light from above and below the plane of interest. By a fairly complex mathematical algorithm, each 2-D slice of the object is stored in the computer as a digital image (Richardson, 2003), which is then processed with an imaging algorithm. It mathematically makes a first estimate as to which 2-D plane light has originated from based on the characteristics of the lens system used. The computer then removes the out-of-focus effects from neighboring slices, then the adjusted image is compared to the original image and the error is determined and used to adjust the estimate. This process is iteratively repeated until a minimal error is obtained, achieving high-resolution fluorescence images. This method also enhances detection of low-level fluorescence. Image processing and analysis is typically done after imaging has been completed, although some systems process the data real-time.

Three-dimensional Cellular Tomography

Three-dimensional cellular tomography is a process for making visually-traced, contour-based 3-D reconstructions and measurements from microscopes. This process is similar to the technique used for X-ray and magnetic resonance imaging discussed in Chapter 7. These systems process images from TEMs, SEMs, confocal microscopes, and light microscopes using stacks of vertically aligned 2-D sections. An algorithm fills in structures and builds a single 3-D image. Several thin sections are combined into artificial thick sections. Tracing can be done within the entire volume or in a series of overlapping thick sections. After tracing, contours are combined and displayed as reconstructed objects using various orientation and display options (stereo, color, rotation, surface rendering, etc.). The contour data are also assigned 3-D coordinates to calculate volumes, surface areas, and perimeters. In addition, low-contrast structures become more apparent and better defined when looking through a volume rather than at individual slices. Manipulations can be done to create a stereoscopic image of the image so it can be viewed in 3-D.

6.3 Cell Orientation

Cell orientation refers to both the orientation of cells within tissue and the orientation of cellular components within the cell. There are a number of reasons to measure the orientation of cells within tissue. Understanding how cells are oriented in relation to one another, and thereby the mechanical properties of natural tissue, is an important aspect in the development of artificial tissue (Tranquillo, 1997). Many types of cells have been found to have specific spatial orientations where the top and bottom of a cell can be identified. Another aspect involves the way in which cells orient themselves and migrate in response to different types of stimuli. An understanding of this process will aid in the growth of artificial tissue.

6.3.1 Orientation Chamber

One way to measure orientation is by using an orientation chamber. One type of orientation chamber tests the ability of cells to orient themselves in a gradient of chemical attractant. The chamber is similar to a hemocytometer (see Chapter 5). The gradient is set up by diffusion from one well to the other and the orientation of cells toward the well containing chemical attractant is scored on the basis of their morphology or by filming their movement.

6.3.2 Video enhanced contrast microscopy

Video enhanced contrast microscopy (VECM), as shown in Figure 6.13, uses a high-resolution video camera and a light microscope that has differential interference contrast (DIC) optics. By using both the analog and digital contrast enhancement as well as background image subtraction, this method is able to observe objects with dimensions an order of magnitude smaller than the resolution limits of the microscope alone (Shotten, 1990). During image subtraction, for example, all identical structures in successive video images can be erased, filtering out any interfering background images. This method involves three steps (Figure 6.14). First the light microscope's DIC optics adjust to just under the saturation level of the video camera. The second step involves reducing the background level by adjusting the black-level control. Then the gain control is adjusted to amplify this dark-low intensity signal. The final step is required because the previous step also captures background blemishes created by lens imperfections and unevenness of illumination. In this step, the specimen free background that has been previously recorded is subtracted from the live image.

6.4 Cell-rolling Velocity

Cell-rolling velocity involves the tracking of cells, particularly blood cells, as they start to tumble or roll as the body responds to infection or foreign matter. This is a parameter that becomes important in the areas of healing and the body's response to biomaterials implanted in the body. When an infection occurs in an area of the body, white blood cells are responsible for attacking the infection. The way the body signals these cells is critical to understanding the healing process. Recruitment of white blood cells from the blood to the area of infection is required. This recruitment comes from cell adhesive molecules on the inside of the blood vessel. These Velcro-like cell adhesion molecules (CAMs) start the rolling process and cause the white blood cells to roll along the endothelium wall of the blood vessel. Because this is the first step in the process of the activation of the body's immune system, it has application in the response to tissue engineering and biomaterials. This is typically measured using fluorescent probes, intensified fluorescence microscopy, cell tracking techniques, and the artificial introduction of CAMs.

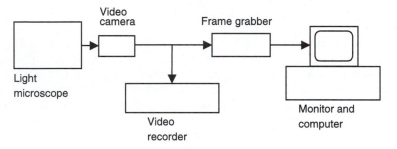

Figure 6.13 Video-enhanced contrast microscope (VECM) system. A light microscope image is recorded by a camera and the image is sent to a recorder and a computer for image processing.

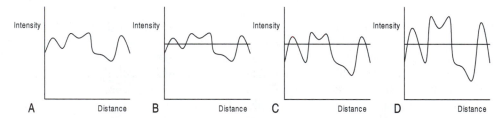

Figure 6.14 (a) Original low-contrast video image. (b) The threshold is set just under the saturation level. (c) Reducing the background level. (d) Gain is adjusted to amplify dark-low intensity signal.

Intensified fluorescence microscopy (IFM) involves the use of a fluorescent specimen, excitation of the specimen with low-level light, and detection of the fluorescence with a microscope and a high-sensitivity video camera. High-sensitivity video cameras use video image tubes with sensitivities about two orders of magnitude greater than conventional video cameras. This configuration was known as video-intensified fluorescence microscopy (VIFM). Recently, these cameras have been replaced with solid-state slow-scan cooled CCD cameras of high resolution. These devices are ideally suited for low-level light imaging and do not suffer from many of the problems associated with the high-intensity video cameras. IFM images do have one major problem—out-of-focus blur. Since the entire field of view of the specimen is illuminated, there are fluorescent emissions from the areas above and below the focal plane causing the out-of-focus blur.

This problem is compensated for in software and is detailed in Section 6.2.7, Image Processing. The actual computation of rolling velocities is typically accomplished after imaging has been completed by using cell-tracking software, which computes the rolling velocities of relative probe locations from the prerecorded images.

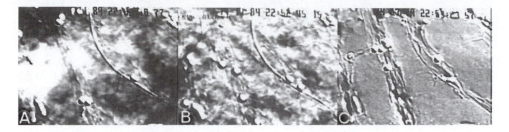

Figure 6.15 (a) An unprocessed photo of cells of the inner epidermis taken through an interference contrast microscope. (b) In the same image with digital contrast enhancement, the single structures become apparent. The background fault remains. (c) Subtraction of the background and resulting with further contrast enhancement.

Figure 6.16 shows a block diagram of an IFM system, which is typically used with the lowest level of excitation light possible so the probe can be traced for a relatively long period of time. This usually gives rise to a relatively low signal-to-noise ratio, which needs to be handled as part of the image processing procedure.

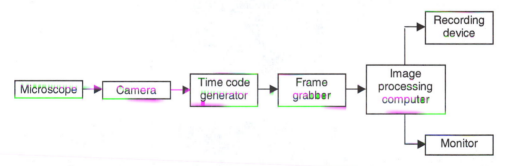

Figure 6.16 Intensified Fluorescence Microscopy system. In this system, the fluorescent signal is received through the microscope by the camera. The video image is typically time stamped and sent to the frame grabber. The frame grabber digitizes the signal and sends it to the computer for image processing. The processed signal is then sent to a monitor for display by the operator and permanently recorded for later playback and analysis.

6.5 Cell Pore Size Determination

Cell pore size is an important parameter for a number of reasons. When a cell secretes hormones, neurotransmitters, or enzymes, a pore is opened in the cell, allowing these molecules to be released (Hoffman, 1997). Also, the first indication that two cells are going to fuse is the formation of a small pore connecting the two cells. The understanding and regulation of these processes are important to the formation and growth of artificial tissue.

6.5.1 TEM

The transmission electron microscope (TEM) (discussed in Chapter 4) is a useful tool for determining pore sizes because of its resolution in the 0.3 to 1 nm range. Electron microscopes did not typically view larger objects well, so it was difficult to view the cell as a whole, locate a point of interest, and zoom in on it. Modern electron microscope systems, however, allow this zooming, and the observation of whole tissues while retaining macromolecular resolution as shown in Figure 6.17. TEMs work in a vacuum, so living cells cannot be observed with this type of microscope. Also, most materials in the cell are transparent to electrons, so they must be stained with dyes. Thus, it is possible that the cellular structure being imaged may be altered by the dye.

Figure 6.17 A TEM image of skeletal muscle cells.

6.5.2 SEM

The scanning electron microscope (SEM) (discussed in Chapter 4) is also useful in cellular analysis because of its ability to show cell surface detail at resolutions in the 3 to 10 nm range and its ability to show 3-D structure as shown in Figure 6.18. The 3-D structure is obtained through the process of Secondary Emission Ion Scanning. As the electron beam of the scanning microscope scans the surface of an object, it can be designed to etch or wear away the outermost atomic layer. The particles are analyzed with each scan of the electron beam. Thus, the outer layer is analyzed on the first scan, and subsequently lower layers analyzed with each additional scan. The data from each layer are then analyzed and reconstructed to produce a 3-D atomic image of the object. Since electrons are relatively small, the etching is sometimes enhanced by bombarding the surface with ions rather than electrons.

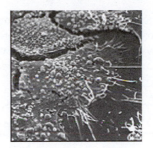

Figure 6.18 A SEM image of stressed liver cells.

6.6 Cell Deformation

Cell deformation involves measuring the mechanical properties of cells and the forces involved in cell activation. If a force is applied to a cell and the deformation is measured, the mechanical properties of the cell can be determined. The deformation of cells, in response to some mechanical force outside the cell, is an area of great interest to biomedical researchers. This process is essential in developing mathematical models of the biomechanical aspects of cellular structure, particularly as they apply to the circulatory system. This research involves *in vitro* biological experiments, mathematical modeling of cellular mechanical behavior, and computer simulation of cell mechanics. Looking at cell deformation, researchers are trying to determine the influence of deformation on the growth of cells, how cells sense deformation at their surface, and how they change their function and structure in response to this deformation. Typically video microscopy is used to monitor, record, and measure the amount of deformation in these experiments. The more difficult engineering task with this parameter is the ability to apply a measurable force to deform the cell.

6.6.1 Micropipet Technique

Several methods can be used to deform cells (Hochmuth, 1990). The micropipet is an important tool for cell deformation studies. This technique involves aspirating part or all of a cell into a glass micropipette as shown in Figure 6.19. A pipet with a diameter of 1 μm is capable of producing a force F of 1 pN. The rigidity and other cell deformation properties can be found as a function of the force on the cell and the displacement of the cell inside the micropipet. By tracking the displacement of the leading edge of the cell as it moves into the pipet, cell deformations as small as 0.1 μm can be measured.

6.6.2 Optical Trapping

Another technique used is called optical tweezers or a laser trap. This system develops an optical gradient generated by a continuous laser. The optical trapping system uses an inverted microscope in differential interference contrast mode, a motorized x–y stage, and two associated lasers. The microscope feeds a video camera and the video is recorded and stored on an optical memory disk recorder. The optical tweezers can be moved throughout the viewing field and measures cell deformations as small as 0.1 μm.

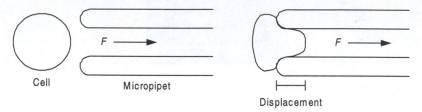

Cell   Micropipet

Displacement

Figure 6.19 In the micropipet technique, force causes a displacement to determine cell deformation properties.

6.7 Cell Shear Stress

When a liquid flows through a vessel of uniform diameter, it adopts a parabolic distribution of velocities from zero at the wall to a maximum at the center. Shear stresses are generated due to regions of the liquid flowing at different speeds. A different velocity of the liquid flowing on either side of a cell creates pressure on the cell (Figure 6.20). This is of particular interest to bioengineers working with the circulatory system. Cell shear stress is defined as the cellular response to this mechanical effect within the body. Endothelial cells produce many different kinds of compounds that control vascular tone, vessel diameter and blood flow rate. The production rate and the release of these compounds are modulated by blood flow and the associated shear stress. Tissue engineers need to understand the effect of physical parameters, such as pressure and shear stress, on tissue formation (Lewis, 1995). For example, bone marrow cells cultivated in a petri dish produce only two of ten types of blood cells. When these cells are cultivated in a bioreactor, a device that simulates some of the physical parameters of the body, all ten types of red blood cells are produced.

There are two problems in dealing with cell shear stress. One is knowing what is going on in the cell when the shear stress is applied. The other is what effect the shear stress has on the cells.

6.7.1 Cone and Plate

One of the problems with shear stress measurements involving the circulatory system is that most methods of generating shear stress require relatively large amounts of blood. However, the cone and plate method solves this problem. With this system, the liquid is placed between a flat horizontal plate and a cone-shaped surface (Eylers, 1992). The cone is rotated at a constant speed, causing the liquid to rotate at speeds that are linearly proportional to their distance from the axis of rotation because the slope of the cone is constant. The shear stress S is the force per unit area of the wall and is defined as

$$S = \eta \frac{dv}{dx} \qquad (6.4)$$

where η is the viscosity of the liquid (0.001 Pa·s for water at 20 °C), and dv/dx is the gradient of the velocity between the cone and the plate. Figure 6.21 shows that with this constant velocity gradient, the shear stress is proportional to the viscosity of the liquid. If the viscosity remains relatively constant, as is the case with blood, the shear stress remains constant. Changing shear stress values can then be accomplished by changing the speed (v) of the rotating cone.

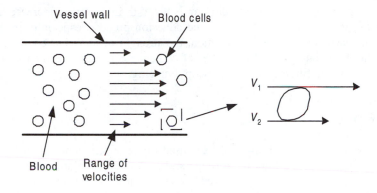

Figure 6.20 Parabolic distribution of velocities and the shear stress it causes on a blood cell.

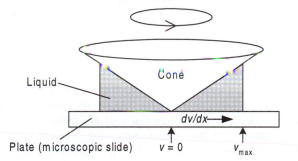

Figure 6.21 Cone and plate system. The cone is immersed in a liquid and then rotated at a constant rate. Since the velocity changes linearly with distance, the shear stress is a function of rotation velocity.

6.7.2 Fluorescent Ratiometric Imaging

Shear stress is reported to induce calcium increase in the cell. Fluorescent ratiometric imaging (see Section 6.13.1) is used to analyze the calcium response with respect to shear stress. The fluorescent image is an image pair acquired at two different wavelengths and obtained by a silicon-intensified target camera. After the signals are sent to an image processing system and calibrated, the images display the calcium concentration within the cells.

6.8 Cell Adhesion

Cell adhesion, like cell rolling velocity, is an important parameter in the healing process, in facilitating the growth of tissue, and in the body's ability to assimilate newly implanted biomaterials.

In some cases reduction of adhesion is the aim, in others an increase is desired. Understanding the mechanisms that start the adhesion process, especially in the bloodstream, are of particular interest. In addition, the strength of adhesive bonds is of prime interest to the tissue engineer. Most studies of cell adhesion use force measurements obtained from video and fluorescence microscopy imaging.

6.8.1 Laser Trap

The optical tweezer, or laser trap, method is used for generating controlled forces in the piconewton range. It uses a light microscope and the force of radiation pressure from an infrared laser. Although piconewton sized forces are quite small, they are sufficient to stop the motion of highly mobile cells or single enzymes. These optical tweezers allow for measurements of subcellular components including torsional stiffness, flexibility of the cell cytoskeleton, and forces generated by enzymes.

6.8.2 Interferometry

Interference microscopy is used for examining cell adhesion. It provides information on the patterns of adhesion, which in turn provides information on the distribution of forces and how the attached areas of cells move. An interferometer splits a light beam into two beams, which are then recombined so they can interfere (Section 4.3). The difference in optical path length between the two beams causes the interference. Changes in optical path length occur with changes in refractive index (changes in density) of the structure on which the beams are focused. Therefore, the image shows different interference patterns in areas with different densities. Interference microscopes are operated in both the reflectance and transmittance modes.

6.8.3 SEM

The SEM uses an intense electron source known as the field emitter, which provides resolution to diameters less than 1 nm (Morton, 1997). This is used to study the cell adhesion molecule. These molecules are found on the surface of leukocytes and are an important mechanism in the body's response to injury.

6.9 Cell Migration

Cell migration is the process by which cells move to a certain area, especially during growth or in response to an abnormal condition in the body. Cell migration is an area of great interest for tissue engineering applications such as tissue growth and biomedical implantation.

6.9.1 4-D Image Processing System

A four-dimensional (4-D) image processing system is used to create a digital movie from time-sequences of 3-D cellular tomographic microscope images. First, a single point image is scanned over a 2-D plane. Vertical slices of the 2-D plane are acquired and a 3-D image is reconstructed. This process must be repeated for each point in time. Then each of these 3-D images is displayed sequentially and the result is an animation of the movement within the image as a function of time. Typically, this animation is recorded on a video recorder for ease of repetitive observation and analysis or is assembled into a computer animation file. Although the animation does not improve the quality of the collected data, it improves the qualitative interpretation of the process.

One problem with this technique is the large amount of storage space and computing power required for the image processing. Cell migration observations typically take 24 to 96 h to complete (Di Francesca and Zerbarini, 1994). Four fields of view, each containing 5 to 20 cells, are collected at 15 to 30 min intervals and the position of each cell is recorded in each frame at every event interval. The data are given to modeling and analysis software to determine where cells moved and how fast they moved there.

6.10 Cell Uptake

Cell uptake is the study of the various types and amounts of biological molecules in the cell. Cell capacitance has been used as an index of secretion in cells. The capacitance of the cell membrane can be measured using techniques similar to the voltage clamp. Fluorescent microscopy is commonly used since fluorescent probes are now available to measure pH, sodium, magnesium, potassium, calcium, and chloride ions.

6.10.1 Nanovid Microscopy

To follow structures within a cell, a probe needs to be sensitive to small amounts of the structure and only label the structure of interest. Submicron colloidal gold particles coupled to antibodies are useful for this purpose (Geerts et al., 1991). These small particles can be detected with a light microscope since the particles scatter a small amount of light, which is detectable using video contrast enhancement microscopy. The particles appear as black dots on a white background and the technique is called nanovid microscopy. The biggest advantage of this technique is its ability to label specific molecules within the cell. Another advantage is the ability for continuous observation without having to worry about the photobleaching effects seen in fluorescent probes. A final advantage is the small size of these particles, which allow diffusion throughout the cytoplasm.

6.10.2 Electroporation

Electroporation is a process that is used to temporarily alter cell membranes by inducing microscopic pores to open in the cell membrane. This allows cell uptake of various biological molecules, and this process involves creating an electric field via a voltage pulse. With the proper pulse shape, the cell completely recovers, the pores reseal, and the cell continues to grow. An advantage of this technique is that it can be used with certain cell types that are affected or killed by chemical methods.

6.11 Cell Protein Secretion

Cell protein secretion involves the movement of proteins across biological membranes and is not well understood. Epithelial cells from certain organs produce and secrete many important proteins that are stimulated by specific hormones and neurotransmitters. Also, cell surface receptor proteins are secreted allowing cells to stick together and form the extracellular matrix. As such, cellular engineers are focusing on the regulation of the external and internal environments of the cell that promote the secretion of specific proteins.

6.11.1 Atomic Force Microscopy

Atomic force microscopy (AFM), as discussed in Chapter 4, allows the magnification of electron microscopy with the ability to work with samples that have not been dehydrated. This resolution allows for observation of individual protein molecules in action. This relatively new technology is being used primarily to provide topological mapping of substances, but is also being used to detect variations in composition, adhesion, and friction (Digital Instruments, 1995). For example, AFM permits study of wear debris a few hundred nanometers in size from ultrahigh molecular weight polyethylene (UHMWPE) used in hip implants.

6.11.2 Fluorescence Lifetime Imaging

In the fluorescence lifetime imaging microscopy (FLIM) method, image contrast in the fluorescence microscope is derived from the fluorescence lifetime at each point in the image. This method is also known as Time-Resolved Fluorescence Microscopy (TRFM) and is analogous to magnetic resonance imaging (MRI) discussed in Chapter 7. It provides chemical information about intracellular ions since the lifetimes of many probes are altered by the presence of certain ions and/or pH. The fluorescence lifetimes of the probes are mostly independent of their concentration and are not normally affected by photobleaching. The most desirable feature of FLIM, however, is that it does not require wavelength specific probes. Therefore, FLIM allows quantitative imaging using visible-wavelength illumination, and there are several methods involved in this technique (Tian and Rodgers, 1990). Figure 6.22 shows the time correlated single photon counting method. In this system, the time difference between the start of an excitation laser light pulse's arrival at the sample and the arrival of the fluorescence photon at the detector is measured. The detector output is recorded in a multichannel analyzer with the intensity at a given channel being proportional to the number of data points (the number of photons) that were collected during that time interval.

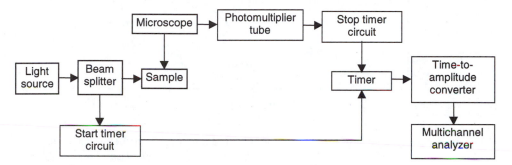

Figure 6.22 Time-correlated single photon counting system block diagram. When a light source provides an excitation pulse, some of the light is deflected to the start timer circuit while the rest illuminates the sample. A single fluorescent photon is detected by a photomultiplier tube, which generates a pulse that stops the timer. The time difference is then converted to an amplitude. The various amplitudes are recorded in the multichannel analyzer and a profile of different time intervals versus number of photons in that interval is displayed.

The other method using this technique simply monitors the fluorescent intensity over time. This measurement is normally achieved with a fast-response photodetector and a sampling oscilloscope for data acquisition. The sample is excited and the fluorescent intensity of the sample is recorded over time.

6.11.3 Fluorescence Recovery after Photobleaching

Fluorescence recovery after photobleaching (FRAP), also known as fluorescence photobleaching recovery (FPR), is a technique used to look at the mobility of plasma membrane proteins within the cell (Peters and Scholz, 1990) by using the effect of photobleaching in a positive way. First, a probe is introduced that binds to plasma membrane proteins. Then, a laser with a low-power beam is focused on a small circular area on the plasma membrane. This causes the probe in that area to fluoresce. The power of the laser is then increased by several orders of magnitude for 0.1 to 1.0 s. This causes the probes in that area to photolyse (convert to a nonfluorescent product). The laser is turned back to the original low level, and initially the area is dark because all the fluorescent probe material no longer fluoresces. With time, plasma membrane proteins from neighboring areas diffuse into the circular area and the fluorescence recovers with time, as in Figure 6.23.

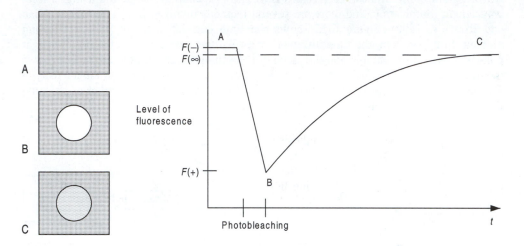

Figure 6.23 A sample graph of fluorescence of proteins during FRAP. (a) Before photobleaching $F(-)$, (b) just after protein photolysed $F(+)$, and (c) long after protein photolysed $F(\infty)$.

If not all the proteins are mobile within the membrane, the percentage of mobile proteins can be calculated using

$$R_{\mathrm{m}} = \frac{F(\infty) - F(+)}{F(-) - F(+)} \qquad (6.5)$$

where R_{m} is the fraction of mobile proteins, $F(-)$, $F(+)$, and $F(\infty)$ are the levels of fluorescence (proportional to light intensity) at times immediately before, after, and long after the area has been photolysed.

6.12 Cell Proliferation

Cell proliferation is the rate at which cells multiply and grow. Many tissue engineering studies depend on the ability to determine the effect of various factors on cell proliferation. These include chemicals and/or conditions that may influence the rate and/or timing of cell proliferation. These studies that help in the understanding of the regulation of cell proliferation are key in the development of new tissue engineering methods and the biomaterials used to support them. This is also an important parameter in developing more successful therapies for such things as the prevention and treatment of cancer and the screening of anticancer drugs. Cell proliferation studies are used when measuring growth factor, drug screening, and the study of toxic, mutagenic, and carcinogenic effects of chemical compounds on cells (Roche Molecular Biochemicals, 2003). The most basic method involves using a microscope counting chamber, or hemocytometer, to count the number of cells in a given area. Density of cells growing on a vessel surface is useful for tissue engineering applications. Density is measured through the use of an ocular grid inserted into an inverted phase contrast microscope. Cell density is recorded and then plotted on a log scale against time in culture. However, more modern techniques are now being used. There are two general methods of measuring cell proliferation in cell populations including measuring DNA synthesis and measuring metabolic activity.

6.12.1 Radiation Monitoring

One method of measuring DNA synthesis involves using a radioactive tracer by adding it to a cell culture. Over time, some of the radioactive material is retained in the DNA of the cells. The cells are washed through filters and the radioactivity of the filters containing the radioactive cells is measured using liquid scintillation counting. Knowing the radiation level of the original tracer in the cell and comparing it to the current radiation level of the cell provides a ratio that is proportional to the number of times the cell has divided.

6.12.2 Fluorescence Microscopy

Fluorescence microscopy techniques are used in a way similar to radiation monitoring. Fluorescent probes are fed to cells, which are given time to incorporate them into their DNA (typically 2 to 24 h). Then, a measurement of the fluorescence level of collected cells can be used to determine the level of proliferation of the cell population.

One other method involves using special salts and detects whether cell proliferation is occurring. The salts react with metabolically active (proliferating) cells to form a red formazan salt, which is easily detected using a colorimetric assay (Chapter 3).

6.13 Cell Differentiation

Differentiation of a cell is defined as the process where cells become structurally and functionally specialized during embryonic development. There are 200 different kinds of cells in our bodies, yet all are developed initially from a single cell. There are many factors that affect cell differentiation, a couple of them being genetics and the environment. Most research on cell differentiation focuses on understanding the mechanisms involved in cell differentiation. This is especially important in tissue engineering. Learning what cell types are involved in tissue regeneration is critical in building tissue. Understanding the mechanisms of cell differentiation aids in the understanding of what a cell needs to regenerate a specific part.

Staining techniques are widely used to visualize those components that are otherwise too difficult to see under an ordinary light microscope because of the lack of color contrast between the object under examination and the background or because of the limited resolving power of the light microscope. In addition, staining techniques are useful in detecting the presence or absence of certain cell components, thus allowing a relatively simple scheme of differentiation or identification of microorganisms. The dye used in staining varies with the cell types.

6.13.1 Fluorescence Ratio Imaging

Fluorescence ratio imaging is used for measuring changes in concentrations of physiological parameters within cells. This is accomplished by taking advantage of the property that fluorescent probes have a wavelength shift when they bind with ions in the cell. A ratio of the intensity of light at the wavelength of emission of the probe itself and the light intensity at the wavelength of emission of the probe that has bound itself to ions in the cell is determined. This ratio of fluorescence intensities at the two wavelengths is then used to calculate the relative concentration of the probe to the concentration of the ion (Bolsover et al., 1993).

First the background must be subtracted from the two signals

$$I'_{\lambda 1} = I_{\lambda 1} - B_{\lambda 1}$$
$$I'_{\lambda 2} = I_{\lambda 2} - B_{\lambda 2}$$

and then the ion concentration ratio (R) is calculated using

$$R = \frac{I'_{\lambda 1} S_{\lambda 2}}{I'_{\lambda 2} S_{\lambda 1}} \tag{6.6}$$

where λ_1 and λ_2 are the two excitation wavelengths, $I_{\lambda 1}$ and $I_{\lambda 2}$ are the fluorescence intensities, $B_{\lambda 1}$ and $B_{\lambda 2}$ are the background light levels, and $S_{\lambda 1}$ and $S_{\lambda 2}$ are the shading corrections that compensate for the small spatial distribution between the two excitation beams. This method has probes that allow ratio imaging of Ca^{2+}, K^+, Mg^{2+}, Cl^- and pH.

6.13.2 Cell Sorter

Flow cytometry systems (Chapter 5) examine cells previously labeled with fluorochromes one at a time and then separate the cells on the basis of their color of fluorescence. Fluorescence activated cell scanning and sorting uses a flow cytometer and a PC workstation for data processing and analysis. The system uses multiple excitation wavelengths in the visible and UV range. Cells may be separated into a wide variety of tissue culture containers. The PC can be used for statistical purposes providing histograms of the various distributions of cells.

6.14 Cell Signaling and Regulation

Cell signaling and regulation involves the complex processes that occur within cells to maintain life (regulation) and trigger new events to occur (signaling). For the biomaterials engineer, the dream is to manufacture material that could (1) sense its environment, (2) integrate the sensed information, and finally (3) take some action based on it. To do this, engineers must study the processes of signaling and regulation within the cell and develop an understanding of how and why they work as they do.

A number of techniques are used for research in this area. NMR provides noninvasive images of cellular structure and chemical composition. This technique is discussed in detail in Chapter 7. Microscope Laser Light Scattering Spectroscopy (MLLSS) is used for motion studies of intercellular macromolecules during diffusion or flow. This method directs laser light into a sample and subsequently detects the scattered light. Fluorescent probes are used to mark specific membrane proteins, which can in turn be microinjected into cells. Their paths through the cell are monitored by computer-aided video systems. With these systems, both quantitative and qualitative observation of membrane flow in live cells can be monitored. In addition, time lapse photography coupled with phase contrast microscopy allows visualization of the cell division process. However, two of the most common methods are described below.

6.14.1 Fluorescence In-Situ Hybridization (FISH)

Fluorescence In-Situ Hybridization (FISH) is a technique for detecting specific DNA sequences in chromosomes, whole cells, and tissue slices. In-situ hybridization is the process of attaching a small fragment of DNA to a specific target strand of DNA or RNA in a preserved cell. Probes are used that only attach to a specific genetic sequence in that cell. Then the point of attachment along the chromosome is identified (Figures 6.24 and 6.25). This method is useful if the identity of a particular gene is known but the chromosome on which it is located is not. FISH identifies the chromosome and the exact spatial location of the gene by exciting the probe and then determining its spatial location. Another application is in the study of embryo development and the mechanisms involved in the sequence of gene selection, activation, and location. Using a CLSM and multiple different colored probes, analysis of fifteen or more probes simultaneously is possible. To do this, the spectrum of the fluorescence emission must be determined since it may contain light that is a mixture of several different colors. This is accomplished with a technique called Fourier spectroscopy (Soenksen et al., 1996). An interferometer splits the fluorescent light emitted by the sample, then the beams are recombined to interfere with one another (Chapter 4). By changing the length of the optical path difference between beams, an intensity versus optical path difference is obtained. This result is sent to a computer where a spectral profile of light intensities for all wavelengths of interest is calculated using a Fourier transform technique. Since the probes used typically have a relatively large span between emission wavelength, the intensity peaks where each probe is fluorescing is readily apparent from the spectral image.

The 3-D organization of specific chromosomes within cells and tissues is also being examined using FISH in combination with laser scanning confocal microscopy and computer assisted 3-D image reconstruction algorithms.

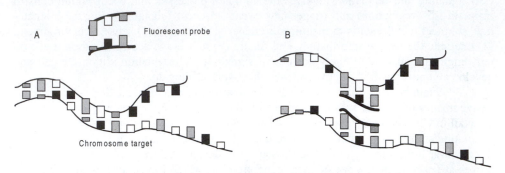

Figure 6.24 (a) The DNA strand is denatured with heat and put in the same culture as the fluorescent probe. (b) The fluorescent probe binds to the target gene.

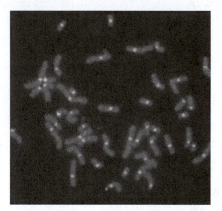

Figure 6.25 Human chromosomes probed and counterstained to produce a FISH image. The lighter area within each chromosome show the fluorescent probe attached to the target gene. (from Detecting Nucleic Acid Hybridization. [Online] www.probes.com/handbook/sections/0805.html)

6.14.2 Radioactive Labeling

A process known as pulse labeling is used to determine the site of molecular synthesis in cells. A radioactive tracer is added to the cellular environment for a short period of time. The radioactive media are then washed or diluted away. The cells are then fixed and autoradiography is used to locate the sites of newly synthesized DNA. The site of utilization of a particular molecule is determined by pulse-chase labeling where cells are again exposed to a radioactive tracer for a short period of time and then washed away. But this time, the cells are allowed to grow for a period of time. When these cells are examined, the radioactivity will have moved to the site of utilization in the cell.

6.15 Problems

6.1   List the size of an atom, molecule, virus, and cell.

6.2   The object length of a microscope is 10 mm. Calculate the focal length for a magnification of 20.

6.3   Find the resolution of a microscope with an index of refraction of 1.2 and a cone angle of 80° that is operated at a wavelength of 770 nm.

6.4   Explain the meaning of the terms brightfield, darkfield, phase contrast and differential interference contrast in the field of microscopy.

6.5   Sketch a phase contrast microscope. Show the light paths and use these to explain how the phase contrast microscope achieves its improvement over a simpler microscope. Explain how the phase contrast microscope is changed to achieve a dark field.

6.6   Explain how a two-dimensional image of incident light is obtained in a solid-state camera.

6.7   Describe the term fluorescent probe as it applies to cell microscopy.

6.8   Explain the reason for using and the basic principles of operation of fluorescence microscopy.

6.9   Describe the effect of photobleaching in fluorescence microscopy.

6.10   The radiation counting system shown in Figure 6.10 has a voltage supply of 100 V. Sketch waveforms of voltage versus time for (1) left of the resistor, (2) right of the resistor, (3) right of the capacitor, when ionization by a radiation particle causes a voltage pulse. Name the electric circuit component that converts the analog voltage at the right of the capacitor to a digital signal suitable for input to a pulse counter.

6.11   A radiation detection system records 350 counts in 20 min. The detector has an efficiency of 0.93. Determine the actual number of decays per minute.

6.12   Explain the reason for using and the basic principles of operation of a confocal laser scanning microscope

6.13   For a confocal laser scanning microscope, for fixed optics, explain how scanning is accomplished in the x, y, and z directions. For nonfixed optics that can deflect the scanning beam, sketch the location of the scanning mechanism. Explain how scanning is accomplished in the x, y, and z directions.

6.14   Explain the reason for using and the basic principles of operation of a two-photon excitation microscope.

6.15   Explain the reason for using and the basic principles of operation of deconvolution image processing.

6.16   Explain the reason for and the basic principles of measuring cell orientation.

6.17   Explain the reason for and the basic principles of measuring cell rolling velocity.

6.18   Explain the reason for and the basic principles of measuring cell pore size determination.

6.19   Explain the reason for and the basic principles of measuring cell deformation.

6.20   Explain the reason for and the basic principles of measuring cell shear stress.

6.21   Calculate the shear stress for water at 20 °C in a cone and plate viscometer with a diameter of 10 cm, a separation gap of 1 mm at the circumference, and rotation at 1 revolution per second. Give units.

6.22   Explain the reason for and the basic principles of measuring cell adhesion.

6.23   Explain the reason for and the basic principles of measuring cell migration.

6.24   Explain the reason for and the basic principles of measuring cell uptake.

6.25   Explain the reason for and the basic principles of measuring cell protein secretion.

6.26   Using fluorescence recovery after photobleaching, calculate the fraction of mobile protein where F(−) = 4, F(+) = 1, F(∞) = 2.

6.27   Explain the reason for and the basic principles of measuring cell proliferation.

6.28   Explain the reason for and the basic principles of measuring cell differentiation.

6.29   Explain the reason for and the basic principles of measuring cell signaling and regulation.

6.16 References

Bolsover, S. R., Silver, R. A. and Whitaker, M. 1993. Ratio imaging measurements of intracellular calcium. In D. Shotton (ed.) *Electronic Light Microscopy, Techniques in Modern Biomedical Microscopy*. New York: Wiley-Liss.

Eylers, J. P. 1992. Liquids. In J. F. Vincent (ed.) *Biomechanics – Materials*. Oxford: Oxford University Press.

Geerts, H., De Brabender, M., Nuydens, R. and Nuyens, R. 1991. The dynamic study of cell surface organization by nanoparticle video microscopy. In R. J. Cherry (ed.) *New Techniques of Optical Microscopy and Microspectrometry*. Boca Raton, FL: CRC Press.

Heidecamp, W. H. 2003. Cell Biology Laboratory Manual. [Online] www.gac.edu/cgi-bin/user/~cellab/phpl?chpts/chpt1/intro1.html.

Hochmuth, R. M. 1990. Cell biomechanics: A brief overview. *Trans. ASME, J. Biomech. Eng.*, **112** (3): 233–4.

Hoffman, M. 2003. Cellular Secretion: It's in the Pits. *Am. Sci.*, [Online] www.sigmaxi.org/amsci/Issues/Sciobs97/Sciobs97-03pits.html

LOCI. 2003. Multiple Photon Excitation Fluorescence Microscopy. [Online] www.loci.wisc.edu/multiphoton/mp.html.

Lewis, R. 1995. Tissue engineering now coming into its own as a scientific field. *The Scientist*. **9**: 15.

Molecular Probes. 2003. *Handbook of Fluorescent Probes and Research Products*. [Online] www.probes.com/handbook/

Paesler, M. A. and Moyer, P. J. 1996. *Near Field Optics: Theory, Instrumentation and Applications*. New York: John Wiley & Sons.

Pawley, J. B. 1990. *Handbook of Biological Confocal Microscopy*. New York: Plenum Press.

Peters, R. and Scholz, M. 1991. Fluorescence photobleaching techniques. In R. J. Cherry (ed.) *New Techniques of Optical Microscopy and Microspectrometry*. Boca Raton, FL: CRC Press.

Photometrics. 2003. CCD Grading. [Online] www.photomet.com/library_enc_grading.shtml

Richardson, M. 2003. Three Dimensional Deconvolution of Microscope Data. Vaytek, Inc. [Online] www.vaytek.com/technical.html.

Roche Molecular Biochemicals 2003. Guide to Cell Proliferation and Apoptosis Methods. [Online] biochem.roche.com/prod_inf/manuals/cell_man/cell_toc.html

Soenksen, D. G., Garini, Y. and Bar-Am, I. 1996. Multicolor FISH using a novel spectral bio-imaging system. In Asakura, T., Farkas, D. L., Leif, R. C., Priezzhev, A. V., Tromberg, B. J. (eds.) *Optical Diagnostics of Living Cells and Biofluids. Proc. SPIE,* **2678**.

Shotton, D. (ed.) 1993. *Electronic Light Microscopy, Techniques in Modern Biomedical Microscopy.* New York: Wiley-Liss.

Shotten, D. M. 1991. Video and opto-digital imaging microscopy. In R. J. Cherry (ed.) *New Techniques of Optical Microscopy and Microspectrometry.* Boca Raton, FL: CRC Press.

Stine, B. 2003. High Resolution Imaging of Biological Samples by Scanning Probe Microscopy. Application Notes, Digital Instruments. [Online] www.di.com/AppNotes/ANMain.html

Tian, R. and Rodgers, M. A. J. 1991. Time-resolved fluorescence microscopy. In R. J. Cherry (ed.) *New Techniques of Optical Microscopy and Microspectrometry.* Boca Raton, FL: CRC Press.

Wright, S. 1993. Introduction to confocal microscopy and three-dimensional reconstruction. In B. Matsumoto (ed.) *Methods in Cell Biology.* San Diego: Academic Press.

Chapter 7

Nervous System

Jang-Zern Tsai

The *nervous system* is defined as all cells, tissues, and organs that regulate the body's response to internal and external stimuli (Costello, 1994). There are many neurological diseases and disorders that can be assessed by the use of medical equipment. Biomedical engineers need to understand how the nervous system works in order to design this equipment.

7.1 Cell Potential

The nervous system is comprised of neuron cells, the conducting elements of the nervous system responsible for transferring information across the body. Only these and muscle cells are able to generate potentials, and therefore are called *excitable cells*. Neurons contain special ion channels that allow the cell to change its membrane potential in response to stimuli the cell receives (Campbell et al., 1999).

7.1.1 Resting Potential

All cells in the body have a cell membrane surrounding them, and across it there is an electric charge referred to as the *resting potential*. This electric impulse is generated by differential ion permeability of the membrane. In the cells, potassium (K^+) channels allow diffusion of K^+ ions out of the cell while sodium (Na^+) ions diffuse into the cell. This Na^+–K^+ pump, which requires ATP to operate, pumps two K^+ ions into the interior of the cell for every three Na^+ ions pumped out. K^+ and Na^+ ions are continuously diffusing across the membrane from where they were just pumped, but at a slower rate. Since there are more K^+ ions inside the cell than outside, a potential exists. The magnitude of the equilibrium membrane resting potential is given, at body temperature, by the Nernst equation

$$V_m = (RT/nF)\ln[K^+]_e/[K^+]_i \qquad (7.1)$$
$$V_m = 0.0615 \log_{10}[K^+]_e/[K^+]_i$$

At 37 °C (body temperature), where V_m is the transmembrane potential in millivolts, R is the universal gas constant, T is absolute temperature in K, n is the valence of the ion, F is the Faraday constant, $[K^+]_e$ is the extracellular potassium concentration, and $[K^+]_i$ is the intracellular potassium concentration. Typically, $[K^+]_e = 4$ mmol/L, $[K^+]_i = 155$ mmol/L, which yields $V_m = -97.7$ mV. However, other ions such as Na^+ and Cl^- have a lesser effect on the resting potential because their permeabilities are smaller, so the actual $V_m = -85.3$ mV (Clark, 1998). The typical resting potential for a neuron is -70 mV (Campbell et al., 1999).

7.1.2 Action Potential

Some cells, such as skin cells, are not excitable, while others, such as nerve and muscle cells, are excitable. When a stimulating electric field acts on an excitable cell, the Na^+ permeability increases, Na^+ enters the cell interior, and the entering positive charge depolarizes (reduces to approximately zero) the transmembrane potential. Later, the K^+ permeability increases and K^+ ions flow out to counter this effect. The Na^+ gates close followed by the K^+ gates, and finally, the resting potential is regenerated (see Figure 7.1). The action potential lasts about 1 ms in nerves and about 100 ms in cardiac muscle. It propagates in nerves at about 60 m/s and carries sensation from the periphery toward the brain via sensory nerves. Through motor nerves, the brain commands muscles to contract. Figure 7.2 shows an invertebrate nerve axon under stimulation of an electric potential high enough to excite the depolarization of the axon (Stein, 1980). We can calculate the action potential propagation velocity $v = d/t$, where $d =$ distance and $t =$ time.

7.2 Brain, EEG, and Evoked Potentials

7.2.1 Anatomy and Function of the Brain

The brainstem is a short extension of the spinal cord and serves three major functions: a connecting link among the cerebral cortex, the spinal cord, and the cerebellum; a center of integration for several visceral functions; and an integration center for various motor reflexes.

The cerebellum is a coordinator in the voluntary (somatic) muscle system and acts in conjunction with the brainstem and cerebral cortex to maintain balance and provide harmonious muscle movements.

Figure 7.3 shows the cerebrum, which occupies a special dominant position in the central nervous system. Within the cerebrum are the conscious functions of the nervous system.

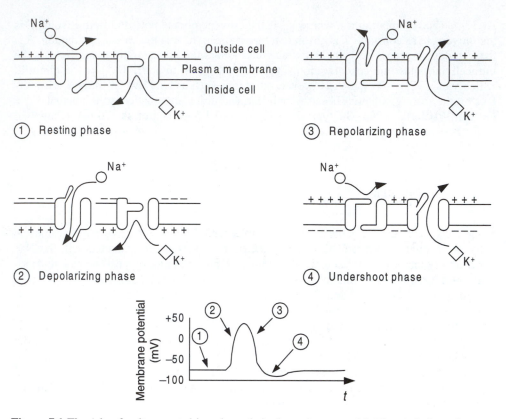

Figure 7.1 The role of voltage-gated ion channels in the action potential. The circled numbers on the action potential correspond to the four diagrams of voltage-gated sodium and potassium channels in a neuron's plasma membrane. (From Campbell, N. A., Reece, J. B. and Mitchell L. G. *Biology*. 5th Edition. Copyright © 1999 by Benjamin/Cummings. Reprinted by permission of Benjamin Cummings.)

7.2.2 Electroencephalography (EEG)

The electroencephalogram (EEG) provides information about the health and function of the brain by detecting electric impulses. EEG can help diagnose conditions such as epilepsy, brain tumors, brain injury, cerebral palsy, stroke, liver, kidney disease, or brain death, and helps physicians find the cause of problems such as headaches, weakness, blackouts, or dizziness (WFUBMC, 2003).

By removing a portion of the skull, it is possible to insert microelectrodes into the brain and record action potentials from single cells. Because this procedure is so invasive, the vast majority of clinical studies are made from electrodes glued onto standard locations on the scalp, as shown in Figure 7.4. These electrodes average the action potentials from large numbers of cells, and therefore do not provide action potentials from single cells. Some localization is possible by using the monopolar connection, which is a recording from a single electrode referenced to the average of all other electrodes, and is also possible by using the bipolar connection, which is a recording between successive pairs of adjacent electrodes. The EEG is typically 100 μV in amplitude with a frequency response of 0.5 to 80 Hz. Sixteen channels are usually recorded simultaneously (Hughes, 1994).

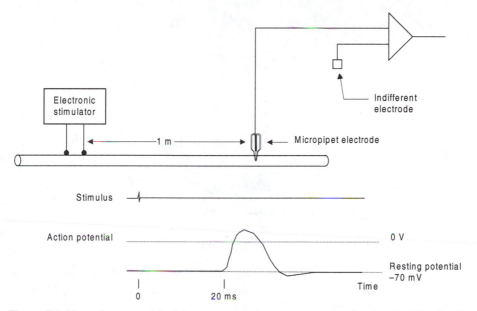

Figure 7.2 The action potential of the nerve axon in response to electric stimulus. The depolarization process first occurs at the stimulation site. The action potential then travels downstream to the recording site where a penetrating micropipet is used to pick up the intracellular potential with respect to an indifferent reference potential.

The EEG is typically recorded with the subject awake and resting. Figure 7.5(b) shows that when the eyes are closed, alpha waves with a frequency of about 10 Hz appear. As the subject becomes drowsy and falls asleep, the frequency decreases and the amplitude increases. During rapid-eye-movement (REM) sleep, frequency increases and amplitude decreases. Alpha waves correspond to when the subject is awake, theta waves indicate sleep, delta waves indicate sleeping well, and finally beta waves indicate REM sleep (Clark, 1998). Each of the four EEG waveforms can be see in Figure 7.5(a).

In people with epilepsy, brain cells create abnormal electricity that causes seizures (AAFP, 2003). Grand mal epilepsy spreads throughout the brain, causes convulsions, and is diagnosed by large amplitude recording from all parts of the scalp. Petit mal epilepsy is identified by a 3 Hz spike-and-dome pattern and may cause an interruption in consciousness. Psychomotor epilepsy is identified by 2 to 4 Hz waves with superimposed 14 Hz waves and causes amnesia or unwanted motor action. Figure 7.5(c) shows various epilepsy waveforms, though other abnormal waveforms also exist that help to diagnose other disease. An EEG with no response for 24 h is an indication of brain death.

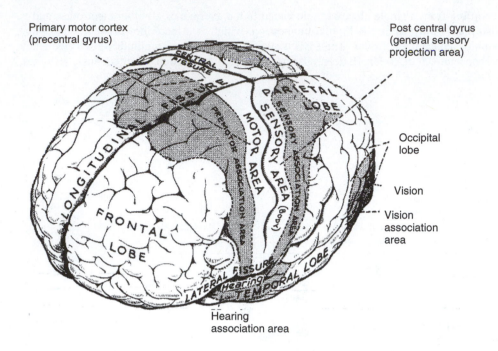

Figure 7.3 The cerebrum contains the frontal, parietal, temporal and occipital lobes. (From McNaught A. B. and Callander, R., *Illustrated Physiology.* 3rd ed. Copyright © 1975 by Churchill Livingstone. Reprinted by permission of Churchill Livingstone.)

7.2.3 Evoked Potentials

An evoked potential is the electric response signal of the brain to a specific stimulus. The response is usually of low amplitude (<5 μV), especially in relation to the background EEG signal (50 to 100 μV), so that stimulus-related cerebral evoked potentials usually cannot be identified in routine EEG recordings. With the use of the signal averaging technique, the background EEG and artifacts can be minimized. For example, a stimulus occurs $N = 100$ times, and each time the response is recorded. All 100 responses at 0 ms are added, all 100 responses at 1 ms are added, etc. The signal averaged output waveform is a plot of the added outputs at 0, 1, 2, etc. ms. Because these responses are synchronized to the stimulus, they add linearly and are N times larger. The interfering noise potentials from the EEG are random and have an amplitude equal to $\sqrt{N}$. Thus the signal-to-noise ratio (SNR) increases by a factor of $N/\sqrt{N} = \sqrt{N}$ (Isley et al., 1993).

Studies of evoked potential are useful in evaluating the functional integrity of sensory pathways. Evoked potentials permit the lesions of these pathways, which may not have any visible symptoms, to be detected and localized. They may be helpful in establishing or supporting a neurological diagnosis, following the course of certain disorders, and assessing sensory function (such as vision or hearing) when normal behavioral or subjective testing is not possible. The electric potentials associated with cognitive processes can also be recorded through the scalp and are helpful in the evaluation of patients with suspected dementia. Evoked potentials provide objectivity, because voluntary patient response is not needed. Clinical abnormalities can be detected by an increase in latency (the delay between the stimulus and the wave response).

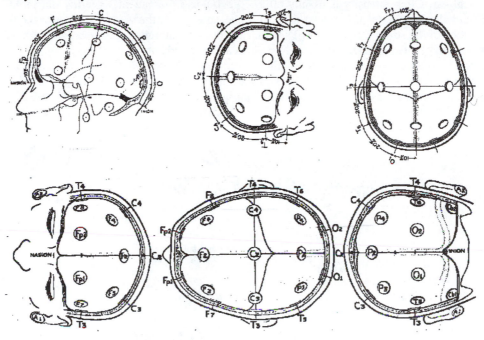

Figure 7.4 The 10–20 electrode system for measuring the EEG. (From H. H. Jasper, The Ten Twenty Electrode System of the International Federation in Electroencephalography and Clinical Neurophysiology, *EEG Journal*, 10 (Appendix), 371–375, 1958. Copyright © by Elsevier. Reprinted by permission of Elsevier.)

Visual Evoked Potentials

During visual-evoked potential (VEP) testing, the subject stares at a central spot on a reversing checkerboard pattern 1 m away. The pattern consists of a counterphase alternation of dark and light squares without any change in total luminance. The reversing rate is about twice per second. The evoked potential generated in the occipital cortex is detected by an electrode placed on the scalp in the midoccipital position with reference to either the linked mastoids or a midfrontal electrode. A recorded VEP is obtained by amplifying, synchronizing, and averaging the responses to 100 such reversals. The procedure is usually done twice to ensure reproducibility, and the two eyes are tested separately. VEP evaluates the visual nervous system from the eyes to the brain. It may help diagnose certain visual or brain disorders (Gevins and Aminoff, 1988) (WFUBMC, 2003).

Brainstem Auditory Evoked Potentials

During brainstem auditory evoked potential (BAEP) testing, rarefaction or condensation clicks at about 10 Hz stimulate one ear of the subject, and white noise masks the contralateral ear. Click intensity is usually 60 to 70 dB above the patient's hearing threshold. The evoked responses are smaller than 1 μV measured from the vertex with reference to the lobe of the stimulated ear. BAEPs within 10 ms of the stimulus are filtered, amplified, and averaged for satisfactory recording. At least two separate results are recorded for each ear to ensure reproducibility. BAEP examination evaluates the nerve pathways from the ears to the brain. It may help uncover the cause of hearing and balance problems as well as other nervous system disorders (Gevins and Aminoff, 1988) (WFUBMC, 2003).

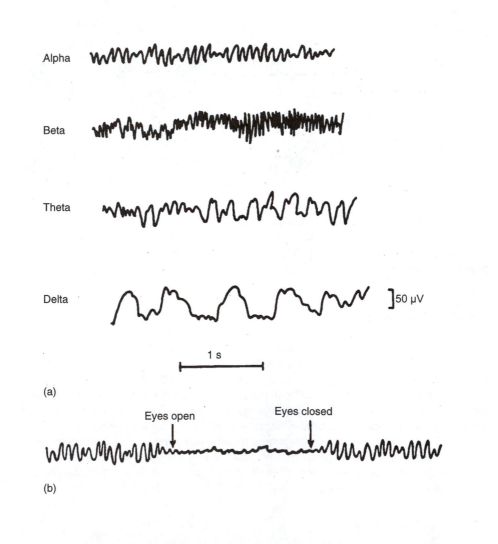

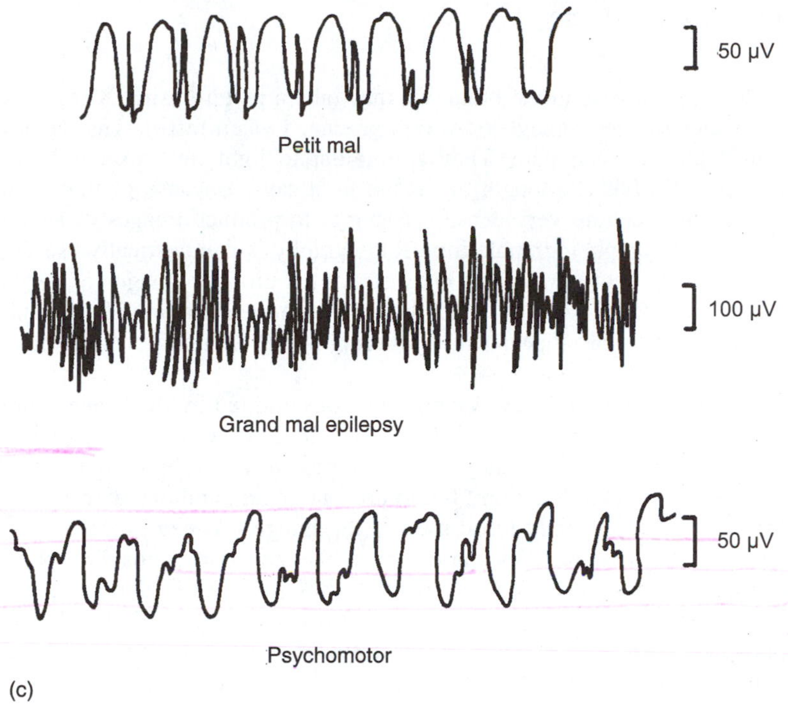

Petit mal

] 50 μV

Grand mal epilepsy

] 100 μV

Psychomotor

] 50 μV

(c)

Figure 7.5 (a) Four types of EEG waves. (b) When the eyes are opened, alpha waves disappear. (c) Different types of epilepsy yield abnormal waveforms. (From Guyton, A. C. *Structure and Function of the Nervous System*, 2nd. Ed. Copyright © 1972 by W. B. Saunders. Reprinted by permission of W. B. Saunders.)

Somatosensory Evoked Potentials

During somatosensory evoked potential (SSEP) testing, peripheral nerves are stimulated with a small electric current through bipolar electrodes on the skin of the subject's arms or legs. Electrodes placed on the subject's scalp and along the spinal cord can detect a series of positive and negative potentials following the peripheral stimulation. The intensity of the stimulus is generally very small. About 1000 responses are averaged to reduce the interference from endogenous cardiac, motor, and EEG activities and exogenous electric noise. The examination is repeated at least once to ensure reproducibility of the results. SSEP assesses the neurological pathways from the nerves in the arms or legs through the spinal cord to the brain. It can also be used to monitor coma patients, monitor patients during surgical procedures, and test hearing in infants and others whose hearing cannot be tested in the standard way (Gevins and Aminoff, 1988) (WFUBMC, 2003).

7.3 Brain Imaging: X ray

In order to diagnose disease in the brain, we may take a picture using X rays, which are very short wavelength electromagnetic waves produced when fast-moving electrons collide with substances in their path. They are similar to light rays, except X-ray wavelengths are only 1/10,000 the length of visible light rays. Because of this short wavelength, X rays can penetrate very dense substances to produce images or shadows that can then be recorded on photographic film. Radiography is diagnostically useful because differences in density between various body structures produce images of varying light and dark intensity on the X-ray film, much like the negative print of a photograph. Dense structures, such as the skull appear white, whereas air-filled areas, such as the lung are black.

Figure 7.6 shows an X-ray system. Electrons emitted by the heated filament are accelerated to the tungsten anode at +100 kV, which emits X rays. An aluminum filter stops low-energy X rays, because they would not pass through the body to form an image. The X rays travel in all directions, but to prevent patient and operator harm, they are shielded by a collimator so only those used in the image proceed. Secondary radiation could fog the film, but is stopped by a grid shaped like Venetian blinds. Phosphor screens emit many light photons for each X-ray photon, thus assisting in darkening the photographic film. To lower X-ray dose, an image intensifier may be used. The X-ray photon strikes a phosphor layer in a vacuum tube. Many light photons stimulate a photocathode to emit many electrons, which are accelerated to strike an output phosphor screen at +25 kV, yielding a good image with low X-ray dose (Siedband, 1998).

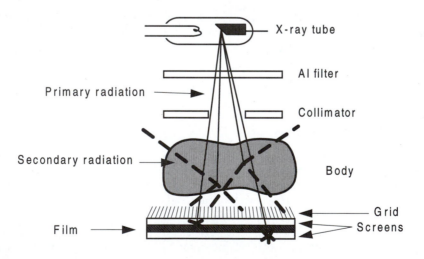

Figure 7.6 The 100 kV X-ray tube generates X rays that form a shadow of the body to expose the film. Unlike a camera, there are no lenses. (From Webster, J. G. (ed.) *Medical Instrumentation: Application and Design.* 3rd ed. Copyright © 1998 by John Wiley & Sons. Reprinted by permission of John Wiley & Sons.)

7.4 Brain Imaging: CT

Figure 7.7 shows a computed tomography (CT) X-ray system. It measures X-ray attenuation of many parallel paths in one plane. Then it repeats this at many angles. The information is processed by computer to yield a two-dimensional image of a slice through the body. Many slices can be used to create three-dimensional images. Figure 7.8 shows how the image may be digitally reconstructed by back-projection, in which the attenuation along each path is uniformly assigned along the path. Thus, if there is high attenuation at one picture element (pixel), the assignment from many angles increases that pixel more than other pixels. Iterative methods guess at the image, measure the resulting attenuation, correct to that measured by X ray, then back project the revised information. Water is used as the benchmark substance, as it has a CT value of zero. Analytic methods use spatial Fourier analysis or convolution techniques. CT can produce cross-sectional images (slices) of anatomic structures without superimposing tissues on each other. Improved CT scanners use multiple simultaneous measurements and X-ray tubes that rotate in 2 s. CT yields images of soft tissue within the brain not possible with conventional X ray, which is largely blocked by the dense skull. The patient is positioned in the center of a donut-like hole, surrounded by the CT scanner. CT is used in the differential diagnosis of intracranial neoplasms (brain tumors), cerebral infarctions (lack of blood supply to the brain), ventricular displacement or enlargement, cortical atrophy (decrease in size of the brain), cerebral aneurysms (enlarged sections of vessels in the brain), intracranial hemorrhage and hematoma (bleeding and subsequent clotting in the brain), and arteriovenous malformation (AVM-malformed blood vessels). However, there do exist limitations of CT as with all imaging devices. The spatial and contrast resolutions limit CT in its ability to distinguish normal from pathogenic tissue. For example, if a tissue (i.e. tumor) in the body has attenuation similar to that of water, it will not be detected.

7.5 Brain Imaging: MRI

Magnetic resonance imaging (MRI) provides two-dimensional images that are slices of ^{1}H isotope concentration in three-dimensional objects. The patient is placed in a very strong z-axis magnetic field of about 2 T. The charged ^{1}H electrons spin and precess like a spinning top at the Larmor frequency of 42.57 MHz/T. The magnetic field is perturbed to produce a small magnetic gradient along the y-axis. Figure 7.9 shows that there is only one slice at a particular Larmour frequency. Surrounding RF coils are pulsed at this frequency, which excites particles in the slice. The magnetic field is quickly perturbed along the x-axis. An RF receiver measures the relaxation along a single line along the z-axis orthogonal to both gradient fields. The z-axis magnetic field can be perturbed to produce additional scan lines. The system is similar to CT and the signals are processed in the same way. The strengths of MRI include: sensitivity to flow (i.e. blood); electronic adjustment of imaging plane (opposed to physical adjustment in CT); no ionizing radiation (only magnetism and radio waves); and high soft tissue resolution (see Figure 7.10). However, some of the weaknesses are contraindications due to implanted medical devices and metal plates in patients, slow image acquisitions, and high cost.

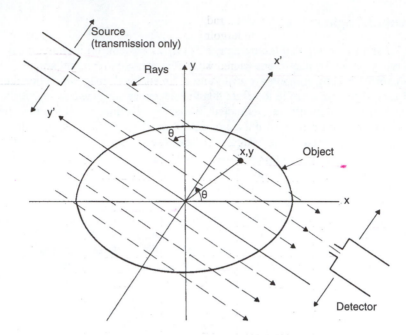

Figure 7.7 In computed tomography, a thin X-ray beam measures the attenuation along a single path. The system is indexed to obtain attenuation in many parallel paths at one angle and then repeated at many angles. The information is processed to yield attenuation at every element in the two-dimensional slice. (From Brooks R. A. and Di Chiro, G. Theory of image reconstruction in computed tomography. *Radiology*, 117: 561–572, 1975. Copyright 1975 © by Radiological Society of North America. Reprinted by permission of Radiological Society of North America.)

7.6 Brain Imaging: Nuclear Imaging

A radionuclide is an atom with an unstable nucleus. In an attempt to reach stability, the radionuclide emits one or more types of radiation, the most common examples being alpha particles, beta particles, and gamma electromagnetic radiation. In general, gamma rays are used for imaging organ systems and provide an indication of how well an organ system functions. Computerized radiation detection equipment, particularly scintillation detectors, detects gamma rays by giving off a light flash (scintillation). The imaging device provides an image of the organ under study and provides information on its size, shape, position, and functional activity. It is used mainly to allow visualization of organs and regions within organs that cannot be seen on a simple X-ray film. Space-occupying lesions, especially tumors, stand out particularly well (Early and Sodee, 1995).

A radiopharmaceutical is administered orally or intravenously to the patient. A sufficient time interval is allowed for the radioactive material to follow its specific metabolic pathway in the body and to concentrate in the specific tissue to be studied. An imaging device records the position and concentration of the radiation that emerged from the radionuclide. In almost all instances, radionuclide imaging exposes the patients to less radiation than would be received undergoing a similar procedure with diagnostic X-ray studies. With a nuclear medicine scan, metastatic disease to the bone can be found 6 months to a year before it can be detected with the usual bone radiograph. Nuclear imaging images function rather than anatomy. A nuclear brain scan is used for the diagnosis of pathologic abnormalities such as tumors, cerebrovascular aneurysms, and hematomas.

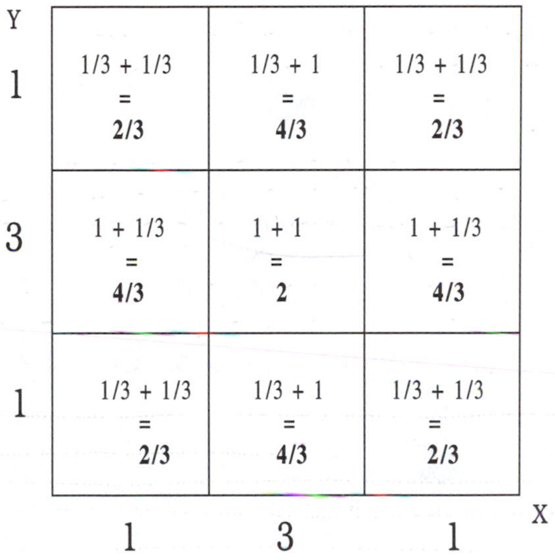

Figure 7.8 Creating images using back-projection. Each row and column signifies a certain path angle. The values of 3 in the center row and column are uniformly back-projected with a value of 1 in each cell. These sum to the maximum value of 2, thus indicating an object in that cell. The values of 1 are uniformly back-projected with a value of 1/3 in each cell. A sum of these two numbers that is significantly greater relative to the others indicates an object.

Gamma Camera

Figure 7.11 shows a gamma camera. Gamma rays from the organ travel in all directions, but the collimator tubes ensure that only radiation from a small region reaches the NaI detector. Light is detected by more than one photomultiplier tube, but the relative amounts received improve spatial resolution. An electronic pulse height analyzer selects events that have the proper gamma-ray energy. The resulting image has spatial resolution of about 1% of the image dimension (Mettler and Guiberteau, 1991).

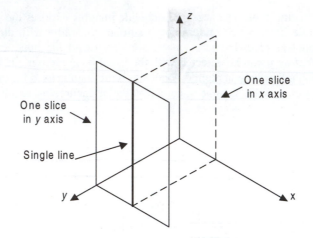

Figure 7.9 In MRI, RF coils excite one slice in the *y*-axis. An RF receiver measures from one slice in the *x*-axis.

7.7 Brain Imaging: Single-Photon Emission Computed Tomography (SPECT)

In single-photon emission computed tomography (SPECT), a scintillation assembly similar to a gamma camera is rotated around the patient. The gamma rays are collected from the patient in a manner similar to CT, but several slices are obtained at the same time. The resulting multiple slices show depth activity in the volume of interest. It is possible to see anomalies not observable with conventional X rays or gamma cameras (Van Heertum and Tikofsky, 1995).

7.8 Brain Imaging: Positron Emission Tomography (PET)

Some isotopes produce positrons that react with electrons to emit two photons at 511 keV in opposite directions. Figure 7.12(a) shows the two detectors on opposite sides of the patient that determine if the two scintillation effects are coincident and have energy levels close to 511 keV. Additional pairs of detectors permit faster operation. Image reconstruction is similar to that of CT, but an advantage of PET is that all of the most common radioisotopes used, ^{15}O, ^{13}N, ^{11}C and ^{18}F, can be compounded as metabolites. For example, CO can be made with ^{11}C. If a portion of the brain is active, increased blood flow carries the isotope to it, where it shows up on the image. Abnormal functioning, tumors, seizures and other anomalies may also be mapped this way. For example, measurement of glucose–fluorodeoxyglucose (FDG) metabolism is used to determine tumor growth. Because small amounts of FDG can be visualized, early tumor detection is possible before structural changes occur or would be detected by MRI or CT.

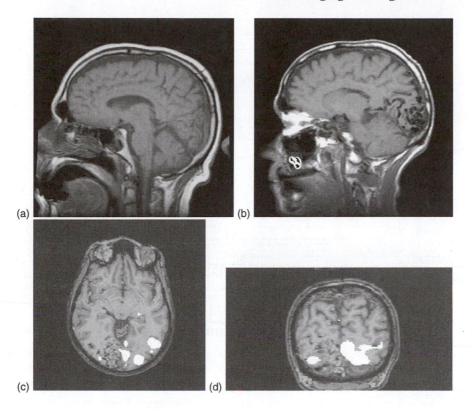

Figure 7.10 MRI images of the brain. (a) A sagittal view of a normal brain. (b) A sagittal view of a patient with an arterio-venous malformation (AVM) which is an explosive growth of the capillary bed. (c) and (d) Brain images from the patient with the AVM in which the visual cortex of the brain was activated using flashing lights while rapid MR imaging was performed. The regions of the brain that were "activated" by the visual stimulus are displayed in white on the images, (Images are from University of Wisconsin-Madison fMRI (functional magnetic resonance imaging) laboratory.)

7.9 Brain Imaging: Biomagnetism

Action potentials within the brain cause ionic currents to flow through cell membranes and between cells. Current flow creates magnetic fields. Magnetic shielding around the subject can remove interference from the earth's magnetic field of 50 µT. A gradiometer (two coils series opposed) can remove the interfering urban noise magnetic fields of 10 to 100 nT from distant sources, yet detect the nearby sources from the brain. The Superconducting Quantum Interference Device (SQUID) magnetometer has a sensitivity of 0.01 pT and can detect the magnetoencephalogram (MEG) of the alpha wave of about 0.1 pT (Malmivuo and Plonsey, 1995); (NATO, 1983).

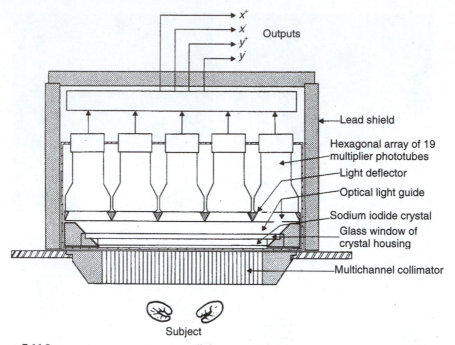

Figure 7.11 In a gamma camera system, radioisotopes emit gamma rays, which are collimated, and strike a NaI crystal, which emits light measured by photomultiplier tubes. (From Hine, G. J. (ed.) *Instrumentation in Nuclear Medicine*. Copyright © 1967 by Academic Press. Reprinted by permission of Academic Press.)

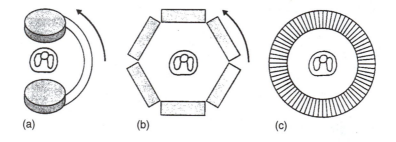

Figure 7.12 In the PET camera, (a) the paired and (b) the hexagonal ring cameras rotate around the patient. (c) The circular ring does not rotate, but may move slightly to fill in the gaps between the detectors. (From Webster, J. G. (ed.) *Medical Instrumentation: Application and Design*. 3rd ed. Copyright © 1998 by John Wiley & Sons. Reprinted by permission of John Wiley & Sons.)

An advantage of the MEG is that the magnetic field is unaffected by the high resistivity of the skull. In fact, no electrodes are required on the scalp. In addition, field distributions of the brain are measured by arrays of 300 or more detectors surrounding the brain (Hari and Lounasmaa, 2000). The combination of functional information, deduced from MEG, with structural information obtained from MRI has become almost routine in most MEG research groups.

7.10 Eye, ERG, EOG, and Visual Field

7.10.1 Electroretinogram (ERG)

When the retina is stimulated with light, an action potential is superimposed on the resting potential. Measuring the electrical response of the retina to flashes of light detects inherited retinal degenerations, congenital (existing from birth) retinal dystrophies, and toxic retinopathies from drugs or chemicals. The electrical response of the retina to a reversing checkerboard stimulus can help distinguish between retinal and optic nerve dysfunction to diagnose macular disease (Vaughan and Asbury, 1995), which is the leading cause of blindness in elderly Americans today and it is also linked to smoking. This disease affects the central area of the retina, causing gradual loss of central vision. The degeneration is caused by the partial breakdown of the insulating layer between the retina and the choroids (the layer of blood vessels behind the retina) (Yahoo! Health, 2003).

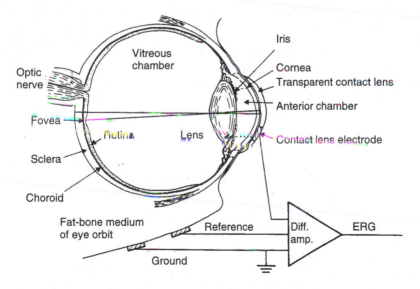

Figure 7.13 The transparent contact lens contains one electrode. The reference electrode is placed on the right temple. (From Webster, J. G. (ed.) *Medical Instrumentation: Application and Design.* 3rd ed. Copyright © 1998 by John Wiley & Sons. Reprinted by permission of John Wiley & Sons.)

The retina has electrical dipole properties that result in the corneal surface being electropositive with respect to the outer scleral surface. Figure 7.13 shows that the record is made by placing an active electrode on the cornea, usually one embedded in a corneal contact lens, with saline solution bridging the gap between the electrode and the cornea, and placing an indifferent electrode on the forehead. An alternative to the contact lens electrode uses electrodes that hook over or rest on the lower eyelid. These electrodes require a large number of responses, but are usually a better option for children than the contact lens electrodes. The entire retina is stimulated with light and the small voltage is amplified and recorded. The retina is stimulated with light after either dark (scotopic) or light (photopic) adaptation (Newell, 1996).

Flash ERG

In flash ERG, the inner and outer nuclear layers of the retina produce a and b waves, respectively (Figure 7.14). Some of the cells that make up the two layers of the retina are the photoreceptors, horizontal cells, bipolar cells, and Muller cells. All of the cells in the two layers give a graduated hyperpolarizing or depolarizing response to light, as compared with the action potentials produced by the retinal ganglion cells. The a wave of the ERG is produced by the hyperpolarizing photoreceptors, and has separate rod and cone components. The b wave is generated from the activity of depolarizing bipolar cells. The a wave represents photoreceptorial activity and the b wave represents postphotoreceptorial activity in the second-order retinal neuronal network, but the sources are still unclear.

To differentiate between rod and cone activity, special conditions can be applied. Since the rods of the eyes are the visual receptors used in dim light, recording from the rod system requires long periods of dark adaptation and one very dim single flash stimulus. Using scotopic conditions and intense white flashes produces a mixed rod and cone response. To obtain cone response, intense light flashes accompanied by an intense light background are required. The differentiation between rod and cone activity is useful in the diagnoses of several diseases including inherited retinal degenerations (mentioned above), which can cause night blindness. The degeneration starts at more peripheral parts of the retina and progressively decreases the field of vision. This disease can be recognized in the later stages by narrowing of the retinal vessels, and scattered clumps of pigment throughout the retina (Encyclopaedia Britannica, 2003).

7.10.2 Electro-Oculogram (EOG)

The resting potential of the front of the retina is electropositive with respect to the back of the retina. This makes the cornea positive compared to the back of the eye. If the eye looks left, this positive cornea makes an electrode to the left of the eye more positive than an electrode to the right of the eye. We can measure the differential voltage of about 100 µV between the two electrodes to determine horizontal direction of gaze. Electrodes above and below the eye can determine the vertical direction of gaze. This electro-oculogram (EOG) is useful for measuring changes in direction of gaze, but cannot measure absolute direction of gaze because variation in skin potential is larger than the EOG signal. This test can be used to detect retinal pigment epithelial dysfunction (Verdon, 2003).

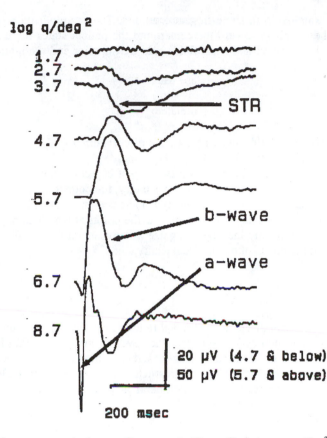

Fig 7.14 The human corneal electroretinogram. A 10 ms flash in quanta/deg^2 yields scotopic threshold respone (STR), the postphotoreceptoral b-wave, and the inner layer a-wave. (From Henkenlively, J. R., and Arden, G. B. (eds.) *Principles and Practice of Clinical Electrophysiology of Vision*. Copyright © 1991 by Mosby Year Book. Reprinted by permission of Mosby Year Book.

7.10.3 Visual Field

The peripheral vision of the retina reduces rapidly away from the fovea centralis (near the center of the retina). This can be assessed by visual field measurement, in which the examinee detects the targets that stimulate the retina at varying distances from the fovea. Several methods are used to measure visual field, including the confrontation test and methods using perimeters, tangent screens, or an Amsler grid (Newell, 1996).

Confrontation Test

For this test the patient fixates their vision straight ahead on the examiner at a distance of 1 m. The patient closes one eye and the examiner puts one hand midway between the patient and the examiner with some fingers extended. The examiner then slowly moves the hand from the periphery toward the center, and the patient states the number of fingers displayed. The temporal and nasal fields, and the superior and the inferior fields are tested in turn.

Perimeters

A perimeter consists of a half bowl with a radius of curvature of 33 cm so that the eye is at the center of rotation of the hemisphere.

In kinetic perimetry, the patient detects a test object of fixed size and illumination moved from a nonsensing area. In static perimetry, the patient detects a test object of constant size and location whose light intensity is increased until it can be seen.

An isopter, the contour line that connects the points at which an object of certain size and certain light intensity may be recognized, represents a line on the macula lutea that passes through the points of equal visual acuity.

Tangent Screen

The evaluation of the central 30° of the visual field is desirable because the photoreceptors of the eye are concentrated in and near the fovea centralis. This can be conducted with a test object from 1 to 50 mm on a tangent screen, which contains concentric circles every 5° and radiating lines 15° or 22.5° apart stitched or marked on black cloth 1 m from the subject. This test demonstrates the peripheral isopter, the blind spot, and various scotomas (blind spots in the visual field).

Amsler Grid

An Amsler grid is formed by intersecting 21 horizontal and 21 vertical dark lines spaced about 5 mm with one dark dot in the middle. It encompasses the central 20° of the visual field when held at the reading distance. The patient tests an eye by fixating it on the central dot and indicates any irregular line section that appears blurred, wavy, curved, discolored, or missing. This may be an indication of leakage or bleeding in the back of the eye causing retinal swelling (Freund, 2003). Amsler grid testing is recommended for patients with macular degeneration or other retinal disorders that threaten central vision. It is advantageous because most patients are able to administer the test themselves, and if a change in condition is noticed, they can come in to see their doctors immediately.

7.10.4 Eye Pressure

Intraocular pressure is an important indication of glaucoma, a disease that can lead to blindness. In a healthy eye, the aqueous humor is constantly produced by the ciliary body in the eye and circulates into the chamber between the iris and the cornea from behind the iris through the pupil. It can drain out of the eye through a network of tissues between the iris and the cornea. If the network of tissues, called the drainage angle, does not function well so that the aqueous humor flows away more slowly than it is produced, or even fails to flow away, the pressure in the eye increases and causes extra pressure on the retinal nerve fibers. Sometimes the extra pressure causes the collapse of tiny blood vessels that nourish the light-sensitive cells of the retina and the optic nerve fiber. The cells and nerve fibers begin to die and vision begins to fade in certain areas (Rich et al., 1996).

For the diagnosis of glaucoma, there are several important tests: Tonometry measures the ocular pressure; gonioscopy observes the anterior chamber angle; ophthalmoscopy evaluates the color and configuration of the cup and neuroretinal rim of the optic disk (where the optical nerve enters the retina and is therefore a blind spot), and perimetry measures visual function in the central field of vision.

Strictly speaking, the intraocular pressure cannot be measured unless we insert a cannula into the eye. Clinically the intraocular pressure is measured indirectly by measuring the ocular tension, which can be determined by the response of the eye to an applied force. Two methods are used to measure ocular tension. Applanation tonometry measures the force required to flatten a standard area of the cornea. Indentation tonometry measures the deformation of the globe in response to a standard weight applied to the cornea.

Figure 7.15 shows the Goldmann applanation tonometer, which measures the force required to flatten the cornea by an area of 3.06 mm in diameter. The flattened area is viewed with a biomicroscope through a split prism after instillation of a topical anesthetic and fluorescein. We calculate the pressure $p = f/a$ where f = force and a = area. The SchiØtz tonometer measures the ease with which the cornea may be indented by the plunger of the instrument. It is less costly and easier to use as compared with applanation-type instrument, but it is not as accurate as the applanation instrument and may be influenced by more factors.

A noncontact applanation tonometer measures intraocular pressure without touching the eye. An air pulse of linearly increasing force flattens the cornea in a few milliseconds. When a beam of light reflects from the flattened surface, a maximum detected signal shuts off the pneumatic pulse and measures the time elapsed, which correlates with pressure.

The normal value of intraocular pressure is about 15 ± 3 mmHg, while an eye with glaucoma possibly has a intraocular pressure around 22 ± 5 mmHg. The intraocular pressure may vary with time by 6 mmHg during a day, so continuous sampling of the eye pressure for several hours is required for adequate evaluation of glaucoma (Newell, 1996).

7.10.5 Ophthalmoscopy

Ophthalmoscopy is used to inspect the interior of the eye. It permits visualization of the optic disk, vessels, retina, choroid (layer of blood vessels behind the retina), and ocular media. Direct ophthalmoscopy and indirect ophthalmoscopy are the two methods used (Newell, 1996).

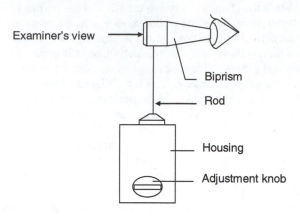

Figure 7.15 The Goldmann applanation tonometer. When the applanation head is moved to the right to rest on the anesthetized cornea, the examiner rotates the adjustment knob to increase the force.

Direct Ophthalmoscopy

In a darkened room, the examiner projects a beam of light from a hand-held ophthalmoscope through the patient's pupil to view an upright image of the retina structure. The ophthalmoscope has rotating lenses on top to magnify a particular area being viewed up to 15 times. With pupillary dilation, about half the fundus (posterior part of the eye) may be seen, while about 15% of the fundus may be seen without pupillary dilation. The resolving power of direct ophthalmoscopy is about 70 μm. Smaller objects, such as capillaries, small hemorrhages, or microaneurysms, cannot be seen.

Indirect Ophthalmoscopy

Indirect ophthalmoscopy, shown in Figure 7.16, is usually performed by using a binocular ophthalmoscope. The patient's retina is illuminated by a light source from headset of the binocular instrument. An inverted image of the fundus magnified about 5 × is formed between the condensing lens and the ophthalmoscope. Prisms within the instrument make it possible to see a stereoscopic image. The entire fundus may be examined by indirect ophthalmoscopy with pupillary dilation and sclera (the white part of the eye) indentation.

Indirect ophthalmoscopy provides a significantly wider field of view than direct ophthalmoscopy, but because of direct ophthalmoscopy's higher resolution both are commonly used during an eye examination (Bickford, 2003). Although indirect ophthalmoscopy only has 200 μm resolving power, it is more advantageous than direct ophthalmoscopy in that the stereoscopic image allows detection and evaluation of minimal elevations of the sensory retina and retina pigment epithelium. These images also allow the only direct view (without performing surgery) of the living network of blood vessels and can help diagnose atherosclerosis, hypertension, diabetes mellitus, and other systemic and eye-specific disorders (Bickford, 2003).

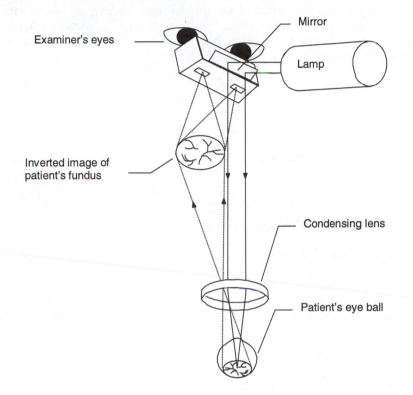

Figure 7.16 In indirect ophthalmoscopy, the examiner stands in front of the patient and holds the condensing lens at an appropriate distance from the patient's eye.

7.11 Ears and Audiometry

7.11.1 The Ears

The ear is a sense organ specialized for hearing and balance functions. Anatomically, it is divided into the external ear, middle ear, and inner ear. The external ear is composed of the auricle, the external auditory canal, and the outermost layer of the tympanic membrane. The middle ear, or tympanic cavity, is an oblong cavity bridged by an ossicular chain consisting of three small bones between the tympanic membrane and the inner ear. The inner ear consists of the cochlea and the labyrinth, which are the end organ receptors for hearing and balance (Cody et al., 1981).

The auditory portion of the ear can be physiologically divided into two parts—a sound-conducting apparatus and an electromechanical transducer (Hawke, 1990). The sound-conducting apparatus consists of the external ear, the tympanic membrane, the ossicular chain, and the labyrinthine fluid. The electromechanical transducer transforms the mechanical energy of sound into electric impulses to be transmitted by the auditory nerves to the auditory cortex of the brain.

7.11.2 Audiometry

Audiometry is used to evaluate hearing pathologies in order to provide diagnostic information and rehabilitation.

Pure Tone Air Conduction Threshold Testing

The audio signal passes through the outer ear, the middle ear, and the inner ear before being further processed by the central auditory system. Pathology of any of these parts of the ear can cause hearing loss. In this test, the subject responds to bursts of single-frequency stimuli presented through calibrated earphones. The signal intensity is adjusted, and the threshold with 50% correct responses is recorded (Laszlo and Chasin, 1988).

Pure Tone Bone Conduction Threshold Testing

A special vibratory transducer is placed on the mastoid process or the forehead, and thus stimulates the inner ear directly through the skull, bypassing the outer and middle ears. If there is a difference between the air conduction and bone conduction responses, this indicates pathology in the outer or middle ear, which may be treatable. Inner ear pathology is difficult to treat (Laszlo and Chasin, 1988).

Speech Discrimination Testing

In speech discrimination testing, the subject listens to lists of single-syllable speech discrimination words presented through earphones, and repeats what he or she hears. The result of this test is scored from 0 to 100% based on the correctness of the subject's answer. In contrast to the pure tone threshold test, which addresses hearing sensitivity, this test assesses the integrity of the entire auditory system's ability to hear clearly and understand speech communication. A low score is related to sensorineural loss, and a higher score may be attributed to normal hearing or conductive hearing loss (Laszlo and Chasin, 1988).

Speech Reception Threshold

This test uses words that are attenuated successively. The patient repeats back two-syllable words that have equal stress. A threshold is determined when the patient repeats 50% of the words correctly (Laszlo and Chasin, 1988).

Impedance Audiometry

In tympanometry, a probe in the ear varies air pressure from +1962 to –1962 Pa, testing middle ear function for the presence of fluid in the middle ear. A stiff middle ear (poor ossicular chain mobility) reduces the mobility of the eardrum, causing a 220 Hz tone to be reflected with little attenuation. A flaccid middle ear system (possible ossicular chain disarticulation) yields high damping and little reflection (Laszlo and Chasin, 1988).

Evoked Response Audiometry (see Section 7.2.3)

Clicks in the ear yield electric potentials measured by an electrode at the vertex (top) of the scalp. An average response computer separates the very small synchronized signals from the random signals from other brain potentials. Early latency evoked potentials (about 10 ms) originate from the brain stem. Middle latency responses (about 30 ms) are responses above the brain stem. Late latency evoked potentials (about 100 ms) evaluate the neurological integrity of the cortical centers.

7.11.3 Otoscopy

The external auditory canal and tympanic membrane, or eardrum, can be inspected with an otoscope, a hand-held instrument with a flashlight, a magnifying glass, and a cone-shaped attachment called an ear speculum. Evidence of injury or congenital (present at birth) malformation is usually obvious. One can find foreign bodies, inflammation, ulcers, lacerations, and tumors by visually inspecting the external auditory canal with an otoscope. This type of ear examination can also detect a rupture or puncture of the eardrum. One may also see scarring, retraction, or bulging of the eardrum. A red or swollen eardrum can indicate an ear infection. In physical diagnosis, the otologist views normal anatomy and pathologic processes in the middle ear through the translucent tympanic membrane (Cody et al, 1981).

7.12 Muscles

7.12.1 Muscle Contraction, Length and Force

About 40% of the human body is skeletal muscles, and another 10% is smooth muscles of internal organs and cardiac muscles from the heart. Here, we are interested in characterizing the function of skeletal muscles. The primary function of skeletal muscles is to generate force, and they are therefore excitable. Thus, skeletal muscles have two fundamental properties: They are excitable (able to respond to stimulus) and contractible (able to produce tension) (Biewener, 1992).

A skeletal muscle consists of numerous fibers with diameters ranging from 10 to 80 μm. Each muscle fiber contains hundreds to thousands of myofibrils (Figure 7.17), each containing about 1500 myosin filaments and 3000 actin filaments lying side by side (Carlson and Wilkie, 1974). Figure 7.18 shows the structure of a single myofibril.

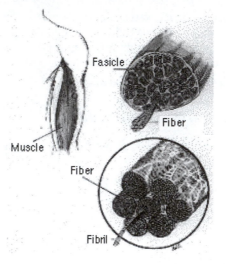

Figure 7.17 The components of a skeletal muscle; the myofiber is the smallest complete contractile system. Each myofiber is composed of many myofibrils. (From Muscle Physiology Laboratory, 2003).

As an action potential travels along a motor nerve to muscle fibers, it initiates an action potential along the muscle fiber membrane, which depolarizes the muscle fiber membrane and travels within the muscle fiber. The subsequent electrochemical reaction within the muscle fiber then initiates attractive forces between the actin and myosin filaments, causing them to slide together. This mechanism produces muscle contraction (Jones and Round, 1990).

Tension is developed in the muscle as it contracts. There are three types of contraction: (1) Isometric, or static, contraction means a muscle contracts without change in its length. (2) Concentric contraction occurs when a load is less than the isometric force produced by the muscle, and the load shortens the muscle. (3) Eccentric contraction occurs when the load is greater than the isometric force and elongates the contracting muscle (Sigel, 1986).

The length–force relation of a muscle can be measured isometrically. With this method, a skeletal muscle is maximally stimulated while its length is held constant. The tension generated by the muscle at various constant lengths is measured to generate the length–force curve (Lieber, 1992). Figure 7.19 shows a typical length–force curve of a skeletal muscle (Ozkaya and Nordin, 1991).

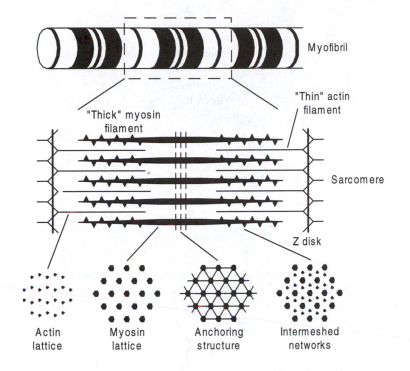

Figure 7.18 The structure of a single myofibril. Each myofibril contains about 1500 myosin filaments and 3000 actin filaments in the pattern shown above. The distance shown between the two Z disks is one sarcomere. During muscle contraction the thin (actin) filaments slide inward past the thick (myosin) filaments, pulling the two Z disks closer together, thus shortening the sarcomere. The shortening of the thousands of sarcomeres in each myofibril causes the myofibrils, the myofibers and then the entire muscle to contract. (From Muscle Physiology Laboratory, 2003).

The amount of force generated by a stimulated muscle depends on how its ends are restrained (Woledge et al., 1985). If neither end of a stimulated muscle is fixed, the muscle shortens at its maximum velocity of 33 cm/s, V_{max}, and no force is generated. If one end of the muscle is fixed and a small force is applied to the other end, the muscle shortens at a steady velocity less than V_{max}. A force of sufficient magnitude, F_0, will prevent the muscle from shortening and isometric contraction occurs within the muscle. The force–velocity relation is described by Hill's equation (Woledge et al., 1985)

$$(F + a)(V + b) = (F_0 + a)b \tag{7.2}$$

where a and b are constants derived experimentally, F is the muscle force during shortening at velocity V, and F_0 is the maximum isometric force that the muscle can produce. Figure 7.20 shows a hypothetical curve for force–velocity relation according to Hill's equation.

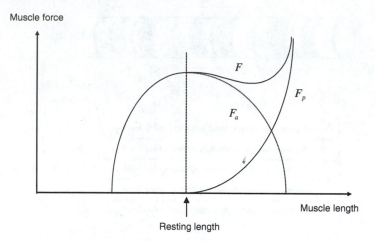

Figure 7.19 The overall force of a muscle, F, is the sum of the active force, F_a, and the passive force, F_p. The active force results from voluntary contraction of the contractile elements of the muscle. The passive force results from elongation of the connective muscle tissue beyond its resting length. No passive force builds up when the muscle is at its resting length or less.

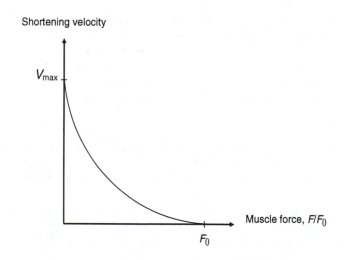

Figure 7.20 The muscle force decreases with increasing shortening velocity.

The force–velocity curve can be obtained experimentally with isotonic measurement (Lieber, 1992). The muscle is allowed to change its length against a constant load during maximal stimulation. The muscle velocity (i.e., shortening velocity, or lengthening velocity) is measured and plotted against the constant load.

7.12.2 Electromyography (EMG)

By placing electrodes into a skeletal muscle, we can monitor the electric activity of the muscle (Keynes and Aidley, 1991). EMG is used to detect primary muscular disorders along with muscular abnormalities caused by other system diseases such as nerve dysfunction (Junge, 1992). Using EMG to study muscle function is also used in various fields such as kinesiology, psychology, and rehabilitation medicine (Basmajian and De Luca, 1985).

The two main types of electrodes for measuring EMG signals from muscles are surface electrodes and inserted electrodes. The inserted type further includes needle electrodes and wire electrodes. Selection of electrode type depends on the particular application and convenience of use (Loeb and Gans, 1986).

The surface electrode consists of silver disks that adhere to the skin. Saline gel or paste is placed between the electrode and the skin to improve the electric contact. The disadvantages of surface electrodes are that they cannot effectively detect signals from muscles deep beneath the skin and that, because of poor selectivity, they cannot eliminate cross-talk from adjacent muscles.

Needle electrodes have a relatively smaller pickup area and, thus, are more suitable for detecting individual motor unit potentials. Needle electrodes can be repositioned in the muscle after they are inserted. Figure 7.21 shows various constructions of needle electrodes (Kimura, 1989).

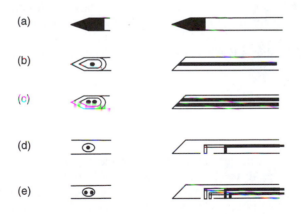

Figure 7.21 Examples of needle electrodes: (a) Monopolar needle with exposed pickup tip. (b) Monopolar configuration with a pickup wire centrally located in a hypodermic needle. (c) Bipolar configuration with two pickup wires in parallel with each other in a hypodermic needle. (d) Monopolar configuration with a wire exposed at the needle's side hole. This can be used to detect the activity of individual muscle fibers. (e) Bipolar configuration with two wires exposed at the side hole of a hypodermic needle. This can be used to detect the motor unit action potential from a large portion of the motor unit territory.

Figure 7.22 shows a wire electrode consisting of two insulated fine wires, 25 to 100 μm in diameter, inserted through the cannula of a hypodermic needle. The distal tips of the wires are deinsulated by about 1 to 2 mm and bent to form staggered hooks. One disadvantage of wire electrodes is that they tend to migrate after insertion into the muscle.

Figure 7.22 A bipolar wire electrode uses a hypodermic needle to insert through the skin into the muscle of interest.

We can obtain the EMG signal simply by placing a surface electrode on the skin enveloping the muscle or by applying an inserted electrode in the muscle. A reference electrode, usually a surface electrode, is placed on a site with minimal electric association to the inserted site. The drawback of this monopolar configuration is that it detects not only the signal from the muscle of interest but also unwanted signals from around the muscle of interest.

In the bipolar configuration, two electrodes with a small distance between each other are placed in the muscle to pick up the local signals within the muscle of interest. A differential amplifier amplifies the signals picked up from the two electrodes with respect to the signal picked up by a reference electrode. Because the interference signals from a distant source are essentially equal in magnitude and phase, as detected by the two electrodes, the common mode rejection capability of the differential amplifier eliminates the unwanted signals.

Figure 7.23 shows EMG waveforms of tongue muscle taken when the subject was speaking some vowels.

7.12.3 Neuromuscular Performance

Nerve cells that control movement emerge from the spinal cord and brainstem. They travel various distances to reach particular muscles. Neuromuscular disorder develops when any part along the path, from the brain to the muscle, does not work properly. If the nerve that controls the muscle is injured and the action potential from the brain cannot reach the muscle, the muscle does not contract normally. On the other hand, if the muscle has some disease, it may not produce normal contraction in response to the action potential from the brain.

For adequate treatment of neuromuscular disorders, it is important to know where the injury occurs and how serious it is. By applying a dc stimulus to the skin over a nerve at various positions in the legs or arms, and recording the time of muscle twitch due to the stimulation with an electrode on the skin over the muscle, one can calculate the conduction velocity of the nerve. The conduction is slowed with injury to the nerve (Ringel, 1987).

EMG is also used for diagnosis of neuromuscular performance. Abnormal electric signals are recorded by the EMG if there is disease in the muscle.

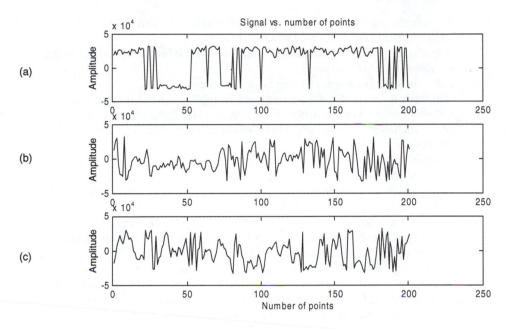

Figure 7.23 EMG signals associated with the speech signal of the letters /ia/iao/ian/iang/iong. The EMG signal measured from (a) the hyoglossus (tongue) muscle, (b) styloglossus muscle, and (c) genioglossus muscle. Wire electrodes were inserted into these tongue muscles to take the EMG measurement.

7.13 Problems

7.1   Calculate the magnitude and sign of the cell membrane potential at body temperature given that the intracellular potassium concentration = 150 millimoles/L and the extracellular potassium concentration = 5 millimoles/L.

7.2   Figure 7.2 shows the change of membrane potential due to an electric stimulus. Suppose that the extracellular potassium concentration keeps constant during the time course, sketch a figure to show the change of the intracellular potassium concentration.

7.3   Calculate the velocity of propagation for the action potential in Figure 7.2.

7.4   Explain the origin of action potentials and how to measure them.

7.5   Explain why and how we measure the electroencephalogram.

7.6   Explain why and how we measure evoked potentials.

7.7   Draw the block diagram of a signal averager (average response computer) for use with evoked potentials and explain how it improves the signal and by what factor.

7.8   Calculate the improvement in signal-to-noise ratio when an average response computer averages 200 recordings.

7.9   Explain the reason for an X-ray filter, collimator, grid, and screen.

7.10   Describe the beam paths used in a CT scanner.

7.11   Explain how magnetic fields are varied to obtain information from a single line of tissue in an MRI scanner.

7.12   Describe the advantages of using nuclear medicine.

7.13   We wish to image the functionally active regions of the thyroid gland, which takes up iodine when functionally active. Explain the procedure and equipment used, including the detector type, localization method, and energy selection method.

7.14   Explain how a gamma camera forms an image.

7.15   Explain why and how we diagnose visual system disease.

7.16   Describe the placement of electrodes to measure horizontal direction of gaze in electro-oculography.

7.17   Explain why and how pressure within the eyeball is measured.

7.18   For the Goldmann applanation tonometer, calculate the force in newtons when measuring a patient with normal ocular pressure.

7.19   Explain why and how we diagnose auditory system disease.

7.20   Explain why and how we measure dimension, force, and electrical activity of skeletal muscle.Design an experiment to determine the force–length relationship of a muscle. Is the method isometric or isotonic?

7.14 References

American Academy of Family Physicans (AAFP). 2003. What You Should Know About Epilepsy. [Online] familydoctor.org/handouts/214.html.

Basmajian, J. V. and De Luca, C. J. 1985. *Muscle Alive: Their Functions Revealed by Electromyography.* 5th ed. Baltimore, MD: Williams & Wilkins.

Bickford, L. 2003. The EyeCare Connection: The EyeCare Reports. [Online] www.eyecarecontacts.com/exam.html.

Biewener, A. A. (ed.) 1992. *Biomechanics-Structures and Systems: A Practical Approach.* New York : IRL Press at Oxford University Press.

Campbell, N. A., Reece, J. B. and Mitchell L. G. 1999. *Biology.* 5th ed. Menlo Park: Benjamin/Cummings.

Carlson, F. D. and Wilkie, D. R. 1974. *Muscle Physiology.* Englewood Cliffs, NJ: Prentice-Hall.

Castello, R. B. (ed.) 1994. *The American Heritage Dictionary.* New York: Dell.

Clark, J. W. 1998. The origin of biopotentials. In Webster, J. G. (ed.) *Medical Instrumentation: Application and Design.* 3rd ed. New York: John Wiley & Sons.

Cody, T. R., Kern, E. B. and Pearson, B. W. 1981. *Diseases of the Ears, Nose, and Throat.* Mayo Foundation.

Early, P. J. and Sodee, D. B. 1995. *Principles and Practice of Nuclear Medicine.* 2nd ed. St. Louis: Mosby.

Encyclopaedia Britannica. 2003. Eye disease: Retinal degeneration. [Online] search.eb.com.

Freund, K. B. 2003. Amsler Grid Testing of Central Vision. [Online] www.vrmny.com/amsler.htm.

Gevins, A. S. and Aminoff, M. J. 1988. Electroencephalography: Brain Electrical Activity. In J. G. Webster (ed.) *Encyclopedia of Medical Devices and Instrumentation*. New York: John Wiley & Sons.

Hari, R., and Lounasmaa, O. V. 2000. Neuromagnetism: tracking the dynamics of the brain. *Physics Today* 33–38, May 2000.

Hine, G. J. 1967. *Instrumentation in Nuclear Medicine*. New York: Academic.

Hughes, J. R. 1994. *EEG in Clinical Practice*. 2nd ed. Boston: Butterworth-Heinemann.

International Conference on Biomagnetism. 1989. *Advances in Biomagnetism*. New York: Plenum Press.

Isley, M. R., Krauss, G. L., Levin, K. H., Litt, B., Shields, R. W. Jr. and Wilbourn, A. J. 1993. *Electromyography/Electroencephalography*. Redmond, WA: SpaceLabs Medical.

Jones, D. A. and Round, J. M. 1990. *Skeletal Muscle in Health and Disease : A Textbook of Muscle Physiology*. Manchester, New York: Manchester University Press.

Junge, D. 1992. *Nerve and Muscle Excitation*. 3rd ed. Sunderland, MA: Sinauer Associates.

Keynes, R. D. and Aidley, D. J. 1991. *Nerve and Muscle*. 2nd ed. Cambridge, New York: Cambridge University Press.

Kimura, J. 1989. *Electrodiagnosis in Diseases of Nerve and Muscle: Principles and Practice*. 2nd ed. Philadelphia: Davis.

Laszlo, C. A. and Chasin, D. J. 1988. Audiometry. In J. G. Webster (ed.) *Encyclopedia of Medical Devices and Instrumentation*. New York: John Wiley & Sons.

Lieber, R. L. 1992. *Skeletal Muscle Structure and Function: Implications for Rehabilitation and Sports Medicine*. Baltimore: Williams & Wilkins.

Loeb, G. E. and Gans, C. 1986. *Electromyography for Experimentalists*. Chicago: University of Chicago Press.

Malmivuo, J. and Plonsey, R. 1995. *Bioelectromagnetism: Principles and Applications of Bioelectric and Biomagnetic Fields*. New York: Oxford University Press.

Mazziota, J. C. and Gilman, S. 1992. *Clinical Brain Imaging: Principles and Applications*. Philadelphia: F. A. Davis Company.

Mettler, F. A. and Guiberteau, M. J. 1991. *Essentials of Nuclear Medicine Imaging*. 3rd ed. Philadelphia: W. B. Saunders.

Muscle Physiology Laboratory. 2003. Myofilament structure. [Online] muscle.ucsd.edu/musintro/fibril.

Muscle Physiology Laboratory. 2003. Fiber types. [Online] muscle.ucsd.edu/musintro/fiber.

NATO Advanced Study Institute on Biomagnetism. 1983. *Biomagnetism: An Interdisciplinary Approach*. New York: Plenum Press.

Newell, F. W. 1996. *Ophthalmology: Principles and Concepts*. 8th ed. St. Louis: Mosby - Year Book.

Ozkaya, N. and Nordin, M. 1991. *Fundamentals of Biomechanics: Equilibrium, Motion, and Deformation*. New York: Van Nostrand Reinhold.

Ringel, S. P. 1987. *Neuromuscular Disorders: A Guide for Patient and Family*. New York: Raven Press.

Ritch, R., Shields, M. B. and Krupin, T. (eds.) 1996. *The Glaucomas*. 2nd ed. St. Louis: Mosby.

Schneck, D. J. 1992. *Mechanics of Muscle*. 2nd ed. New York: New York University Press.

Siedband, M. P. 1998. Medical imaging systems. In Webster, J. G. (ed.), *Medical Instrumentation: Application and Design*. 3rd ed. New York: John Wiley & Sons.

Siegel, I. M. 1986. *Muscle and its Diseases: An Outline Primer of Basic Science and Clinical Method*. Chicago: Year Book Medical Publishers.

Stein, R. B. (ed.) 1980. *Nerve and Muscle: Membranes, Cells, and Systems*. New York: Plenum Press.

Van Heertum, R. L. and Tikofsky, R. S. (eds.) 1995. *Cerebral SPECT Imaging*. 2nd ed. New York: Raven Press.

Vaughan, D. G. and Asbury, T. 1999. *General Ophthalmology*. 15th ed. Appleton & Lange.

Verdon, W. 2003. Electrodiagnostic and Vision Functions Clinic. [Online] spectacle.berkeley.edu/ucbso/vfc/

WFUBMC. 2003. Wake Forest University Baptist Medical Center. Diagnostic Neurology Department. EEG Laboratory. [Online]
www.wfubmc.edu/neurology/department/diagneuro/eeglab.html

WFUBMC. 2003. Wake Forest University Baptist Medical Center. Diagnostic Neurology Department. Evoked Potential Laboratory. [Online]
www.wfubmc.edu/neurology/department/diagneuro/ep.html

Webster, J. G. (ed.) 1998. *Medical Instrumentation: Application and Design*. 3rd ed. New York: John Wiley & Sons.

Woledge, R. C., Curtin, N. A. and Homsher, E. 1985. *Energetic Aspects of Muscle Contraction*. London; Orlando: Academic Press.

Yahoo! Health. 2003. Macular degeneration. [Online]
health.yahoo.com/health/encyclopedia/001000/0.html.

Chapter 8

Heart and Circulation

Supan Tungjitkusolmun

Cardiovascular disease is the leading cause of death in the United States. There are over one million heart attacks per year and over 600,000 deaths, of which 300,000 die before reaching the hospital. One important trend that has emerged is the emphasis on prevention. Today, most people are aware of the things they can control that prevent heart disease (diet, not smoking, exercise, blood pressure, etc.). Attention to these areas helps reduce the chance of getting the disease but does not eliminate it completely. Early detection and treatment of the disease continue to be key areas of emphasis in the medical community. The level of sophistication in dealing with cardiovascular diseases has taken great strides in the past two decades and has been greatly aided by advances in technology. New diagnostic techniques have been developed and old ones improved, providing added sensitivity and specificity. It is now very common for primary care physicians to conduct tests in their offices to detect cardiovascular diseases.

It is very important for biomedical engineers to understand the basic principles of medical instruments. This chapter presents several important cardiovascular variables and the most common methods by which they are measured, as well as an introduction to the physiology of the heart and blood vessels.

The human heart and blood vessels are a transportation system that delivers essential materials to all cells of the body and carries away the waste products of metabolism. The heart serves as a four-chamber mechanical pump. The two chambers on the right side of the heart send deoxygenated blood to the lungs via the pulmonary arteries (pulmonary circulation). The other two chambers, on the left side of the heart, supply oxygenated blood through the arteries, which branch in an orderly fashion until they reach the capillary beds of the organs (systemic circulation). Here, there is an exchange of nutrients, waste products and dissolved gases. The blood then flows from the capillaries into the veins, which lead back to the right atrium. Hormones are also transported by the cardiovascular system.

8.1 Cardiac Anatomy and Physiology

Figure 8.1 illustrates important features of the human heart. The heart, about the size of a clenched fist, is located in the thoracic (chest) cavity between the sternum (breastbone) and the vertebrae (backbone). The heart is divided into right and left halves and has four chambers—an upper and lower chamber within each half. The upper chambers, the *atria*, receive blood returning to the heart and transfer it to the lower chambers, the *ventricles*, which pump the blood from the heart. Figure 8.2 shows a simplified circulatory system. Blood returning from the systemic circulation enters the right atrium via large veins known as the *venae cavae*. The blood entering the right atrium has returned from the body tissues, where O_2 has been extracted from it and CO_2 has been added to it by the cells of the body. This partially deoxygenated blood flows from the right atrium into the right ventricle, which pumps it out through the pulmonary artery to the lungs. Thus, the right side of the heart pumps blood into the pulmonary circulation. Within the lungs the blood loses its CO_2 and picks up a fresh supply of O_2 before being returned to the left atrium via the pulmonary veins. This highly oxygenated blood returning to the left atrium subsequently flows into the left ventricle, the pumping chamber that propels the blood to all body systems except the lungs; that is the left side of the heart pumps blood into the systemic circulation. The large artery carrying blood away from the left ventricle is the *aorta*.

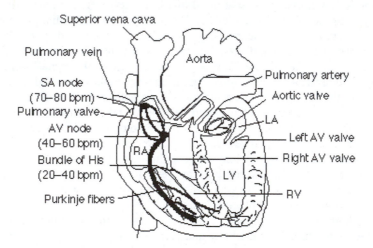

Figure 8.1 Basic structure of the heart. RA is the right atrium, RV is the right ventricle; LA is the left atrium, and LV is the left ventricle. Basic pacing rates are shown.

8.1.1 Heart Valves

Blood flows through the heart in one fixed direction—from veins to atria to ventricles to arteries (see Figure 8.2). The presence of four one-way heart valves aides the unidirectional flow of blood, and the valves are positioned so that they open and close passively due to pressure differences, similar to a one-way door. The backward gradient can force the valve to close but cannot force it to swing open in the opposite direction.

Two of the heart valves, the right and left *atrioventricular* (AV) valves, are positioned between the atrium and the ventricle on the right and left sides, respectively (see

Figure 8.1). These valves allow blood to flow from the atria into the ventricles during ventricular filling (when atrial pressure exceeds ventricular pressure), but prevent the backflow of blood from the ventricles into the atria during ventricular emptying (when ventricular pressure greatly exceeds atrial pressure). If the rising ventricular pressure did not force the AV valves to close as the ventricles contracted to empty, much of the blood would inefficiently be forced back into the atria and veins instead of being pumped into the arteries. The right AV valve is also called the *tricuspid* valve because it consists of three cusps, or leaflets. Likewise, the left AV valve, which consists of two leaflets, is often called the *bicuspid* valve or, alternatively, the *mitral* valve.

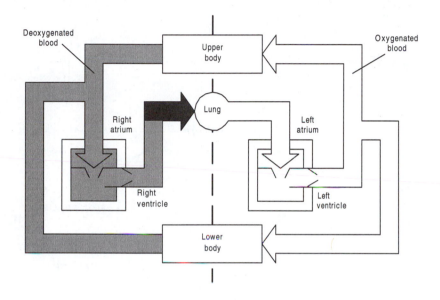

Figure 8.2 The simplified circulatory system. The blood is delivered from the right ventricle to the lung. The oxygenated blood from the lung is then returned to the left atrium before being sent throughout the body from the left ventricle. Deoxygenated blood from the body flows back to the right atrium and the cycle repeats.

The other two heart valves, the *aortic* and *pulmonary* valves, are located at the junction where the major arteries leave the ventricles. They are also known as *semilunar* valves because they are composed of three leaflets, each resembling a shallow half-moon shaped pocket. These valves are forced open when the left and right ventricular pressures exceed the pressure in the aorta and pulmonary arteries (i.e. during ventricular contraction and emptying). The valves close when the ventricles relax and ventricular pressures fall below the aortic and pulmonary artery pressures. The closed valves prevent blood from flowing from the arteries back into the ventricles from which it has just been pumped. The semilunar valves are prevented from swinging back into the ventricles by the anatomical structure and positioning of the cusps (Sherwood, 2001).

8.1.2 Cardiac Cycle

A cardiac cycle contains a sequence of events during one heart beat. The ventricles contract during systole and fill during diastole. In an average human at rest, the entire cardiac cycle takes about 0.8 s (giving a pulse of about 75 beats/min). Heart rate is the number of heartbeats per minute. We can easily measure our own heart rate by counting the pulsations of arteries in the wrist (radial artery) or neck (carotid artery). Figure 8.3 shows the synchronization of the electrocardiogram (see Section 8.2.2) and the pressures in the left ventricle.

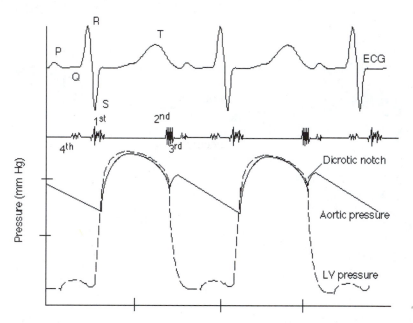

Figure 8.3 In the top figure, the electrocardiogram (ECG) initiates the cardiac cycle. The cardiac sounds (Section 8.5) are also shown. The bottom figure shows that ejection occurs when the pressure in the left ventricle exceeds that in the arteries.

During diastole, the pressure in the left ventricle is low and less than that of the arteries, thus, the aortic valve is closed. The blood from the left atrium flows into the left ventricle. The left atrium subsequently contracts due to the excitation from the pacemaker cells in the heart (see Section 8.1.3) and further fills the left ventricle. The ventricles contract, which increases the pressure and closes the mitral valve. The pressure increases until it exceeds the aortic pressure, the aortic valve opens, and blood flows into the aorta. The blood continues to flow from the left ventricle into the aorta as long as the ventricular pressure is greater than the arterial pressure, as shown in Figure 8.3. However, if the venticular pressure is much greater than the aortic pressure, there is a problem with the valve and it may be stenotic (narrow). The dicrotic notch results when the aortic valve slams shut and the AV valve is still closed. Table 8.1 summarizes the events of the cardiac cycle.

Table 8.1 Duration and characteristics of each major event in the cardiac cycle.

Event	Characteristics	Duration at 75 bpm (0.8 s cycle)
Atrial diastole Ventricular diastole	AV valves opened. Semilunar valves close. Ventricular filling.	0.4 s
Atrial systole Ventricular diastole	AV valves open. Semilunar valves closed. Ventricular filling.	0.1 s
Atrial diastole Ventricular systole	AV valves closed. Semilunar valves open. Blood pumped into aorta and pulmonary artery	0.3 s

The aortic pulse contour changes with aging due to changes in aortic compliance (distensibility). Reflection of the systolic pulse from the distal aorta distorts the aortic contour and creates two systolic shoulders. A pulse from a younger person with a more compliant aorta is rounded. That from an older person with decreased aortic compliance has a reflected wave, which causes a more peaked pulse and higher pressure (hypertension) (O'Rourke, 1995).

Currently, there are numerous electric heart rate monitoring systems that continuously measure heart rate during rest and exercise. They rely on sensing electrodes placed on the skin surface, which measure the electric frequency of the heart, or the electrocardiogram (Section 8.2.2). Some models, like the ear clip, use a photosensor to measure change in tissue light absorption caused by the pulsatile increase in arteriole volume, which is then converted into a beat-per-minute readout.

8.1.3 Cardiac Excitation and Control

Figure 8.1 shows the *sinoatrial node* (SA node), which serves as the heart's pacemaker, located in the wall of the right atrium, near the point where the inferior vena cava enters the heart. It is composed of specialized muscle tissue that combines characteristics of both muscle and nerve (see Chapter 7). Nodal tissue contracts like muscle, thus it generates electric impulses similar to those found in nervous tissue. This tissue is self-excitable, and can contract without any signal from the nervous system. Each time the SA node contracts, it initiates a wave of excitation that travels through the wall of the heart at a rate of approximately 1 m/s (Berne, 1981). The impulse spreads rapidly, and the two atria contract simultaneously. Cardiac muscle cells are electrically coupled by the intercalated disks between adjacent cells. At the bottom of the wall separating the two atria is another patch of nodal tissue, the *atrioventricular* (AV) *node*. The atria and ventricles are separated by nonconductive tissue except for the AV node. When the wave of excitation reaches the AV node, it is delayed for about 0.1 s (AV-nodal delay), which ensures that the atria will contract before the ventricles. After this delay, the signal to contract is conducted to the apex of the ventricles along the bundle of His, and the wave of excitation then spreads upward through the ventricular walls via the Purkinje fibers (Davies et al., 1983).

Simple tests of the functions of the heart include monitoring the heart rate, the blood pressure, the cardiac rhythm, and the heart sounds. More sophisticated diagnoses can be performed by studies of the electrocardiogram, the echocardiogram, the stress test, Holter monitoring, and angiography. These techniques will be discussed in the subsequent sections.

8.2 Cardiac Biopotentials

8.2.1 Electrodes

Surface recording electrodes convert the ionic current within the body to electron current in the metal wires, which is then sent to a recorder. Electrodes are generally made from metal. The silver/silver chloride (Ag/AgCl) electrode is electrochemically stable and is thus most widely used (Brown et al., 1999). An electrode is usually constructed by electrolyzing a silver plate as an anode in an aqueous solution of sodium chloride to form a film of AgCl on the surface of the silver. The reaction is $Ag + Cl^- \rightarrow AgCl + e^-$. Often a transparent electrolyte gel containing chloride ions as the principle anion is used. The gel also allows for good contact between the skin–electrode interface. Figure 8.4 shows a disposable surface electrode. Tobuchi et al. (1981) discuss several techniques used to manufacture Ag/AgCl electrodes. Other metals such as nickel-plated brass can be used as reusable ECG electrodes but they have larger motion artifacts.

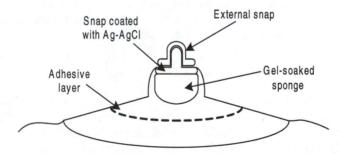
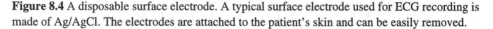

Figure 8.4 A disposable surface electrode. A typical surface electrode used for ECG recording is made of Ag/AgCl. The electrodes are attached to the patient's skin and can be easily removed.

Motion artifacts (error caused by movement of the electrodes) contribute a large error to signals from surface electrodes. Artifact may be caused by stretching of the skin, which changes the voltage across the skin. This can be minimized by abrading the skin with sandpaper, which short circuits the voltage. There are several types of electrodes: suction electrodes, floating metal body-surface electrodes, and dry electrodes (Neuman, 1998a). Other types of electrodes, such as microelectrodes, are discussed in Chapter 7.

8.2.2 Electrocardiogram

Contraction of cardiac muscle during the heart cycle produces electric currents within the thorax. Voltage drop across the resistive tissue can be detected by electrodes placed on the skin and recorded as an *electrocardiogram* (ECG, sometimes called EKG). Figure 8.3 shows components of an electrocardiogram (P, Q, R, S, and T waves).

The electrocardiogram contains information on the electric rhythm, electric conduction, muscle mass, presence of arrhythmia (irregular heart beat), ischemia (lack of blood flow) or infarction, even electric disturbance and drug effects on the heart (see Figure 8.5). The P wave in the ECG represents the contraction of the atria. From Figure 8.3, the R wave—also called the QRS complex—of the ECG corresponds to the contraction of the ventricles. The T wave is due to repolarization (re-establishing the electric potential) of the ventricles. The atria repolarize also, but this event is covered up by the QRS complex. The downslope of the T wave is referred to as the vulnerable period. It is during this time of the cardiac cycle that a premature impulse may arrive and fibrillation can occur. Fibrillation is an irregular contraction of cardiac muscle, which results in an ineffective propulsion of blood. When ventricular fibrillation occurs a normal rhythm can be re-established by passing a strong electric current that puts the entire myocardium in a refractory state in which no impulse can occur. Thus, "resetting" the heart. Direct current defibrillation shock has been found to be more effective than alternating current shock (Berne, 1981).

The ECG recording instrument is called an *electrocardiograph*. Figure 8.6 shows the block diagram of a typical electrocardiograph. The leads attached to the surface electrodes from the patient are fed to the resistors and switch network in order to select the desired potential difference between electrodes. The input signals are amplified and converted from analog signals to digital signals before being processed by a microprocessor. Several algorithms within the microprocessor perform QRS detection and filtering. Clinicians can view the results on the screen of a microcomputer or print a hard copy of the ECG signals for further analysis. Onoda (1994) shows the complex system of a modern electrocardiograph, which provides the physician with valuable information on various physical characteristics of the heart, the extent and progress of damage incurred by the heart and the influence of drugs on the patient.

Clinical electrocardiography would not be practical without the differential amplifier. The differential amplifier rejects ground-referred interferences. Figure 8.7 shows an ECG amplifier with a dc-coupled instrumentation amplifier that has a gain of 25. The high-pass filter blocks dc offset and the last stage is a noninverting amplifier that has a gain of 34 (Neuman, 1998b). Chapter 2 provides more information on the differential amplifier.

Connecting each lead of the ECG to a buffer amplifier prevents loading error caused by the added resistance of the skin. The buffer amplifier offers a very high input impedance and unity gain. If a differential amplifier or inverting amplifier were used without a buffer, the gain would be affected (see Section 2.2.5).

12-Lead ECG

The electric potential generated by the heart appears throughout the surface of the body. We can measure the potential differences between surface electrodes on the body. Different pairs of electrodes at different locations generally yield different results because of the spatial dependence of the electric field of the heart (Neuman, 1998b).

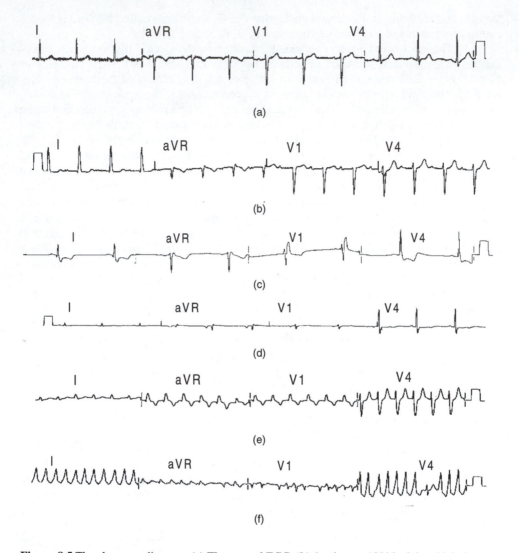

Figure 8.5 The electrocardiogram. (a) The normal ECG. (b) 1st degree AV block in which the delay from the P wave to the Q wave is lengthened. (c) acute inferior myocardial infarction (lack of blood flow to heart muscle, which causes tissue to die), in which the S–T sement is depressed. (d) right atrial hypertrophy (increase in muscle mass of the atria), in which V4 is large. (e) ventricular tachycardia (faster than normal heart rate) with clear AV dissociation. (f) Wolff–Parkinson–White syndrome with atrial fibrillation.

Physicians attach several surface electrodes and record a few beats of the 12-lead diagnostic ECG (see Figure 8.6). The 12-lead ECG monitors the heart's electrical activity from 12 different angles. By selecting different pairs of electrodes, we can obtain ECGs that have different shapes. Common locations for the electrodes are the limbs—right arm, left arm, right leg, and left leg. A pair of electrodes, or combination of several electrodes through a resistive network that gives an equivalent pair, is referred to as a *lead*. The most widely used bedside ECG diagnostic system records 12 different potential differences, or ECG leads (Tompkins, 1993).

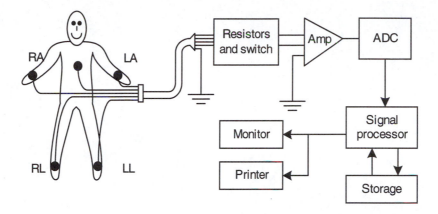

Figure 8.6 Block diagram of an electrocardiograph. The normal locations for surface electrodes are right arm (RA), right leg (RL), left arm (LA), and left leg (LL). Physicians usually attach several electrodes on the chest of the patients as well.

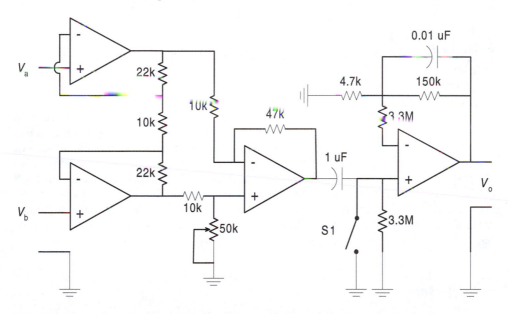

Figure 8.7 A circuit of an ECG amplifier. The instrumentation amplifier, located on the left of the circuit provides a high input impedance and has a gain of 25 in the dc-coupled stages. (From Webster, J. G. (ed.) *Medical Instrumentation: Application and Design*. 3rd ed. Copyright © 1998 by John Wiley & Sons. Reprinted by permission of John Wiley & Sons.)

Figure 8.8 shows the Einthoven triangle. The three leads shown are lead I, right arm (RA) to left arm (LA); lead II, right arm (RA) to left leg (LL); and lead III, LA to LL. The scalar signal on each lead of Einthoven's triangle can be represented as a voltage source, thus we can write Kirchhoff's voltage law for three leads as

$$I - II + III = 0 \qquad (8.1)$$

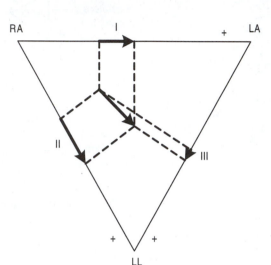

Figure 8.8 Einthoven's triangle. Lead I is from RA to LA, lead II is from RA to LL, and lead III is from LA to LL.

Three additional leads, called the *unipolar leads*, or augmented limb leads, are routinely used in taking diagnostic ECGs. These leads are based on signals obtained from more than one pair of electrodes. Unipolar leads consist of the potential appearing on one electrode taken with respect to an equivalent reference electrode, which is the average of the signals from two or more electrodes. The three different unipolar leads are positive electrodes placed on the right arm, augmented right (aVR), the left arm, augmented left (aVL), and the left foot, augmented foot (aVF). The other leads that are frequently measured in clinical ECGs are the *precordial* leads. Physicians place an electrode at various anatomically defined positions on the chest wall (Neuman, 1998b). Required frequency response is 0.05 to 150 Hz.

ECG Monitoring

The ECG may be monitored continuously by a recorder or telemetry when the patient is in emergency care, in a coronary care unit, an intensive care unit, or during stress tests. In these cases, only one lead, usually lead II, is monitored on a display. The system automatically monitors rhythm disturbances and gives an alarm when the heart rate is too slow or too fast. To avoid motion artifacts and muscle noise the frequency response is reduced to 0.5 to 40 Hz.

Ambulatory (Holter) Monitor

An ambulatory monitor, also called a Holter monitor, is a portable electrocardiograph, which can be worn for up to 24 h to monitor heart abnormalities. The monitor records the heart's activity through electrodes placed on the chest of a patient. The electric impulses are then transmitted to an amplifier, which records them on a small recorder for later review by a physician. The recording provides a doctor with important information about the heart and its rhythm and can help identify the cause of such normally encountered symptoms as chest pain, palpitations (abnormally rapid pulsation in the heart), or dizziness.

Ambulatory electrocardiography has been available since the early 1950s, when Dr. Holter introduced his portable electrocardiograph to the medical community. Since that time, these devices have decreased in size and weight, but have increased in sophistication. The first devices had to be worn in a backpack, and despite their large size and weight, could only review the heart along one axis. Modern devices can look at the heart along several different axes, and their weight and size are such that they do not interfere with the patient's daily activities.

The electrodes are applied to the chest. The number of wires and electrodes applied depends upon the number of leads the physician feels should be monitored—usually one or two. After the electrodes are securely attached to the chest, the excess wire is taped to the skin to prevent accidental disconnection. After the technician checks the system to make sure that it is working properly, the monitor is connected to an electrocardiogram recorder to determine if the ambulatory electrocardiogram provides a high quality tracing, free of excessive artifact or electric interference. The recorder is then tested and a battery inserted into the device. Patients can wear the recorder either on their belt or over their shoulder.

Two types of Holter monitors are generally used. The first type is the *tape-based system*, which is composed of three components: a monitor with a cassette tape, which records the ECG signals; a PC scanning device that scans at 60 times real time and uses a data-reduction processor; and a printer. The patient wears the monitor, which records an analog recording on the cassette tape. When the monitoring session is finished, the cassette tape is put into the data-reduction processor and data are digitally stored on the hard drive of the PC. From the PC, a report is generated, reviewed, and printed. These reports are in two or three channel format depending upon the instrumentation. Tape-based systems are usually preferred by physicians who need numerous recorders because these are less expensive than digital recorders.

The second type of monitor is the *digital solid state system*. This system is usually made up of two components, the monitor and the printer. In some cases, a PC is used to visualize the information on a screen. The patient wears the solid-state monitor, which digitizes, stores, and analyzes the ECG signal in real time. In a tape-based system, the computer performs the function of digitizing, storage and analysis. Digital Holters have no moving parts or tapes. They produce a two- or three-channel report depending upon the instrument and are generally more expensive than individual tape-based monitors because of their capabilities.

ECG Telemetry System

Medical personnel usually monitor one lead of the patient's ECG using a bedside monitor in the critical care unit (CCU) of a hospital. As the patient recovers, the doctor may feel comfortable with the patient moving around but may want to maintain ECG monitoring. A bedside monitor limits the movement of the patient to the length of the ECG leads. With a telemetry system, a nurse attaches the patient leads to the patient in the same way as for the bedside monitor. Instead of the patient leads feeding back directly to the bedside monitor, they feed to a portable transmitter, which resembles a portable radio and is small enough for the patient to carry. The transmitter sends the patient's ECG by modulating it with a radio-frequency (RF) carrier, much in the same way that a cordless telephone works. Manufacturers may transmit the information using frequency modulation (FM), amplitude modulation (AM), or even pulse code modulation (PCM) depending upon design. Antenna systems pick up the RF signal from the transmitter and send it to the receiver, which removes the RF signal from the patient's ECG and sends the resulting signal to the monitor for analysis. The signal strength of the transmitter and tuning of the system limit the distance a patient can move away from the antenna system. In the ECG telemetry system, the receiver alarms on high and low ECG limits the user can adjust, or possibly some arrhythmia conditions (Reinhard and Fowler, 1987).

Exercise Stress Testing

Exercise stress testing is an evaluation of the patient's cardiovascular system using one lead of an ECG, monitor, treadmill, and blood pressure unit. It is done to screen patients for heart disease and help predict or unmask potential coronary problems.

The procedures for stress testing are as follows. The technician first abrades the skin and applies stress electrodes to the patient's chest. Then, a baseline ECG and blood pressure are taken. The exercise protocol (there are numerous protocols, which gradually increase both speed and elevation of the treadmill) is started. During exercise, the physician monitors the patient's ECG, heart rate, blood pressure, and physical symptoms. S–T segment depression may indicate cardiac ischemia due to coronary arterial occlusion. When the test ends, ECG and blood pressure measurements are done for documentation of the heart rate at maximum effort or submaximum effort. It is important to continue monitoring the patient's ECG, heart rate, and blood pressure during recovery, because some problems are uncovered during that phase.

8.2.3 Electrograms

Electrograms are electric signals recorded from within the heart. Electrograms are measured by introducing a catheter electrode into the veins. The catheter is then placed into the heart under X-ray guidance, which is usually a biplane cardiac imaging system with a C-arm, allowing multidirectional biplane projections with extreme angulation capabilities. We can use electrograms to study abnormal heart rhythms under controlled situations to diagnose the specific problem with the heart's electric system. Electrophysiologists can obtain information on the propagation of excitation from the SA node to the ventricles with the catheter.

There are certain groups of patients who have problems that can be corrected by radio-frequency ablation—the use of RF energy via a catheter for lesion making to modify the conduction pathway of the heart (Huang and Wilber, 2000). One example is patients with refractory arrhythmias of the upper chamber who need complete interruption of the AV conduction pathway. Electrophysiologists usually monitor the abnormal conduction of the heart by applying three ablation catheters, each having three recording electrodes and one ablation electrode at the tip. The catheters are placed at different locations, such as the SA node, the AV node, and in the ventricle. By observing the timing of the excitation activity at various locations, the electrophysiologist can determine the locations of unwanted conduction pathways. They are then heated and destroyed with RF energy.

Any invasive procedure that requires catheters to be inserted in the body includes some risk. However, the risk is quite small and the electrophysiology study is relatively safe. One possible side effect is that some patients may develop bleeding at the site of catheter insertion into the vessels.

8.3 Cardiac Pressures

The state of the heart and heart valves can be assessed by measurement of cardiac pressures. The sensors used are designed so that the application of pressure modifies the electric properties of some component by stretching or otherwise deforming it. Sensors with this property, called strain gages are very important in biomedical instrumentation design. Section 8.3.1 describes the fundamentals of strain gages. Section 8.3.2 describes a catheter that is widely used for invasive measurements of cardiovascular variables.

Figure 8.9 shows a typical system for measuring cardiac pressures. A pressure port is located at the tip of the catheter. A syringe temporarily inflates a balloon to guide the catheter into the pulmonary artery. Slow infusion prevents a clot from forming at the pressure port.

8.3.1 Strain Gage

Pressure measurement systems in use today are essentially all electric strain gages and apply the principle of the Wheatstone bridge, as described in Sections 2.1.6 and 9.11.3. Older pressure sensors had unbonded strain gage wires mounted under stress between a frame and movable armature so that preload was greater than any expected external compressive load. This was necessary to avoid putting the wires in compression. Thus blood pressure caused diaphragm movement, which caused resistance change, which caused a voltage change.

Modern pressure sensors use an etched silicon diaphragm embedded with strain gages formed by diffusing impurities into it. The strain gage electric resistance changes with diaphragm deformation, which responds to pressure applied. Unfortunately the resistance of the silicon also changes with temperature, but the effect of this change may be minimized by diffusing four strain gages into the diaphragm. These then form the four arms of a resistance bridge. Two of the strain gages are placed close to the center of the diaphragm, where their resistance increases with applied pressure, and the other two close to the periphery, where their resistance decreases with pressure. However, if the temperature changes, all four resistances change by the same percentage and this does not change the output from the resistance bridge.

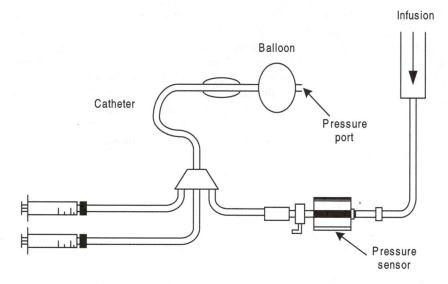

Figure 8.9. A system for cardiac pressure and flow measurement. The pressure is usually measured with a catheter placed in the right side of the heart. An external strain gage pressure sensor is also shown. (Adapted from Orth, J. L. 1995. System for calculating compliance and cardiac hemodynamic parameters, US Patent, 5,423,323.)

8.3.2 Catheter

In the cardiac catheterization laboratory, catheters are inserted into the chambers of the heart to measure pressures, flows, and oxygen saturation to determine if valve replacement is required. Catheters can also inject radiopaque dye for X-ray fluoroscopy, which can image vessel narrowing to determine if vessel replacement is required.

A catheter is a flexible tube for insertion into a narrow opening such as the blood vessels, so that fluids may be introduced or removed. Figure 8.9 shows a Swan–Ganz catheter. The first person who passed a catheter into the human heart was Werner Forsmann, a medical student at Eberswalde, Germany. In 1929, at the age of 25, he passed a 65 cm catheter through one of his left antecubital veins, guiding it by fluoroscopy until it entered his right atrium. Forsmann then climbed up the stairs to the Radiology Department where the catheter position was documented.

Catheters are inserted through sheaths (the layers of connective tissue that envelop structures such as nerves, arteries, tendon, and muscle) into the arteries and the veins of the body then pass up to the heart. The pressures within different areas of the heart are measured. Two approaches may be used to apply a catheter to the heart.

The brachial (*in the arm*) *approach* usually utilizes cutdown on the brachial artery (an artery that extends from the auxillary artery, down the side and inner surface of the upper arm to the elbow, where it divides into the radial and ulnar arteries) and vasilic vein at the elbow. The direct brachial approach may have advantages in a very obese patient, in whom the percutaneous femoral technique may be technically difficult and bleeding hard to control after catheter removal.

The indirect femoral approach. The percutaneous femoral approach has its own set of advantages. Arteriotomy and arterial repair are not required; it can be performed repeatedly in the same patient at intervals, whereas the brachial approach can rarely be repeated more than two or three times with safety.

Cardiac pressures are usually measured by one of several ways. In the *catheter-type* system, the blood pressure communicates with the pressure sensor via a fluid-filled catheter. The sensor is outside the body. With an appropriate length and diameter, accurate pressure readings can be obtained.

With the *catheter-tip* sensor, the elastic unit is placed in a blood vessel, thus the elastic member is in direct contact with the blood. This arrangement avoids the damping and resonance sometimes encountered with catheter-type systems and is more accurate than the catheter-type sensors (Grossman and Baim, 1991).

8.4 Cardiac Output

Cardiac output is a measure of the well being and performance of the heart. The maintenance of blood flow commensurate with the metabolic needs of the body is a fundamental requirement of human life. In the absence of major disease of the vascular tree, the maintenance of appropriate blood flow to the body depends largely upon the heart's ability to pump blood in the forward direction. The quantity of blood delivered to the systemic circulation per unit time is termed the cardiac output, generally expressed in liters/minute (L/min).

Cardiac output of a normal patient is directly correlated with the size of the body. Most investigators use total body surface area as the standardizing variable. The ratio of the cardiac output to area is called the *cardiac index.* Table 8.2 lists several variables related to cardiac output.

$$\text{Cardiac output (CO)} = \text{heart rate (HR)} \times \text{stroke volume (SV)} \qquad (8.2)$$

The ejection fraction is defined as

$$\text{Ejection fraction} = \frac{\text{Stroke volume}}{\text{End - diastolic volume}} \qquad (8.3)$$

Table 8.2 Some physiological variables. The data presented in this table are the average values of a group of subjects.

Variables	Mean (±SD)
Weight (kg)	70
Cardiac output (mL/s)	110
Heart rate (min^{-1})	76
Mean velocity, ascending aorta (mm/s)	16
LV end-diastolic volume (mL)	125 (±31)
LV end-systolic volume (mL)	42 (±17)
LV stroke volume (mL)	82 (±20)
LV ejection fraction	0.67 (±0.11)
LV mean wall thickness (mm)	10.9 (±2.0)

End-diastolic volume is the amount of blood in the ventricle at the end of diastole. End-systolic volume is the volume of blood remaining in the ventricle at the end of systole when ejection is complete.

The stroke volume, the volume of ejected blood from the ventricles, is about 80 mL/beat. The average resting heart rate of human is about 70 beats/min. Thus, cardiac output is about $80 \times 70 = 5,600$ mL/min $= 5.6$ L/min. Cardiac output is regulated by changes in both HR and SV. Heavy exercise increases both HR and SV, and CO can increase to as high as 25 L/min.

8.4.1 Fick Method

In 1870, Fick stated that if the concentration of O_2 in arterial and mixed venous blood is known, and the rate of inhalation of O_2 is also known, the cardiac output, CO, can be computed from the formula (Webster, 1998)

$$CO = \frac{dm/dt}{C_a - C_v} \tag{8.4}$$

where dm/dt is the consumption of O_2 (L/min), C_a is the arterial concentration of O_2 (L/L), and C_v is the venous concentration of O_2 (L/L).

The Fick technique is invasive, that is, the O_2 concentrations must be measured in the vicinity of the lungs. The rate of inhalation or exhalation of gas is measured using the spirometer as discussed in Chapter 9. C_a can be measured by obtaining a sample from any artery (except the pulmonary artery) and C_v can be measured in the pulmonary artery. The blood-O_2 analyzer is used to find the concentrations of O_2 in blood.

The Fick method has been widely used, and its accuracy, when carefully performed, is approximately ±5%. The principal limitation is the need for a steady state; output and pulmonary gas exchanged must remain constant while samples are collected, which in practice means a period of several minutes. For that reason, rapid changes in the circulation during exercise or other events cannot be studied by this technique.

Example 8.1 A patient's O_2 concentration, meaused in the pulmonary artery, is 0.12 L/L. The O_2 concentration measured in the patient's aorta is 0.19 L/L. A spirometer is used to obtain the patient's O_2 consumption rate 0.250 L/min. Calculate the patient's cardiac output.

The patient's O_2 concentrations and consumtion are known: $C_a = 0.19$ L/L, $C_v = 0.12$ L/L and $dm/dt = 0.250$ L/min. From Eq. (8.4)

$$CO = \frac{dm/dt}{C_a - C_v} = \frac{0.250 \text{ L/min}}{0.19 \text{ L/L} - 0.12 \text{ L/L}} = 3.57 \text{ L/min}$$

8.4.2 Thermodilution

Thermodilution is considered a gold standard for measuring cardiac output. A bolus of 10 mL, 0.9% cool physiological saline is injected into the venous circulation in the right atrium. The drop in temperature in the arterial circulation is usually measured using a thermistor attached to a catheter in the pulmonary artery. The cardiac monitor calculates the cardiac output of the patient by measuring the time the bolus takes to travel through the heart's right side. The nurse inserts a thermodilution pulmonary artery catheter, also known as the Swan–Ganz catheter, as shown in Figure 8.9, into the patient, usually at the internal jugular vein or subclavian vein (in the neck area). This catheter has a small balloon that once inflated naturally floats toward the heart to the pulmonary artery.

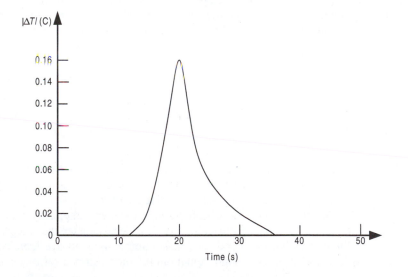

Figure 8.10 The relationship of the temperature gradient and time. (Adapted from Baker, P. D., Orr, J., Westenskow, D. R. and Johnson, R. W. Method and apparatus for producing thermodilution cardiac output measurements utilizing a neural network, US Patent, 5,579,778.)

The cardiac output can be calculated from (Webster, 1998)

$$CO = \frac{|Q|}{\rho_b c_b \int_0^{t_1} |\Delta T_b(t)| \, dt} \quad (\text{m}^3/\text{s}) \tag{8.5}$$

where Q is the heat injected in joules, ρ_b is the density of the blood in kg/m^3, c_b is the specific heat of the blood in J/(kg·K), and ΔT_b is the temperature gradient function as shown in Figure 8.10. The area under the temperature versus time curve is inversely proportional to the blood flow or cardiac output. Therefore, healthy patients with normal cardiac output will have less area under temperature versus time curve than unhealthy patients with some type of coronary insufficiency. If cardiac output is reduced, the bolus takes longer to pass through the heart.

Example 8.2 Given the following data: blood density = 1060 kg/m^3, heat capacity of the blood = 3640 J/(kg · K), amount of heat injected = 1.25 J, and the area under the curve in Figure 8.10 is 1.0 s·°C, calculate the cardiac output.

From Eq. (8.5), the cardiac output can be computed as

$$CO = \frac{1.25 \, \text{J}}{1060 \, \text{kg/m}^3 \times 3640 \, \text{J/(kg} \cdot °\text{C)} \times (1.0 \, \text{s} \cdot °\text{C})} = 3 \times 10^{-7} \, \text{m}^3/\text{s}. \tag{8.6}$$

The disadvantage of this method is its invasiveness. Also, the lower temperature bolus can be rapidly warmed to body temperature. Repeat measurements are easier to perform than for the dye dilution technique (see below). However, it is more difficult to calculate the size of the temperature perturbation than it is to measure the quantity of injected dye. Another disadvantage of the thermodilution method is that the injecting catheter acts as a heat exchanger between the blood and the cold saline. The cold saline left at the end of the injection can further cool the blood. However, the error is not too significant.

8.4.3 Dye Dilution

The principle of this technique is similar to that of thermodilution. A small quantity of dye is injected into the venous blood stream through a catheter tip at a region very close to the heart. The dye passes through the heart, is mixed with blood, then appears in the arterial circulation. Arterial blood is drawn off through another catheter by a motorized syringe and passes at a constant rate through an optical densitometer that measures the concentration of the dye within the blood. The most commonly used dye is indocyanine green, which has low toxicity and does not remain in the blood stream for extended periods of time.

The disadvantages of the dye dilution technique are: it is not completely nontoxic and the dye cannot be removed immediately from the blood stream, thus repeat measurements are difficult to perform (Cohn and Wayne, 1982).

8.4.4 Radionuclide Angiography

External detectors are used to record activity from short-lived radiopharmaceuticals. This is a procedure that yields a high degree of contrast between the blood and the surrounding structures. There are two classes of radionuclides:

1. Positron Emission Tomography (PET), which uses substances that release positrons, which travel a short distance, then react with electrons to emit pairs of photons on paths 180° apart. This technique is excellent for measuring the extent of myocardial ischemia and is highly specific for the diagnosis of coronary artery disease (Marwick 1995).

2. Single Photon Emission Computerized Tomography (SPECT), which is a method based on radionuclides such as ^{99m}Tc. The imaging device in SPECT is usually a gamma-ray camera. Diagnostic accuracy and adjunctive assessment of ventricular function make SPECT the currently favored stress imaging radioisotope technique (Merz 1997). Chapter 7 discusses the principles of radionuclides.

8.4.5 Echocardiography

Sound of frequencies greater than 20 kHz, the highest frequency that a human ear is able to detect, are called ultrasound. The application of the ultrasonic imaging technique to the heart is called *echocardiography*, a noninvasive technique. The transducer contains a piezoelectric crystal, which transmits and receives ultrasound resulting in images. Echocardiography can measure the size of the heart, functions of the heart, the blood flow through the heart (when combined with the Doppler technique), and the movement of the heart wall. It can help identify tumors and clots in the heart as well as structural abnormalities of the heart wall, valves, and the blood vessels going in and out of the heart.

Cardiac diagnosis uses high-frequency ultrasound (1.6 to 2.25 MHz). The sound waves are both emitted and received by the same sensor and are reflected at interfaces between media of different acoustic impedances. Motion mode, or *M mode*, echocardiograms are generated by means of a single ultrasonic beam traversing cardiac structures with definable anatomic shapes (chamber walls, interventricular septum) or characteristic motions (cardiac valves). The echocardiograms are then recorded on an oscilloscope, where the time-of-flight of reflected sound occupies the ordinate and slower time the abscissa. Advances in echocardiographic equipment including real-time cross-sectional scanning (two-dimensional echocardiography), Doppler echocardiography, and contrast echocardiography have broadened the role of echocardiography in the diagnosis of a variety of cardiac diseases.

M-mode Echocardiography

With this method, the ultrasonic sensor is placed on the anterior chest. The beam passes through the right ventricle, interventricular septum, left ventricular cavity, posterior left ventricular wall, pericardium and lung. Figure 8.11 shows that these structures are portrayed on the ordinate and time is displayed on the abscissa. Additional information is obtained by sweeping the sensor in an arc between the base and apex of the heart. By proper positioning of the sensor, all four cardiac valves, both ventricles, and the left atrium can be visualized (Bom, 1977). M-mode echocardiography provides good axial resolution and is therefore superior in qualitative diagnoses such as pericardial thickening, small effusion, and constrictive hemodynamics (Cikes and Ernst, 1983). However, because it lacks spatial geometry, most potential information from using many cardiac cross-sections is unavailable (Cikes and Ernst, 1983).

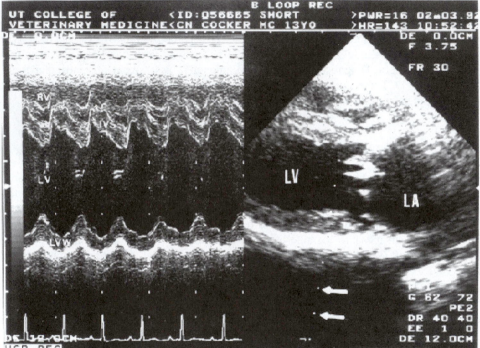

Figure 8.11 Simultaneous recording of motion mode (M-mode) and two dimensional echocardiograms. The arrows on the right image indicates the position of the ultrasound beam from which the M-mode recording was made. LVW = left ventricular wall, LV = left ventricle, LA = left atrium, RV = right ventricle (From College of Veterinary Medicine, Univ. of Tennessee. 2003. M-Mode Echocardiography [Online] www.vet.utk.edu/)

Two-Dimensional Echocardiography (2-D Echo)

M-mode echocardiography provides extremely useful information about cardiac structures, but cannot depict lateral motion (motion perpendicular to the ultrasonic beam). In two-dimensional echocardiography, the ultrasonic beam angle is swept very rapidly and can provide pictures of lateral motion and cardiac shape not available by M-mode echocardiography alone. Images recorded in 2-D echo are displayed on videotape. A variety of views are possible but those commonly employed are the long axis view, the cross-sectional view, and the four-chamber view (Figure 8.12). Two-dimensional echocardiography is superior to M-mode echocardiography in evaluating left ventricular function, left ventricular aneurysm, and intracavitary thrombus. Table 8.2 shows the relative advantages of M-mode and 2-D echocardiography.

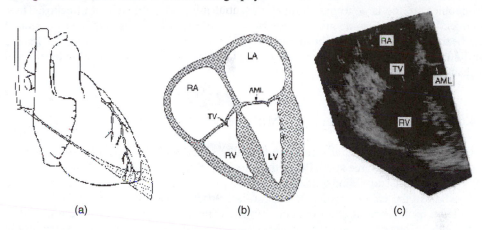

(a)   (b)   (c)

Figure 8.12 2-D echocardiography of the long axis view of the right ventricle (RV): (a) the ultrasonic beam angle through the heart, (b) the cross-sectional diagram of the image and (c) the actual 2-D image. TV = tricuspid valve, AML = anterior mitral leaflet. (Adapted from Rafferty, T. 1992. Transesophageal two-dimensional echocardiography: www.gasnet.org/echomanual/html/2-d_echo.html).

Table 8.2 Relative advantages of echocardiographic examination techniques. (Adapted from Roelandt, 1983)

M-mode echocardiography	Two-dimensional echocardiography
Excellent time resolution	Anatomical relationships
Accurate dimensional measurements	Shape information
Timing of events against other parameters	Lateral vectors of motion
Easy storage and retrieval	Easier to understand

Doppler Echocardiography

Doppler echocardiography employs the Doppler principle, which is discussed in Section 8.8.3. When the ultrasonic wave is reflected by a moving object, the frequency of the reflected beam is shifted. Doppler echocardiography was originally employed to evaluate blood flow in peripheral arteries and veins; however, Doppler is now combined with standard cardiac ultrasound. Doppler studies can measure the transvalvular pressure gradient in patients with aortic, pulmonic, tricuspid, and mitral stenosis. There is a good correlation between the gradient measured by Doppler and that obtained with catheterization.

Contrast Echocardiography

Contrast echocardiography is based on the fact that the injection of almost any liquid through a small catheter produces a cloud of echoes on the M-mode echocardiogram. The most commonly used dye for contrast echocardiography is indocyanine green. The microbubbles are too large to pass through the capillaries and yield a cloud of echos. Thus, an injection of indocyanine green on the right side of the heart remains in the right side of the heart unless there is a right-to-left shunt. We measure the cardiac output by tracing the intravascular bubbles with a high degree of sensitivity by ultrasound (Meltzer and Roelandt, 1982).

8.4.6 Magnetic Resonance Imaging (MRI)

MRI provides higher spatial resolution than ultrasound and the capability of measuring the velocities in the three spatial directions for any plane through the heart (Ingels et al., 1996). In contrast to radionuclide angiography, MRI utilizes no ionizing radiation and can diagnose the characteristics of the cardiac muscle. Also, the stroke volume can be calculated by integrating the flow curve in the main pulmonary artery during a cardiac cycle. MRI offers excellent anatomical detail of the cardiovascular system, particularly in congenial heart disease and in diseases of the aorta (Underwood, 1992).

8.5 Cardiac Sounds

Cardiac sounds shown in Figure 8.3 are associated with the movement of blood during the cardiac cycle. Table 8.3 gives descriptions of the origins of various heart sounds. The first heart sound is low-pitched, soft, and relatively long. It is associated with the close of the AV valves. The opening of the valves does not produce any sound. The sounds are caused by vibrations within the walls of the ventricles and major arteries during valve closure, not by the valves snapping shut. The first heart sound signals the onset of systole. The second heart sound has a higher pitch, is shorter and sharper. It is associated with the closure of the semilunar valves that occurs at the onset of ventricular relaxation as the left and right ventricular pressures fall below the aortic and pulmonary artery pressures, respectively. Thus the second heart sound signals the onset of ventricular diastole. Murmurs are vibrations caused by turbulence in the blood moving rapidly through the heart. Table 8.4 lists the types of murmurs and what they represent.

Table 8.3 The heart sounds. The first and second heart sounds are most prominent.

Sound	Origin
1st sound	Closure of mitral and tricuspid valves
2nd sound	Closure of aortic and pulmonary valves
3rd sound	Rapid ventricular filling in early diastole
4th sound	Ventricular filling due to atrial contraction

Table 8.4 Timing of murmurs. For example, if the physician hears the first heart sound, a swishing sound, and then the second heart sound, the patient likely suffers from AV valve insufficiency.

Characteristic	Type of murmur	Valve disorder
Systolic murmur	Whistling	Stenotic semilunar valve
1st HS–murmur–2nd HS	Swishing	Insufficient AV valve
Diastolic murmur	Whistling	Stenotic AV valve
2nd HS–murmur–1st HS	Swishing	Insufficient semilunar valve

A stenotic valve is a stiff, narrowed valve that does not open completely. Thus, blood must be forced through the constricted opening at tremendous velocity, resulting in turbulence that produces an abnormal whistling sound similar to the sound produced when you force air rapidly through narrowed lips to whistle. An insufficient valve is one that cannot close completely, usually because the valve edges are scarred and do not fit together properly. Turbulence is produced when blood flows backward through the insufficient valve and collides with blood moving in the opposite direction, creating a swishing or gurgling murmur (Sherwood, 2001). Improved diagnosis can be obtained from heart sound audiograms, where the magnitude of each frequency component is displayed.

8.5.1 Stethoscopes

A stethoscope is used to transmit heart sounds from the chest wall to the human ear. High-frequency sounds, or murmurs, are easier to hear with the *diaphragm*. The *bell*, which should be applied lightly to the chest, transmits low-frequency sounds more effectively—for example, the diastolic murmur of mitral stenosis, and third and fourth heart sounds. Murmurs arise from turbulent blood flow and are characterized by their timing, quality, and intensity. Intensity is graded from 1 (just audible) to 6 (audible without a stethoscope).

Figure 8.13 shows a diagram of a stethoscope which incorporates both the bell and diaphragm modes into the same chestpiece. Using this particular model (Mohrin, 1995), we can interchange between the bell and diaphragm modes by simply pressing the chestpiece against a patient's body and twisting the bell housing so that the two diaphragms (shown in Figure 8.13) adjust their openings. If the two openings of the diaphragms coincide with each other, the stethoscope is operating in the bell mode. If there is no through opening, it is in the diaphragm mode. Thus, physicians are able to change between the bell and diaphragm modes without having to remove the bell housing from the patient's body.

8.5.2 Microphones

Microphones in use today are either crystal microphones (piezoelectric effect) or dynamic microphones (Faraday's principle) (Peura and Webster, 1998). *Piezoelectric sensors* are used as microphones since they are in the form of a thin film. They are very useful if one is interested in detecting surface vibrations of an object, such as the heart sound. When a force, f, is applied to a polarized crystal, the resulting mechanical deformation gives rise to an electric charge Q. Microphones turn an acoustical pressure into a voltage.

$$Q = kf = Kx \tag{8.7}$$

where k is the piezoelectric constant in C/N, K is the proportionality constant in C/m, and x is the deflecting distance of the piezoelectric sensor.

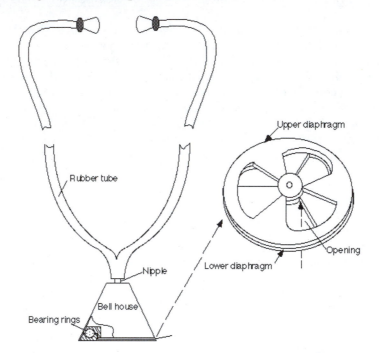

Figure 8.13 A stethoscope with bell and diaphragm modes. (Adapted from Mohrin, C. M., 1995. Stethoscope. US Patent, 5,389,747.)

Figure 8.14 illustrates a charge amplifier. Chapter 2 gives the voltage across an initially uncharged capacitor, C, by

$$v = \frac{1}{C}\int_0^{t_1} i\,dt = \frac{1}{C}\int_0^{t_1} K\frac{dx}{dt}\,dt = K\frac{x}{C} \tag{8.8}$$

where t_1 is the integrating time and i is the current flowing through the capacitor, C. A charge amplifier acts as a high-pass filter, and Figure 8.15 shows the time constant τ is

$$\tau = RC \tag{8.9}$$

Hence it only passes frequencies higher than the corner frequency $f_c = 1/(2\pi RC)$.

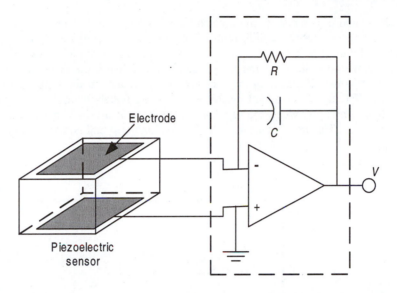

Figure 8.14 The piezoelectric sensor generates charge, which is transferred to the capacitor, C, by the charge amplifier. Feedback resistor R causes the capacitor voltage to decay to zero.

8.6 Myocardial Viability

The assessment of myocardial viability is one of the newest and most rapidly expanding areas of clinical and experimental research in cardiology. A number of diagnostic techniques have been developed to help the clinician determine with a good degree of accuracy whether myocardial viability is present or absent in dyssynergic (failure of muscle coordination) regions of the heart.

Several methods have been used to assess the presence or absence of myocardial viability. These methods depend on the demonstration of myocardial perfusion (presence of oxygen and nutrients), myocardial metabolism, or a response to a stimulus. The methods that use the demonstration of perfusion as the basis for the presence of viability include positron emission tomography (PET), thallium-201 imaging, and myocardial contrast echocardiography.

8.6.1 Positron Emission Tomography (PET)

Positron Emission Tomography (PET) uses isotopes that emit photons 180° apart (i.e. in opposite directions). It has been successfully used to assess myocardial viability. This technique has the distinct advantage of allowing for the assessment of both myocardial perfusion and metabolism. This is particularly important since severely ischemic but viable myocardium might show almost no flow, despite the presence of myocardial metabolism. Under normal circumstances, the myocardium uses fatty acids for its metabolism. However, when blood flow slows and oxygen supply decreases, anaerobic metabolism develops, and consequently, the myocardium begins using glucose instead of fatty acids as a source of energy. A high ratio of regional glucose utilization to myocardial flow has been found to be a reliable sign of severely ischemic but viable muscle.

Several studies have demonstrated that PET is a better predictor of myocardial viability than thallium-201 scintigraphy. This probably relates to the improved imaging resolution of PET and to its added benefit of assessing metabolism. The advantage of PET over thallium is lost when sophisticated quantitative analysis of perfusion by thallium is undertaken. The disadvantages of PET are its high cost and nonportability.

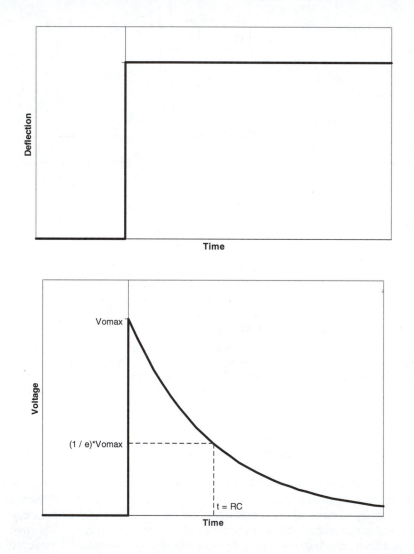

Figure 8.15 The charge amplifier responds to a step input with an output that decays to zero with a time constant $\tau = RC$.

8.6.2 Thallium-201 Imaging

Thallium, a member of the heavy metal group of elements, had little biologic use other than as a rodenticide until its application in myocardial imaging in the early 1970s. Although a heavy metal capable of causing typical heavy metal poisoning at high doses, doses of 5,000 to 10,000 times less than that associated with clinical poisoning can be employed for myocardial imaging. The biologic properties of thallium at these low concentrations closely resemble those of potassium.

Thallium-201 imaging of the myocardium is one of the more difficult nuclear medicine procedures from the standpoint of technique. The relatively low photon abundance and low energy of the imaged photons require excellent equipment and technique to obtain high-quality images. The cost of Thallium-201 and the time required for exercise testing and imaging makes the repetition of technically inadequate studies difficult (Ritchie et al., 1978).

8.6.3 Myocardial Contrast Echocardiography

The principle of this technique is discussed in Section 8.4.5.

8.7 Circulation

The three kinds of blood vessels are arteries, veins, and capillaries, which in the human body have been estimated to extend a total distance of 100,000 km. Arteries carry blood away from the heart to organs throughout the body. Within these organs, arteries branch into arterioles, tiny vessels that give rise to the capillaries, which form networks of microscopic vessels that infiltrate each tissue. It is across the thin walls of capillaries that chemicals are exchanged between the blood and the interstitial fluid surrounding the cells. At the downstream end, capillaries rejoin to form venules, and these small vessels converge into veins. Veins return blood to the heart. Notice that arteries and veins are distinguished by the direction in which they carry blood, not by the quality of the blood they contain. Not all arteries carry oxygenated blood, and not all veins carry blood depleted of oxygen. However, all arteries do carry blood from the heart to capillaries, and only veins return blood to the heart from capillaries. The arterial walls are muscular, thus the arteries are able to control the blood flow through constriction and dilation. The size of the capillary is 10 μm in diameter, and is where the gaseous exchange of CO_2 and O_2 occurs.

Peripheral vascular diseases are a group of distinct diseases or syndromes involving the arterial, venous, or lymphatic system (Strandness et al., 1986). The major vascular disease is coronary artery disease, which kills more than a million people every year, and afflicts more than 5 million others with disabling symptoms. The development of various noninvasive tests for diagnosis of arterial disease continues to improve rapidly. These tests enable us to diagnose and follow the course of arteriosclerosis (abnormal condition of an artery) much more accurately than was possible in the past. The Doppler ultrasonic method is the most reliable method of evaluating the severity of lower extremity ischemia and monitoring postoperative results. A promising new development is Duplex scanning (Section 8.11.1) of the peripheral arteries, combining Doppler spectral analysis of blood flow with B-mode ultrasound imaging. Using the Duplex scan to localize the stenosis and to accurately estimate its severity may allow selected vascular reconstructions to be done in the future without pre-operative arteriography.

Venography remains the standard for the diagnosis of deep venous thrombosis (blood clots), but does have significant limitations. In addition to the cost and discomfort, some venograms are not diagnostic because of technical reasons. Likewise, invasive techniques have become useful in the long-term evaluation of treated and untreated peripheral vascular disease. The increasing sophistication of instrumentation has provided not only physiological information but also anatomic depiction of intraluminal lesions by ultrasonic imaging. Direct arteriography has also seen new developments with the advent of digital subtraction angiography. While venous digital subtraction angiography has not proven to be as clinically useful as originally hoped, application of digital subtraction techniques to direct arteriography has permitted significant reduction in amounts of contrast agent required. By combining digital subtraction techniques with direct arterial angiography, patients at high risk of renal failure can be studied safely while combined arterial systems, such as abdominal and carotid vessels, can be simultaneously studied in patients with normal renal function. As for the future, magnetic resonance imaging (MRI) holds promise as an arterial diagnostic tool, particularly with large vessels. However, it appears unlikely that it will replace ultrasound and contrast arteriography.

8.8 Blood Flow

Physical factors that influence blood flow are pressure and resistance. The flow through arterial grafts is measured at the time of surgery to ensure that the graft has been successfully inserted. The flow in peripheral blood vessels is measured as an aid in the diagnosis of peripheral vascular disease.

The flow rate, F, of a single vessel is the volume of blood moving past a fixed point per unit time and can be calculated from

$$F = \Delta P/R, \text{ (mL/min).} \tag{8.10}$$

Pressure, P, is usually measured in mmHg, and R is the resistance.

$$R = 8L\eta/\pi r^4, \tag{8.11}$$

where L = length, η = viscosity, r = radius. Thus,

$$F = \Delta P \pi r^4/(8L\eta) \text{ (Poiseuille's law).} \tag{8.12}$$

ΔP is the difference between the mean arterial blood pressure (MABP) and right atrial pressure. Since the right atrial pressure is about zero, ΔP = MABP. R is the resistance through all vascular beds in parallel. Fung (1997) provides further information on the biomechanics of the blood vessels.

For detecting or measuring blood flow, ultrasonic Doppler flowmetry has become the main technique of noninvasive investigation, both in the clinical and research laboratory. The continuous wave (CW) and pulsed varieties each have their advocates, the former being easier to use while the latter affords more precise flow interrogation. Although the audible signal suffices for many clinical applications, recordings are required for others and for most research endeavors. Recordings made with the zero-crossing detector are adequate to some extent for real-time frequency spectrum analyzers. These devices not only depict the velocity flow envelope with precision but also allow analysis of the power spectrum at each frequency. Increased velocity and disturbances in the pattern of blood flow, revealed by these processing methods, are good indicators of vascular disease. Measurement of volume flow requires knowledge of the cross-sectional area of the vessel (which can be obtained from the B-mode image) and the mean of the velocity profile. Since velocity profiles are seldom completely axisymmetric, precise flow measurement has been difficult to achieve with either the CW or pulsed-Doppler.

Laser Doppler methods are applicable only to the cutaneous tissues and are difficult to quantify. At present their role remains uncertain. The use of magnetic resonance imaging (MRI) to measure blood flow, though promising, is in its infancy. Positron emission tomography (PET) has proved useful in the evaluation of local blood flow, especially in the brain, but is not widely available and remains prohibitively expensive. The transcutaneous electromagetic flowmeter seems to have little merit. Monoitoring the clearance of radionuclides, a semi-invasive technique for evaluting blood flow in isolated areas, was initially greeted with enthusiasm; however, the washout curves proved to be difficult to interpret, and the results inconsistent. At present isotope clearance is employed only for special clinical studies or research work.

8.8.1 Dilution

This technique was discussed in Section 8.4.3. Applying a similar approach, we can determine the blood flow.

8.8.2 Electromagnetic Flowmeter

This technique utilizes Faraday's law of magnetic induction, which states that a voltage is induced in a conductor that is moving in a magnetic field. It can measure pulsatile flow. The blood is an electrical conductor, therefore voltage will be developed when the blood flows through a magnetic field. Figure 8.16 illustrates the principle of an electromagnetic flowmeter. If the magnetic field, the direction of motion and the induced voltage are mutually at right angles (orthogonal), then

$$e = Blu \hspace{4cm} (8.13)$$

The induced voltage e is proportional to the velocity of the conductor u (averaged over the vessel lumen and walls), the strength of the magnetic flux density B, and the length between electrodes, l.

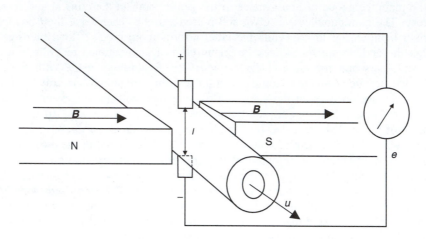

Figure 8.16 Principle of an electromagnetic flowmeter.

Small-size probes can be made using an iron-cored electromagnet to increase the signal strength. Air-core coils can be used for the larger probes to reduce the problems of hysteresis in the magnetic cores. The electrodes make contact with the outer wall of the blood vessel.

Dc Excitation

If the generation of the magnetic field is done with dc, the induced voltage at the electrodes is also dc. The induced voltage is small compared to surface potentials generated at the electrodes. It is difficult to distinguish changes in blood flow from changes in the surface potentials due to polarization at the electrodes, and electrodes may detect other signals such as electric impulses from the heart. Thus, dc excitation is not used commercially.

Ac Excitation

If the generation of the magnetic field is done with ac then the induced voltage at the electrodes is alternating. The surface potentials at the electrodes are removed by capacitive coupling. The ac excitation causes a different problem than dc—the change in magnetic field intensity causes an induced current in the conductor, which is proportional to the rate of change of the magnetic field intensity. The induced current cannot be separated from the desired signal at the electrodes. One of the solutions is to measure the voltage from the electrodes when the magnetic field intensity is not changing.

The electromagnetic flowmeter measures the average of the blood velocity across the lumen of the vessel. The outside diameter of the vessel is constrained to be the same as the internal diameter of the probe, so the volume flow through it can be calculated. In practice, this is not an ideal assumption. Most blood vessels display laminar flow—flow is greatest in the center of the vessel and slowest near the walls. It is not uniformly sensitive to the same blood velocity at different positions across the lumen of the vessel. Alterations in the velocity profile across the vessel therefore alter the measured mean velocity, giving rise to error. Fortunately, this induced error is not large enough to cause unacceptable measurements.

8.8.3 Ultrasonic Flowmeter

Ultrasonic flowmetry is a commonly used technique for measuring blood velocity in the peripheral arteries. It can measure pulsatile flow. A piezoelectric sensor is used to convert from electric to acoustic signals (see Section 8.5.2). Figure 8.17 shows a system for blood flow measurement using an ultrasonic flowmeter. The Doppler effect shifts the frequency of the ultrasonic beam from the oscillator when it intercepts the moving blood cells by an amount Δf, which can be computed from

$$\Delta f = \frac{2u f_0 \cos\theta}{c},$$
(8.14)

where f_0 is the fundamental frequency of an ultrasonic wave from the source, traveling at velocity c through the blood. The ultrasonic wave intercepts a stream of moving blood with velocity u, crossing the ultrasound beam at an angle θ, to produce the frequency shift Δf. The factor of 2 occurs because the Doppler shift arises both on absorption of the sound by the moving blood particles and between the transmitting blood cell and the receiving transducer.

The Doppler probe has a piezoelectric ultrasound transducer integrated into the tip. The crystal on the tip sends and receives ultrasound waves. The timing of the sending and receiving velocity allows the ultrasonic flowmeter to measure blood flow velocities. The returning signal is transmitted, in real time, to the display device, which shows the average velocity of all the red cells within the sample volume. When the sample area is constant, the peak velocity is accurately tracked in the center of the artery and the key parameters remain relatively positionally insensitive and reliable.

Computerized parameters of intracoronary flow velocity, including peak and mean diastolic and systolic velocities, diastolic and systolic velocity integrals, mean total velocities, and the total velocity integral are automatically analyzed.

8.8.4 Laser-Doppler Flowmeter

The Doppler ultrasonic flowmeter can assess blood flow from the Doppler shift that sound waves experience when they travel through a large blood vessel. However, the ultrasonic flowmeter is not able to measure the blood flow in small blood vessels, or microcirculation. In contrast, this is possible if laser light is used instead of sound (Shepherd and Öberg, 1990). Light is capable of measuring the velocities of red blood cells even at the relatively slow speeds with which they move through capillaries. Laser-Doppler blood flowmetry thus utilizes the short wavelengths (extremely high frequencies) of visible and infrared light to measure the blood flow velocity in a similar manner as the Doppler ultrasound blood flowmeter. The spectral purity of the laser makes it practical to detect the slight frequency shifts produced by the interactions between photons and moving red blood cells.

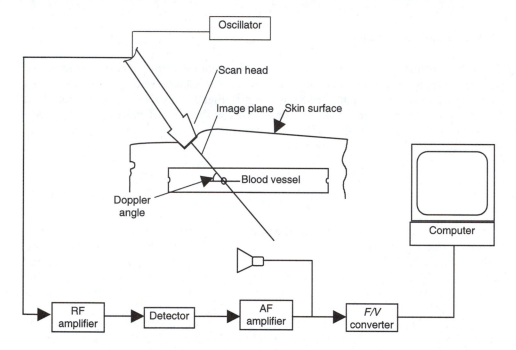

Figure 8.17 Ultrasonic flowmeter. The sensor at the scan head transmits the signal from the oscillator and receives the reflected wave from the blood cells. The RF (radio-frequency) amplifier amplifies the received signal and the carrier frequency, then AF (audio-frequency) signal is produced by a detector. (Adapted from Picot, P. A. and Fenster, A. 1996. Ultrasonic blood volume flow rate meter. US Patent, 5,505,204.)

Figure 8.18 shows the principle of laser-Doppler measurement. Laser light is transmitted to the blood vessels and reflected to a detector. Light that intercepts the moving red blood cells is Doppler-shifted. Light in tissue is scattered by stationary tissue and it reaches the detector without experiencing a Doppler shift (Shepherd and Öberg, 1990).

A laser-Doppler catheter velocimeter can be used to measure intravascular blood flow velocity. This can be achieved easily because optical fibers have minute diameters and permit high resolution. However, catheters disturb the flow of blood in their neighborhood, and because light is multiply scattered in passing through the blood, combining the high spatial and temporal resolution of laser-Doppler velocimetry with catheter delivery is not straightforward. At the present time, the development of instruments for the practical measurement of blood flow velocity through a catheter is an area of active research.

The major advantage of the laser-Doppler flowmeter is the ability to measure blood flow in regions other than the limbs. Another advantage is its frequency response. Not only can the laser-Doppler flowmeter be used to examine variations in blood flow through the cardiac cycle, but also, on a somewhat longer time scale, slower rhythms can be observed.

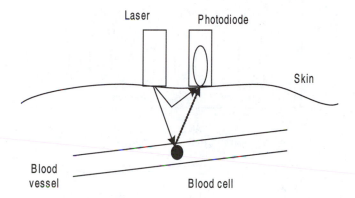

Figure 8.18 Laser-Doppler flowmetry. Light that intercepts the red blood cells experiences a Doppler frequency shift.

8.9 Blood Pressure

Arterial pressure is an essential check on the performance of the heart. The systolic and diastolic pressures not only provide those maximum and minimum values that are an important aspect of the pulsations, but also permit a rough estimate of mean pressure in large arteries. Normal blood pressure is considered to be 120/80 mmHg. Mean blood flow is essentially the same at all cross sections of the arterial tree, in the total population of peripheral arterioles as in the ascending aorta. Mean pressure, however, falls continually from the ascending aorta to the terminus of the vena cava, an indication that the energy imparted to the blood by ventricular ejection is gradually dissipated as it travels through the vascular system. The decrement in pressure per unit length of the system, or vascular resistance, is greatest in the microcirculation.

Blood pressures measured in the distal portions of the extremities are especially useful for identifying the presence of hemodynamically significant disease. The degree of circulatory impairment is usually reflected by the relative or absolute reduction in resting pressure. Stress testing increases the sensitivity of pressure measurement, permitting the recognition of clinically significant disease in limbs in which resting pressures are borderline normal. Unfortunately, pressure measurements are not always reliable.

8.9.1 Indirect Measurement

Figure 8.19 shows a sphygmomanometer, which indirectly measures blood pressure. An inflatable cuff formed by cloth containing an inflatable bladder encircles the arm. A hand or electric pump and a pressure gage are used. A pressure release valve allows controlled release of pressure within the cuff.

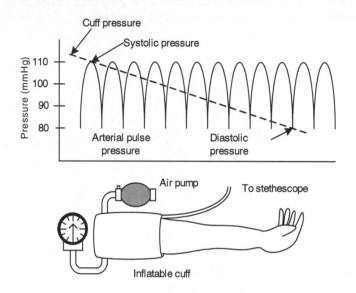

Figure 8.19 The sphygmomanometer detects arterial opening and closing that occurs between systolic and diastolic pressures.

The auscultatory method of measuring blood pressure refers to listening to sounds produced by blood turbulence within the artery. The occlusive cuff is pumped up to about 180 mmHg, then bled off at about 3 mmHg/s. When the arterial pressure exceeds the cuff pressure, blood squirts through the partially occluded artery and creates turbulence, which creates Korotkoff sounds (Figure 8.20). A stethoscope placed distal to the cuff over the brachial artery detects the tapping noise that signals systolic pressure at about 120 mmHg. When the cuff pressure decreases below about 80 mmHg, the artery remains open the entire cycle. The sounds disappear, which indicates diastolic pressure. A piezoelectric sensor placed within the cuff replaces the stethoscope for automatic detection.

There are several automatic noninvasive blood pressure monitoring devices available commercially. A typical automatic noninvasive blood pressure monitor is composed of an inflatable cuff with a microphone. The signal from the microphone is then passed to a filter. Generally, the cuff method gives systolic pressures within ±2 mmHg of the values obtained by direct arterial puncture and diastolic pressure within ±4 mmHg. Pressure in the cuff is sometimes measured with a mercury manometer, eliminating the need for calibration.

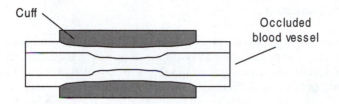

Figure 8.20 The pressure of the cuff occludes the blood vessel. When the arterial pressure is greater than the pressure applied by the cuff, Korotkoff sounds are created and blood pressure can be measured.

The oscillometric method uses the same cuff. When the artery changes blood volume under the cuff, this changes air volume within the cuff, which changes pressure within the cuff. These pressure fluctuations can be seen when using an aneroid or mercury manometer and the cuff pressure is between systolic and diastolic. Alternatively a pressure sensor (piezoelectric sensor) can detect these fluctuations in an automatic instrument. Because the pressure sensor is within the instrument, no separate sensor is within the cuff and only a single air tube leads to the cuff.

Figure 8.21 shows the indirect measurement of blood pressure. S_o is the point where cuff-pressure oscillations start to increase. A_s is the amplitude corresponding to auscultatory systolic pressure and A_d is the amplitude corresponding to auscultatory diastolic pressure. A_m is the maximal oscillation amplitude, which signals mean pressure.

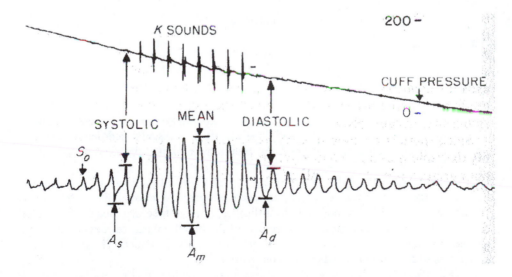

Figure 8.21 Top: Cuff pressure with superimposed Korotkoff sounds, which appear between systolic and diastolic pressures. Bottom: the oscillometric method detects when the amplified cuff pressure pulsations exceed about 30% of maximal pulsations. (From Geddes, L. A. 1984. *Cardiovascular devices and their applications.* Copyright © by John Wiley & Sons. Reprinted by permission of John Wiley & Sons.)

8.9.2 Direct Measurement

Arterial and venous blood pressure can be measured by inserting a catheter into the blood vessel and maneuvering it until the end is at the site at which the blood pressure is to be measured. In this case, the diaphragm of the sensor is mounted at the end of the catheter outside the body (similar to Section 8.2). Catheters for use in the artery are usually thin, short, and highly flexible. A catheter can be inserted into the artery inside a needle. The alternative method is to use a catheter-tip sensor where the miniature diaphragm is at the tip inside the vessel. Although accurate, these methods are very invasive.

8.9.3 Arterial tonometry

Arterial tonometry is a noninvasive technique for monitoring the arterial blood pressure in a continuous manner. A linear array of pressure sensors is pressed against the radial artery so that at least one sensor is directly over the lumen. Pressure is increased from low to high so that measurements are made when the artery is half collapsed. By selecting the maximal peak-to-peak pressure reading from all sensors, the arterial pressure is determined in the same way as for the applanation tomometer (Section 7.10.4). Zorn et al. (1997) compared the Colin Pilot 9200 tonometric blood pressure measurements with intra-arterial blood pressure measurements. Tonometric values were slightly less than the intra-arterial pressure measurements; the mean difference for systolic blood pressure was 2.24 ± 8.7 mmHg, and diastolic pressure was 0.26 ± 8.88 mmHg.

8.10 Vessel Distension

Blood vessels can be distended (stretched) by raising intravascular pressure or decreased in radius by lowering it because they are elastic. The difference between the pressures inside and outside the vessel—the transmural pressure—is one of the factors that controls the radius. The contraction or relaxation of vascular smooth muscle alters vessel diameter by changing the elasticity of the wall and the elasticity of a vascular bed determines how much of the blood volume is accommodated within that region at the existing local pressure.

The distribution of the blood volume among different parts of the system is determined by the relation between local pressure and vascular distensibility. Pressure is high in arteries, but they are not very distensible and the volume of blood in the arterial tree is relatively small. In contrast, a large part of the blood volume resides in the veins, even though their pressure is low, because they are readily distended. The distension of an arterial wall can be calculated from

$$\text{Distension} (\%) = \frac{\Delta d}{d_{\mathrm{D}}} \times 100 \tag{8.15}$$

where Δd is the diameter change and d_{D} is the diastolic diameter (Nakatani et al., 1995).

8.10.1 Intravascular Ultrasound

Intravascular ultrasound provides both *in vitro* and *in vivo* two-dimensional visualization of arteries in real time. Figure 8.22 shows that a rotating ultrasonic transducer illuminates the walls. Reflections from the inner and outer wall surfaces yield accurate determination of artery luminal dimension, cross-sectional area, wall thickness, and wall morphology (Forestieri and Spratt, 1995).

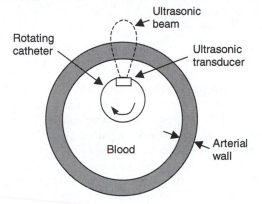

Figure 8.22 A catheter is inserted through the blood vessels. A rotating ultrasonic transducer is attached at its tip and illuminates the walls.

8.10.2 Angiography

Peripheral angiograms are most commonly done to test the arteries that supply blood to the head and neck or the abdomen and legs. Because arteries do not show up well on ordinary X rays, angiograms utilize a *contrast agent* containing iodine, which is injected into the arteries to make them visible on radiographs. A catheter is placed into the artery, via the groin, and is manipulated by the physician into the artery requiring study. Once the catheter is in place, contrast agent, or dye is injected through the catheter into the arteries and a series of X rays taken. The dye allows the doctor to see the inside of the patient's artery and determine how well blood is moving through the vessel.

In digital subtraction angiography (DSA), a first image is taken without contrast agent, then a second image is taken with contrast agent. The two images are digitized, then subtracted, yielding no image in all regions without contrast agent. Vessels containing contrast agent show very well, so less contrast agent is required.

8.11 Vessel Volume Flow

One application of vessel volume flow measurement is the prediction of stenosis of both common and internal carotid arteries by monitoring common carotid blood flow. Generally, an adequate degree of stenosis (usually accepted as 50% diameter reduction), results in a measurably decreased volume flow. Vessel volume flow measurements may also be applied in the diagnosis and treatment of vascular malformations. Specifically, the measurement of volume flow may help to distinguish arteriovenous malformations and fistulae, which are high-flow lesions, from venous malformations, which are low-flow lesions. Moreover, volume flow measurements provide a quantitative way both of assessing blood steals (deviation of normal flow) and of evaluating the effectiveness of embolization (mass of foreign matter in the blood vessel) therapy. Renal dialysis patients may also benefit from these measurements.

8.11.1 Flow Visualization

Continuous-wave or pulsed-Doppler flow-mapping devices produces a static image on an oscilloscope screen that corresponds to a projection of the lumen of the vessel underlying the probe. The frequency of the Doppler signal is color-coded to indicate regions of increased velocity (stenotic areas), and provided a pulsed-Doppler is used, the image can be processed by a microcomputer to depict the outline of the vessel in a plane or produce a histogram of its cross-sectional area. Although B-mode ultrasonic scanning allows large vessels, such as the abdominal aorta, to be visualized, it is inappropriate for the routine study of smaller vessels. With the development of real-time B-mode scanners, it is now practical to examine virtually all vessels of interest, even those with diameters as small as 2.0 mm.

A major advance in flow visualization is the combination of B-mode scanning and the pulsed-Doppler velocity detector in a single instrument, called the Duplex scanner. The two technologies are complementary. Doppler flow patterns help distinguish between arteries and veins, and the image permits more accurate positioning of the Doppler sample volume. One instrument superimposes a real-time, color-coded Doppler flow-map on a real-time B-mode image, thus incorporating the advantages of both modalities.

By using a gamma camera to trace the passage of an isotope through the veins of the extremities, one can obtain crude, but sometimes clinically revealing flow-maps.

8.11.2 Ultrasonic Sonography

To estimate the volume flow rate of blood through an artery, pulsed Doppler ultrasound measures the velocity of the blood at 16 or more discrete locations across the vessel diameter. From the velocity measurement, and a measurement of the diameter of the vessel, we estimate the volume of blood flowing through the vessel. Other techniques, such as color M-mode, directly measure the one-dimensional velocity profile and are also used.

A clinical color Doppler ultrasound instrument is used with a position and orientation sensing device, and a computer with video digitizer to acquire blood velocity measurements in two dimensions across an entire blood vessel lumen. Subsequently, the volume flow rate can be determined by integrating the two-dimensional velocity profile over the vessel lumen area. The volume flow rate measurements are then displayed to an operator in real time as a scrolling graph.

8.11.3 Video Microscopy

With the developments in video and computer technology over the past two decades, we are now able to see structures that are very fine, movements that were too fast too be seen, and images that were too dim or too noisy to be seen. Generally, the components used in video microscopy are a camera, mixer, recorder, processor, analyzer, and monitor. It also utilizes advances in image-processing (Inoué, 1986). Chapter 6 discusses the principles of video microscopy.

8.12 Problems

8.1   Sketch the anatomy of the heart chambers and valves and show the direction of blood flow.

8.2   Explain why the wall of the left ventricle is thicker and more muscular then the wall of the right ventricle.

8.3   Describe the relation between the arterial pressure and the left ventricular pressure during a cardiac cycle.

8.4   Explain why the AV node delays the wave of excitation.

8.5   Draw a typical ECG waveform, label all waves, and explain what is happening within the heart muscle during each wave.

8.6   Explain why different pairs of electrodes are used on specific locations on the body for an ECG.

8.7   An ECG has a scalar magnitude of 0.8 mV on lead I and a scalar magnitude of 0.2 mV on lead III, calculate the scalar magnitude of lead II.

8.8   Given the stroke volume = 80 mL, LV end-diastolic volume = 51 mL , and the heart rate = 80 bpm, calculate the LV ejection fraction and the cardiac output.

8.9   Use the Fick method to calculate the cardiac output (in mL/min) of a patient whose arterial concentration of O_2 is 0.20 L/L, venous concentration of O_2 is 0.13 L/L, and the O_2 consumption is 0.3 L/min.

8.10   Calculate the cardiac output using the thermodilution method from the following data: $\int_0^{t_1} |\Delta T_b| \, dt = 3\,\text{s}\cdot{}^\circ\text{C}$, $Q = 1250$ J, $\rho_b = 1060$ kg/m^3, and $c_b = 3640$ J/(kg· K).

8.11   Explain why and how we measure cardiac output by the thermal and dye dilution techniques and compare the two methods.

8.12   A Doppler-ultrasonic blood flowmeter has a carrier frequency of 7.5 MHz, with a transducer angle of 30°. The velocity of the sound is 1500 m/s, and the audio frequency is 12.6 kHz. Calculate the blood velocity.

8.13   Describe the cardiac sounds, their origins, and the time they occur in the cardiac cycle.

8.14   An artery has a diameter of 3 mm. The viscosity of the blood is 0.0028 Pa · s at 37 °C, and the mean arterial blood pressure is 90 mmHg at one end and 0 mmHg at the other end. Calculate the blood flow rate in mL/min, assuming that the length of the artery equals 0.5 m.

8.15   Sketch an electromagnetic flowmeter and explain its principle of operation.

8.16   Sketch an ultrasonic flowmeter and explain its principle of operation.

8.17   Draw the block diagram for and describe the automatic indirect auscultatory peripheral blood pressure measurement system.

8.18   Draw the block diagram for the automatic indirect oscillometric (not auscultatory) peripheral blood pressure measurement system. Show all pneumatic connections between parts. Show location of all sensors and explain how they work. Describe the measurement cycle. Sketch the resulting waveforms. Explain how the signal processing identifies significant pressures.

8.19   Explain why and how we measure direct cardiac pressure using a catheter (catheter-type and catheter tip).

8.20   Explain why and how we measure vessel pressure using arterial tonometry.

8.21   Explain why and how we measure vessel pressure distensibility.

8.22   Explain why and how we measure vessel volume flow.

8.13 References

Berne, R. and Levy, M. 1981. *Cardiovascular Physiology*. 4th ed. St. Louis: Mosby.

Bom, N. (ed.) 1977. *Echocardiography: With Doppler Applications and Real Time Imaging*. The Hague: Martin Nijhoff Medical Division.

Brown, B. H., Smallwood, R. H., Barber, D. C., Lawford, P. V., and Hose D. R. 1999. *Medical Physics and Biomedical Engineering*. Bristol, UK: IOP Publishing.

Burch, G. and Winsor, T. 1972. *A Primer of Electrocardiography*. 6th ed. Philadelphia: Lea and Febiger.

Cikes, I. and Ernst, A. 1983. New aspects of echocardiography for the diagnosis and treatment of pericardial disease. In J. Roelandt (ed.) *The Practice of M-mode and Two-dimensional Echocardiography*. Boston: Martinus Nijhof.

Cohn, P. F and Wayne, J. 1982. *Diagnostic Methods in Clinical Cardiology*. Boston: Little, Brown.

College of Veterinary Medicine, University of Tennessee. 2000. M-Mode Echocardiography [Online] www.vet.utk.edu/.

Davies, M. J., Anderson, R. H. and Becker, A. E. 1983. *The Conduction System of the Heart*. London: Butterworth.

Forestieri, S. F. and Spratt, R. S. 1995. *Angiography using ultrasound*. US Patent, 5,394,874.

Fung, Y. C. 1997. *Biomechanics: Circulation*. 2nd ed. New York: Springer-Verlag.

Grossman, W. and Baim, D. S. 1991. *Cardiac Catheterization. Angiography and Intervention*, 4th ed. Philadelphia: Lea & Febiger.

Huang, S. K. S., and Wilber, D. J. (eds.) 2000. *Radiofrequency Catheter Ablation of Cardiac Arrhythmias: Basic Concepts and Clinical Applications*, 2nd ed. Armonk, NY: Futura.

Ingels, N. B. Jr., Daughters, G. T., Baan, J., Covell, J. W., Reneman, R. S. and Yin, F. C. 1996. *Systolic and Diastolic Function of the Heart*. Amsterdam: IOS Press.

Inoué, S. 1986. *Video Microscopy*. New York: Plenum Press.

Marwick, T. H., Shan, K., Go, R. T., MacIntyre, W. J. and Lauer, M. S. 1995. Use of positron emission tomography for prediction of perioperative and late cardiac events before vascular surgery. *Amer. Heart J*. **130** (6): 1196-202.

Meltzer, R. S. and Roelandt, J. (eds.) 1982. *Contrast Echocardiography*. Hague: Martinus Nijhoff.

Merz, C. N. and Berman, P. S. 1997. Imaging techniques for coronary artery disease: current status and future directions. *Clinical Cardiology*. **20** (6): 526-32.

Mohrin, C. M. 1995. *Stethoscope*. US Patent, 5,389,747.

Moran, P. R. 1982. A flow zeugmatographic interlace for NMR imaging in humans. *Magn. Reson. Imaging* **1**: 197–203.

Nakatani, S., Yamagishi, M., Tamai, J., Goto, Y., Umeno, T., Kawaguchi, A., Yutani, C. and Miyatake, K. 1995. Assessment of coronary artery distensibility by intravascular ultrasound: application of simultaneous measurements of luminal area and pressure. *Circulation* **91**: 2904–10.

Neuman, M. R. 1998a. Biopotential electrodes. In J. G. Webster (ed.) *Medical Instrumentation: Application and Design*. 3rd ed. New York: John Wiley & Sons.

Neuman, M. R. 1998b. Biopotential amplifiers. In J. G. Webster (ed.) *Medical Instrumentation: Application and Design*. 3rd ed. New York: John Wiley & Sons.

Onoda, M. 1994. *Electrocardiograph system*. US Patent, 5,284,151.

O'Rourke, M. 1995. Mechanical principles in arterial disease. *Hypertension* **26**: 2–9

Orth, J. L. 1995. *System for calculating compliance and cardiac hemodynamic parameters*. US Patent, 5,423,323.

Peura, R. A. and Webster, J. G. 1998. Basic sensors and principles. In J. G. Webster (ed.) *Medical Instrumentation: Application and Design*. 3rd ed. New York: John Wiley & Sons.

Picot, P. A. and Fenster, A. 1996. *Ultrasonic blood volume flow rate meter*. US Patent, 5,505,204.

Rafferty, T. 1992. Transesophageal Two-Dimensional Echocardiography [Online] www.gasnet.org/echomanual/html/2-d_echo..html.

Reinhard, C. J. and Fowler, K. A. 1987. *Telemetry system and method for transmission of ECG signals with heart pacer signals and loose lead detection*. US patent, 4,658,831.

Ritchie, J. L., Hamilton, G. W. and Wackers, F. J. T. (eds.) 1978. *Thallium–201 Myocardial Imaging*. New York: Raven Press.

Shepherd, A. P. and Öberg, P. Å. (eds.), 1990. *Laser-Doppler Blood Flowmetry*. Boston: Kluwer Academic Publishers.

Sherwood, L. 2001. *Human Physiology: From Cells to Systems*. 4th ed. Pacific Grove, CA: Brooks/Cole.

Strandness, D. E., Didisheim, P., Clowes, A. W. and Watson, J. T. (eds.) 1986. *Vascular Diseases: Current Research and Clinical Applications*. Orlando: Grune & Stratton.

Tobuchi, K., Kato, T. and Namba, K., 1981. *Silver-silver chloride electrode*. US Patent, 4,270,543.

Togawa, T., Tamura, T. and Öberg, P. Å. 1997. *Biomedical Transducers and Instruments*. Boca Raton FL: CRC Press.

Tompkins, W. J. (ed.) 1993. *Biomedical Digital Signal Processing*. Englewood Cliffs NJ: Prentice Hall.

Underwood, R. 1992. Magnetic resonance imaging of the cardiovascular system. In J. Reiber and E. Ernst (eds.) *Cardiovascular Nuclear Medicine and MRI*. Dordrecht: Kluwer Academic.

Webster, J. G. 1998. Measurement of flow and volume of blood. In J. G. Webster (ed.) *Medical Instrumentation: Application and Design*. 3rd ed. New York: John Wiley & Sons.

Zorn, E. A., Wilson, M. B., Angel, J. J., Zanella, J. and Alpert, B. S. 1997. Validation of an automated arterial tonometry monitor using Association for the Advancement of Medical Instrumentation standards. *Blood Press. Monit.* **2** (4): 185–8.

Chapter 9

Lung, Kidney, Bone, and Skin

Shilpa Sawale

This chapter describes four major organs in the body: lung, kidney, bone, and skin. Each of these organs has its own properties and functions: the lungs help respiration, the kidneys help clean the blood, bone supports the body, and skin protects the body. Measurements are performed to confirm whether each of these organs is functioning properly, and also to measure some of their properties.

9.1 Lung

The exchange of gases in any biological system is called respiration. To sustain life, the human body needs oxygen, which is utilized in cells with other essential nutrients during the metabolic oxidation process. Carbon dioxide is a by-product of cellular metabolism. The hemoglobin in the blood is the dominant transport mechanism by which oxygen is brought to cells. Carbon dioxide produced by cells is dissolved in the blood plasma and carried off for disposal via the lungs. The most important function of the lungs is to supply tissue with adequate oxygen and to remove excess carbon dioxide.

To properly exchange gases during inspiration, alveolar pressure must be less than the atmospheric pressure. There are two ways of producing the pressure difference necessary for inspiratory flow: (1) the alveolar pressure can be reduced below atmospheric. This is called natural or negative pressure breathing; (2) the atmospheric pressure can be raised above normal and therefore above the resting alveolar pressure. This is called positive pressure breathing.

Normal breathing is accomplished by active contraction of inspiratory muscles, which enlarges the thorax. This further lowers intrathoracic pressure, which normally is less than atmospheric pressure and hence, the air at atmospheric pressure flows through the nose, mouth and trachea to the lungs.

The process of bulk gas transport into and out of the lungs and the diffusion of gases across the alveolar membrane is known as pulmonary function. Tests performed to determine parameters of system efficiency are called Pulmonary Function Tests (PFTs) (Feinberg, 1986). PFTs do not tell us the ultimate answer; for example, finding a decreased lung volume reveals restriction, but not the cause of it. However, in several circumstances PFTs are a useful and powerful tool: diagnosing a lung disease and finding the extent of the abnormality, following a patient during the course of a disease to determine the efficiency of the treatment or the need for supplemental oxygen and mechanical ventilation, determining if preoperative patients can withstand the surgery, assessing disability, and deciding whether an individual can perform an occupational task requiring a certain work load (Petrini, 1988).

9.2 Pulmonary Volume

The most commonly used measures of the mechanical status of the ventilatory system are the absolute volume and changes of volume of gas space in the lungs achieved during various breathing maneuvers. Pulmonary volumes have been divided according to physiological limits (Petrini, 1988). The common terms used to describe various capacities (capacities refer to quantities that include more than one volume) are as follows:

1. Total Lung Capacity (TLC) is the largest volume to which the subject's lung can be voluntarily expanded.
2. Residual Volume (RV) is the smallest volume to which the subject can slowly deflate his or her lung.
3. Functional Residual Capacity (FRC) is the resting lung volume achieved at the end of normal expiration. It is the point where inward pulling forces of the lung and outward pulling forces of the chest wall are equal in magnitude.
4. Tidal Volume (TV) is the volume inspired and expired during normal breathing.

All of the lung capacities are based on these four parameters (Eqs. (9.1) to (9.4)). IRV is the inspiratory reserve volume. ERV is the expiratory reserve volume.

$$\text{Inspiratory capacity (IC)} = \text{IRV} + \text{TV} \qquad (9.1)$$

$$\text{Functional residual capacity (FRC)} = \text{RV} + \text{ERV} \qquad (9.2)$$

$$\text{Vital capacity (VC)} = \text{IC} + \text{ERV} \qquad (9.3)$$

$$\text{Total lung capacity (TLC)} = \text{VC} + \text{RV} \qquad (9.4)$$

In dealing with volumes of gases, the conditions under which the values are reported must be well defined and carefully controlled because gases undergo large changes under different thermodynamic conditions. These volumes and specific capacities, represented in Figure 9.1, have led to the development of specific tests to quantify the status of the pulmonary system.

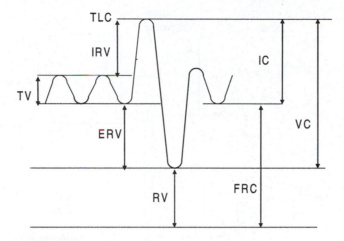

Figure 9.1 The spirometer measures lung capacities and lung volumes. Because the subject cannot make the lung volume equal to zero, the spirometer cannot measure RV and FRC.

9.2.1 Measurement of Pulmonary Volume

The measurement of changes in the lung volume have been approached in two different ways. The first involves measurement of the gas through the airway opening. This is referred to as spirometry. The second method is to measure changes in the volume of the gas space within the body using plethysmographic techniques.

Spirometer

The spirometer was one of the early instruments used for estimating respiratory volumes. It collects the gas passing through the airway opening and computes the volume it had occupied within the lungs. The tests performed by the spirometer can be classified into two major groups: single-breath tests and multiple-breath tests. There are three types of tests under the category of single-breath tests: tests that measure expired volume only, tests that measure expired volume in a unit time, and tests that measure expired volume/time. Maximal Voluntary Ventilation (MVV) is measured in the category of multiple-breath tests (Feinberg, 1986).

A spirometer is an expandable compartment consisting of a movable, statically counterbalanced, rigid chamber or bell, a stationary base, and a dynamic seal between them, often made of water. Changes in internal volume of the spirometer are proportional to the displacement of the bell. This motion can be observed on a calibrated scale or a rotating drum (kymograph) as shown in Figure 9.2.

The mouthpiece of the spirometer is placed in the mouth of the subject whose nose is blocked. As the gas moves into and out of the spirometer, the pressure of the gas in the spirometer changes, causing the bell to move.

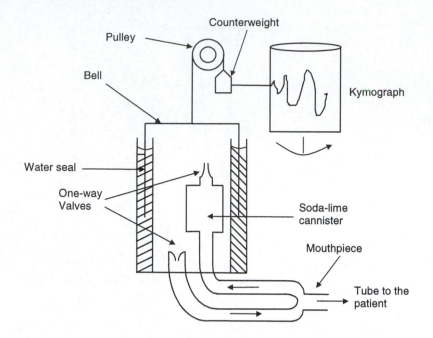

Figure 9.2 In the water sealed spirometer, expired CO_2 is removed in the soda-lime cannister.

The system can be modeled as two gas compartments connected such that the number of moles of gas lost by the lungs through the airway opening is equal and opposite to the number gained by the spirometer. For rebreathing experiments, most spirometer systems have a chemical absorber (soda lime) to prevent build up of carbon dioxide. With the exception of water vapor in a saturated mixture, all gases encountered during routine respiratory experiments obey the ideal-gas law during changes of state:

$$P = \frac{N}{V} RT = \rho RT \tag{9.5}$$

where P is the pressure of the ideal gas, R is the universal gas constant, T is the absolute temperature, and ρ is the mole density, which for a well-mixed compartment equals the ratio of moles of gas N in the compartment to the compartment volume V.

Spirometers should be able to measure at least 7 liters of gas with 3% accuracy. They should be linear, have stability (no drift), low inertia, low resistance, and no hysteresis. During slow breathing maneuvers, spirometers yield the output shown in Figure 9.1, from which impared vital capacity can indicate disease.

Over a few minutes, the O_2 in the spirometer is consumed. The expired CO_2 is absorbed by the soda lime cannister. Thus the bell slowly decends. The average slope of the spirometer output yields the O_2 consumption vs. time, as required for basal metabolism tests and the Fick method of cardiac output measurement.

The soda lime cannister adds pneumatic resistance, so is removed when measuring maximal voluntary ventilation. The subject breathes as hard as possible for 15 s and the peak-to-peak sum of all breaths multiplied by four yields the volume breathed per minute.

Forced expiratory volume (FEV) measures airway restriction. The subject inhales as much as possible, then blows out hard. A normal subject can exhale 83% of vital capacity in 1 s whereas a subject with restricted airways takes longer.

Body Plethysmograph

Because alveolar pressure cannot be measured directly, an indirect method has been developed using a device known as a total body plethysmograph. This device is a volumetric displacement box into which the patient is placed. The box is closed and sealed so that no gas can enter or escape except through the connection leading to a patient mouthpiece or mask. When the outlet connection is attached to a spirometer, pneumotachometer, or respiratory function computer, a wide variety of measurements can be easily made. It is used clinically not only to evaluate the absolute volume and volume changes of the lungs, but also to provide a continuous estimate of alveolar pressure, from which airway resistance can be found. Figure 9.3 shows a total body plethysmograph.

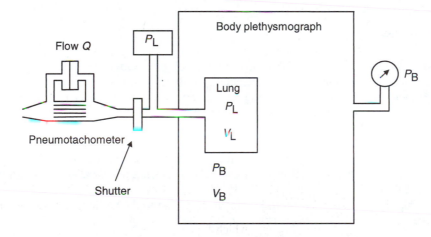

Figure 9.3 The total body plethysmograph measures lung volume with the shutter closed and the airway resistance via a pneumotachometer with the shutter open.

In order to compute the alveolar pressure using plethysmograph pressure response to thoracic volume changes, the Volume of Thoracic Gas (VTG) must be known. A patient sits inside the plethysmograph with a pneumotachometer flow meter in his or her mouth. Both mouth and box pressures are displayed on the *x* and *y* axes of the oscilloscope for one breath cycle against a total mechanical obstruction to airflow. This is accomplished using a solenoid operated shutter to block the pneumotachometer. When a patient breathes against a closed shutter, the air flow is zero, and thus the pressure measured at the mouth is the same as the alveolar pressure (Feinberg, 1986).

The tissues of the body are composed of liquids and solids and can be considered incompressible when compared with the gas in the lungs. During breathing movements, the tissues change shape, volume does not change. Also because the volume of the plethysmograph is constant, changes in the volume of the plethysmograph are zero. Consequently the change in the volume of the gas in the lungs ΔV_L is equal and opposite to the change in the volume of the gas space in the box ΔV_B.

During compression, $\Delta V_L = -\Delta V_B$. There is a large change in pressure of the lung ΔP_L, while there is a small change in pressure in the box ΔP_B. As a first order approximation, $V_L/V_B = -\Delta P_B/\Delta P_L$. The partial pressure of water vapor in the lung must be subtracted from the total pressure to yield compressible dry gas. The lung is isothermal whereas the box is adiabatic. All this can be calibrated by plotting $\Delta P_B/\Delta P_L$ during breathing with the shutter closed, from which we learn V_L (Primiano, 1998). After calibration, we can measure ΔP_B and use it to predict ΔP_L. Then we can open the shutter to obtain the airway resistance

$$R_{AW} = \Delta P_L/\Delta Q \tag{9.6}$$

where Q is the flow measured by the pneumotachometer.

Impedance Plethysmography

Impedance plethysmography is the measurement of physiologic volume changes by measuring electric impedance. We can calculate the fixed impedance from Eq. (2.8).

Each time the heart ejects blood into the arteries, their diameter increases about 7%. Because the arteries have a regular cylindrical shape, we can calculate their longitudinal change in impedance. We shall follow Swanson's derivation to develop the equations used in impedance plethysmography. The derivation is based on three assumptions: the expansion of vessels is uniform (this may not be the case in diseased vessels), the resistivity of the blood does not change, and lines of current are parallel to the arteries (Webster, 1998).

The shunting impedance of the blood, Z_b, is due to the additional blood volume ΔV that increases the cross-sectional area ΔA.

$$Z_b = \frac{\rho_b L}{\Delta A} \tag{9.7}$$

ΔV can be calculated using the following equation:

$$\Delta V = \frac{-\rho_b L^2 \Delta Z}{Z^2} \tag{9.8}$$

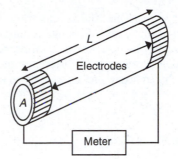

Figure 9.4 A model for two electrode impedance plethysmography for cylindrical vessels.

If the assumptions are valid, the above equation shows that we can calculate ΔV from ρ_b and other quantities that are easily measured.

However, the lung does not have a regular shape, so we cannot use equations to predict volume changes. In respiratory impedance plethysmography, respiration is electrically monitored by measuring changes in chest impedance. The electric impedance of the thoracic cavity changes with breathing movements and can be sensed in order to monitor ventilatory activity. Ventilation produces changes in the shape and content of the chest that are electrically observable as small impedance variations. A 100 kHz impedance measurement between electrodes on each side of the chest yields a waveform that follows ventilation. Electrodes can be easily attached to a segment of the chest and the resulting impedance of the chest measured. As resistivity increases in response to increased air in the chest, the impedance of the chest increases. It is useful for monitoring breathing of infants as in infant apnea monitors.

Inductance Plethysmography

Respiratory inductive plethysmography continuously monitors the motions of the chest wall that are associated with changes in thoracic volume. The variable inductance sensor uses a single loop of wire attached in a zigzag pattern to its own compliant belt. It is excited by a low-level radio-frequency signal. Changes in the loop's cross-sectional area produce corresponding changes in self-inductance. An output is obtained proportional to the local cross-sectional area of the segment of the chest wall that is encircled by the loop, after demodulation. Usually two separate instruments are used, one around the chest and the other around the abdomen. Their sum is a more accurate estimate of lung volume that that obtained from a single loop. It is used during sleep studies to diagnose sleep apnea (Kryger, 1994).

9.3 Pulmonary Flow

When the lung changes volume during breathing, a mass of gas is transported through the airway opening by convective flow. Measurement of the variables associated with the movement of this gas is very important. The volume flow rate and the time integral of volume flow rate are used to estimate rate of change of lung volume and changes of lung volume. Convective flow occurs as a result of a difference in pressure between two points. From the relationship between pressure difference and volume-flow rate through a system, measurement of the difference in pressure yields an estimate of flow.

9.3.1 Measurement of Pulmonary Flow

Pneumotachometer

The most commonly used instrument for measurement of flow is the pneumotachometer. This instrument is designed to measure respiratory gas flow during inspiration and expiration. It is based on the pneumatic equivalent of Ohm's law:

$$Q(\text{flow}) = \frac{\Delta P \text{ (difference in pressure)}}{R \text{ (fixed resistance of pneumotachometer)}} \tag{9.9}$$

The flow resistors have approximately linear pressure–flow relationships. Flow-resistance pneumotachometers are easy to use and can distinguish the direction of alternating flows. They also have sufficient accuracy, sensitivity, linearity, and frequency response for most clinical applications. Even though other flow resistance elements are incorporated in pneumotachometers, the most common are one or more fine mesh screens placed perpendicular to flow or a tightly packed bundle of capillary tubes or channels with its axis parallel to flow. These devices exhibit a linear pressure drop–flow relationship for a wide range of steady flows, with the pressure drop nearly in phase with the flow.

This element is mounted in a conduit of circular cross section. The pressure drop is measured across the resistance element at the wall of the conduit, and the pressure difference is measured by a differential pressure sensor and is used as a measure of air-flow through it.

The prevention of water vapor condensation in the pneumotachometer is very important because the capillary tubes and the screen pores are easily blocked by liquid water droplets, which decreases the effective cross-sectional area of the flow element and causes a change in resistance. Also, as water condenses, the concentration of the gas mixture changes. To avoid these problems, a common practice is to heat the pneumotachometer element using various techniques. Usually a pneumotachometer is provided with an electric resistance heater. Also, the screen of the pneumotachometer can be heated by passing a current through it, heated wires can be placed inside, or heating tape or other electric heat source can be wrapped around any conduit that carries expired gas (Primiano, 1998).

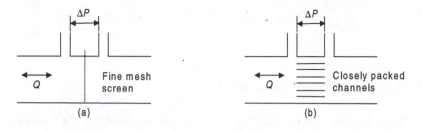

Figure 9.5 A pneumotachometer measures flow from a pressure drop ΔP across resistance elements such as (a) a fine mesh screen or (b) capillary tubes or channels.

9.4 Pulmonary Diffusion

The amount of gas transferred from the lung to blood is a function of the pressure gradient and the diffusion capacity (D_L) if the gas is diffusion limited. The equation is

$$-V_g = D_L(P_{Ag} - P_{cg}) \tag{9.10}$$

where V_g is the rate of transfer of gas (g), from alveoli (A), to the capillaries (c), and P is the pressure. D_L is the diffusing capacity in mL/(min · mmHg) and includes the parameters of the surface area of gas transfer, the thickness of the membrane across which the gas is transferred, and the properties of the gas such as the molecular weight and solubility. Since D_L is measured from the gas side, V_g is negative since the gas is moving away from the lungs (Primiano, 1998).

Carbon dioxide diffuses across the alveolar membrane much more easily than oxygen, so a diffusion defect affects oxygen transfer first. Because of this and because of the important role oxygen plays in sustaining life, it is important to evaluate a diffusion capacity for oxygen. However, obtaining the P_aO_2 requires a sample of arterial blood. To avoid this, clinicians use carbon monoxide as the tracer gas, as its properties are quite close to those of oxygen, and so its diffusion capacity provides a reasonable estimate of the diffusion capacity of oxygen. In addition, it also has affinity for hemoglobin at low concentrations; hence, all carbon monoxide that enters the blood chemically combines with the hemoglobin in the red blood cells. One of the many methods that have been used to find out the diffusion capacity of carbon monoxide involves a single breath technique. The subject inspires a mixture of air, 0.3% (or less) carbon monoxide and helium (approximately 10%) from RV to TLC. The subject holds his breath at TLC for about 10 s and then forcefully exhales down to RV. Although it requires subject cooperation, it can be performed quickly and repeated easily. Also, it does not require samples of arterial blood flow. The computation of D_LCO from measurements made during this single breath technique are based on a one compartment model of the lung. If a well-mixed alveolar compartment is filled with a mixture of gases containing some initial alveolar fraction F_ACO, then during breath holding with the airway open, the CO diffuses into the blood in the pulmonary capillaries, and the alveolar F_ACO decreases exponentially with time (Primiano, 1998).

$$\hat{F}_A CO(t_2) = \hat{F}_{\dot{A}} CO(t_1) \times \exp\left[-\frac{D_L CO(P_{atm} - P_A H_2 O)(t_2 - t_1)}{V_A}\right] \qquad (9.11)$$

9.5 Pulmonary Airway Resistance

Resistance to air flow is determined by the same factors governing the flow of low viscosity fluid in tubes. Resistance to air flow is a calculated quantity instead of a directly measured one. The equation used is as follows

$$Pulmonary\ resistance = \frac{Mouth\ pressure - Intrathoracic\ pressure}{Flow} \qquad (9.12)$$

A noninvasive method of measuring airway resistance is the body plethysmograph. An invasive method for measuring intrathoracic or intra-esophageal pressure involves passing an air-filled catheter with a small latex balloon on its end through the nose into the esophagus. The balloon should be located such that pressure changes due to motions of other organs competing for space in the thoracic cavity are minimized. A region below the upper third of the thoracic esophagus gives low cardiac interference and provides pressure variations that correspond well in magnitude and phase with directly measured representative changes in pleural pressure. The correspondence decreases as lung volume approaches the minimum achievable volume (RV). The frequency response of an esophageal balloon pressure-measurement system depends on the mechanical properties and dimensions of the pressure sensor, the catheter, the balloon, and the gas within the system. The use of helium instead of air can extend the usable frequency range of these systems (Primiano, 1998).

9.6 Kidney

Kidneys are paired bean-shaped organs lying on either side of the spine in the upper part of the abdomen. Each kidney is connected to the aorta and vena cava by a single renal artery and vein. By means of these blood vessels, the kidneys filter about 1200 mL/min of blood. Each kidney consists of approximately a million microscopic units called nephrons, which are made up of two components: a glomerulus and a tubule. The key separation functions of the kidney are: elimination of water-soluble nitrogenous end-products of protein metabolism, maintainence of electrolyte balance in body fluids and elimination of excess electrolytes, contribution to the obligatory water loss, discharge excess water in the urine, and maintenance of the acid–base balance in body fluids and tissues. Normal renal function is to remove water, electrolyte, and soluble waste products from the blood stream. The kidneys also provide regulatory mechanisms for the control of volume, osmolality, electrolyte and nonelectrolyte composition, and pH of the body fluids and tissues. Figure 9.6 shows the general anatomy of the kidney.

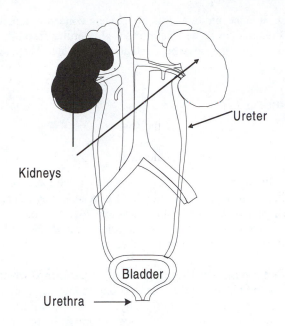

Ureter

Kidneys

Bladder

Urethra →

Figure 9.6 The kidneys excrete urine through the ureters to the bladder, where it is voided through the urethra.

9.7 Kidney Clearance

9.7.1 Creatinine

The glomerulus acts as a filter and separates water and crystalloids from the blood plasma, while retaining blood cells and proteins. Nonselective glomerular filtration is followed by the selective reabsorption in the renal tubes. Of the 180 L of water filtered each day, perhaps 1 to 2 L are excreted. Likewise, all glucose, amino acids, and the small amount of protein filtered together with most of the sodium, chloride, bicarbonate and calcium is reabsorbed. The reabsorption of these ions is precisely regulated to return to the body the exact quantity required to maintain the composition of the extracellular fluid. Creatinine, a byproduct of phosphate metabolism in muscle, is frequently used as a measure of GFR (Glomerular Filtration Rate) because it is freely filtered and not reabsorbed: the quantity crossing the glomerular filter is thus same as the steady state excretion of urine. For creatinine, the clearance (or volume of plasma that contains the quantity excreted in unit time) is the same as GFR, and is expressed in L/day (Gregory, 1988).

$$\text{GFR} \approx C = \frac{UV}{P} \qquad (9.13)$$

where U is the urinary concentration, P is the plasma concentration, V is the urine flow rate, and C is the creatine clearance. The normal glomerular filtration rate averages 120 mL/min, about 116 mL/min is reabsorbed, and about 4 mL/min is excreted as urine.

9.8 Kidney Imaging

9.8.1 Pyelogram

The main components of the urinary tract have either a tubular or cavitational structure and their delineation with radio-opaque medium is readily possible. In this method the radioactive dye is injected into the urethral passage. After a fixed amount of time has passed, the X ray is taken. A satisfactory pyelogram demonstrates the kidney's position, size, shape, its cup-shaped cavity, and its functional ability to concentrate the opaque medium.

The one shot Intravenous Pyelogram (IVP) is performed when there is suspected or potential injury to the kidney or the ureter. If one kidney is injured, it is important to know if the other kidney is functioning normally. The one shot intravenous pyelogram is a very good way to determine this. The presence of foreign bodies, retroperitonial hematomas, and spine fractures can also be determined by the one shot IVP. In this technique 1 mL/kg of nonionic radiocontrast agent is usually injected as a rapid intravenous bolus. Five minutes after the injection of contrast agent, an anteroposterior radiograph of the abdomen is obtained. In a hypotensive patient, the one shot IVP may not be helpful, as there may be no secretion of the contrast due to decreased renal blood flow.

9.9 Hemodialysis

Dialysis is necessary when the GFR falls below 5% to 10% of normal. There are two types of renal failure: acute (days or weeks) and chronic (months or years). Chronic renal failure is a group of conditions in which renal function is slowly eroded, from which recovery is very unusual, while in acute renal failure recovery is within a few days or weeks. Thus dialysis is a temporary measure in acute renal failure but in chronic renal failure it is needed for life or until a kidney is successfully transplanted. Hemodialysis is a process in which blood is continuously circulated in contact with a permeable membrane, while a large volume of balanced electrolyte solution circulates on the other side of the membrane. Diffusion of dissolved substances from one stream to the other removes molecules that are in excess in the blood and replaces those for which there is a deficiency. Three processes occur simultaneously during dialysis—osmosis, diffusion, and ultrafiltration. Osmosis is movement of fluid across membranes from a lower concentration of solutes to a higher concentration of solutes. Diffusion is the passage of particles through a semipermeable membrane from higher concentration to a lower concentration. Ultrafiltration is the bulk flow of solution brought about by hydrostatic pressure across the semipermeable membrane.

The amount of fluid removed from the blood each hour is calculated using the formula

$$mL/h = TMP \times KUF \tag{9.14}$$

where TMP is the Transmembrane Pressure (mmHg) and KUF is the Coefficient of Ultrafiltration of the dialyser ((mL/h)/mmHg). Transmembrane pressure is the arithmetic difference between the pressure on the blood side of the dialyser (usually positive) and the dialysate pressure (usually negative). The negative pressure is calculated using the formula

$$NP = TMP - \frac{AP + VP}{2} \tag{9.15}$$

where NP is the negative pressure, AP is the arterial pressure, and VP is the venous pressure. The transmembrane pressure can be calculated using the formula:

$$TMP = \frac{\textit{Total fluid to be removed} \div \textit{hours on dialysis}}{\textit{KUF of a dialyser}} \tag{9.16}$$

To perform dialysis, an artificial kidney is used, which is commonly known as a hemodialyser, dialyser, or filter. This dialyser can be a coil type, parallel plate type, or hollow fiber type. In the hollow fiber type of dialyser, which is the most commonly used dialyser, the blood flows through the hollow fibers and the dialysate streams past the outside of the fibers. The blood and the dialysate flow counter current to maximize efficiency of solute transfer. Figure 9.7 shows a typical dialyser.

Hemodialysis membranes vary in chemical composition, transport properties, and biocompatibity. These membranes are fabricated from three classes of biomaterials: regenerated cellulose, modified cellulose, and synthetics. Chapter 4 describes some details of biomaterials. Cuprophan (cupramonium cellulose), cellulose acetate, polyacrylonitrile, polymethyl methacrylate polycarbonate, and polysulphate are some other commonly used materials to prepare dialysis membranes.

Figure 9.8 shows a schematic of a hemodialysis machine (Gregory, 1988). The blood from the patient is passed into the dialyser with the help of a roller blood pump. Blood clots rapidly when it comes in contact with most foreign substances, to avoid clotting an anticoagulation agent is added. Heparin is the most commonly used anticoagulation agent. The blood passes through the dialyser where the diffusion takes place between the dialysate and the blood. Clean blood that comes out of the dialyser is then injected into the patient. The blood is taken out from the arterial side of the patient and is injected into the venous side. A venous bubble trap is inserted in the circuit to make sure that the blood going back to the patient does not have an air bubble in it. An ultrasonic sensor or photocell is used for bubble detection. The ultrasonic sensor monitors any change in sound transmission and the photocell monitors any change in light transmission. Whenever there is air on the venous side, the signals being transmitted are interrupted and trigger an alarm. The pressure on the venous side is also monitored using a pressure gage.

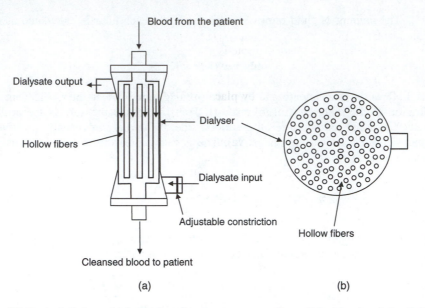

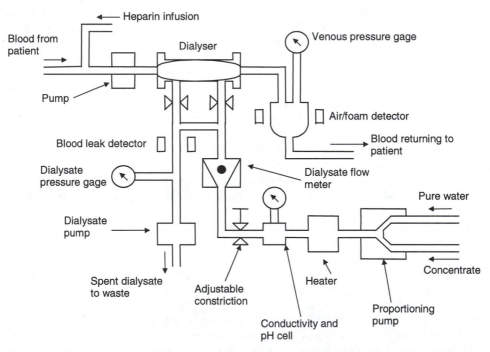

Figure 9.7 Typical dialyser (a) indicating the counter current flow of the blood and the dialysate. (b) Cross-sectional view of the dialyser.

Figure 9.8 Blood from an artery flows past a membrane in the hemodialysis machine, then returns to a vein.

Proportioning pumps first mix the dialysate concentrate with water until a desired concentration is reached. The dialysate temperature, conductivity, pH, and flow are monitored and controlled. The dialysate is warmed by a heater to near normal body temperature. The high temperatures may cause hemolysis (rupture of RBCs). The negative pressure of the dialysate is usually generated by having the dialysate pump draw the dialysate through the dialyser, and by placing an adjustable constriction upstream of the dialyser, as shown in Figure 9.7. The pH of the dialysate must match the acid/base balance of the patient's blood. The dialysate is passed through the dialyser in the opposite direction to blood flow, then passed through a blood leak detector to check if there has been a rupture of the membrane and if any blood has leaked into the dialysate circuit side.

The blood leak detector uses a beam of light that shines through the used dialysate and into a photocell. Normally the dialysate is clear and lets light pass through. Even a tiny amount of blood causes a break between the light beam and the photocell. Any break in the transmission of the light beam triggers an alarm. The spent dialysate is then drained off.

9.10 Peritoneal Dialysis

Each organ in the abdominal or peritoneal cavity is contained within a visceral peritoneum membrane. The cavity itself is lined with a parietal peritoneum membrane. The linings are thin layers of flattened mesothelial cells. The average adult peritoneal cavity has a surface area of 1 to 2 m^2. Peritoneal dialysis takes advantage of the enclosed volume. During the process the peritoneal cavity is filled with dialysate through a surgically implanted catheter. Through diffusion, urea, potassium, excess electrolytes, and other wastes cross the peritoneal membrane into the dialysate. The diffusion continues until an equilibrium concentration is reached between the plasma and the dialysate. The dialysate is periodically drained and replaced with fresh dialysate to continue the process. In addition to waste removal, the kidneys maintain the body's water level. For excess water to diffuse across the membrane the dialysate must be made hypertonic by adding a high concentration of glucose to it. Glucose can readily cross the membrane and be absorbed by the body while water filters across the membrane into the dialysate. The patient may absorb 200 to 500 kcal during peritoneal dialysis.

Peritoneal dialysis provides the patient with added independence. The patient controls the process. Unlike hemodialysis, they are not required to spend 4 to 6 h three times per week in a clinic. However, during the process the patient must be connected to a control system to weigh, warm, and refresh the dialysate. The process is typically performed three times a week and may take 8 to 10 h. To improve patient independence, manufacturers of systems aim to provide a low-cost, portable system. Many patients prefer an automated system that performs the dialysis while they sleep. Figure 9.9 shows a peritoneal dialysis system.

To provide a mobile system, the control hardware, dialysate supply, and spent dialysate are commonly mounted on a wheeled stand. The dialysate supply is normally positioned high on the stand, above the abdomen, to provide a gravity flow. The spent dialysate collection is positioned at the bottom of the stand for same reason. Most systems are equipped with peristaltic pumps to control flow. Traveling or portable systems may exclusively use gravity to provide a smaller, lighter system. The dialysate is warmed by the control electronics prior to flowing through the catheter and into the peritoneal cavity, to prevent thermal shock. After approximately 30 min, the dialysate is pumped out of the cavity and into a collection bag. The weight of the dialysate supply and spent dialysate is monitored to determine the amount of fluid and waste removed from the body.

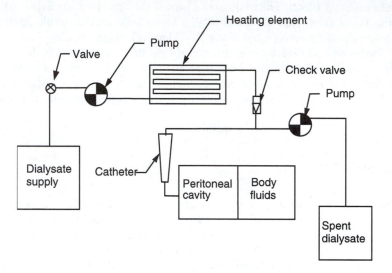

Figure 9.9 A simple schematic of a peritoneal dialysis system.

The key measurement in the peritoneal dialysis process is the weight of the fluids. It is not practical to attempt to measure the change in concentration of the dialysate. Therefore, the amount of fluid pumped into the peritoneal cavity and the amount of fluid removed are measured as the means of monitoring the amount of water and waste diffused from the body fluids. The circuit in Figure 9.10 shows a possible means of measuring the fluid weight.

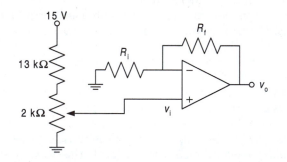

Figure 9.10 Dialysate weight measuring circuit.

The weight of the fluid determines the position of the wiper on a potentiometer. The fluid may be suspended from a spring-loaded mechanism that varies the wiper position. In this case, the value of the variable 2 kΩ resistor is dependent upon the weight of the fluid and has been installed in a voltage divider powered by the 15 V power supply to the op amp. A change in the resistance varies the input voltage, v_i, to the noninverting op amp circuit. The input voltage is amplified by $(R_f + R_i)/R_i$ (see Section 2.2). The amplification is limited by the 12 V saturation of the op amp. Acceptable resistance values for R_i and R_f are 1 kΩ and 5 kΩ, respectively. The maximum value of v_i is 2 V, which results in a 12 V output voltage. The output voltage can be sent to an analog-to-digital converter and then to a CPU for data storage and processing.

The greater the osmotic pressure difference across the membrane, the more diffusion occurs. As the pressure decreases, less diffusion occurs. The concentration of body fluid waste is exponentially removed. It is important to understand that it is this relationship that is used to determine the frequency with which the dialysate should be refreshed. If peritoneal dialysis can accurately be simplified into the model shown in Figure 9.9 (two volumes of fluid separated by a membrane), a relationship can be developed to determine the rate at which the body fluid wastes or solutes diffuse across the membrane.

The concentrations of the solutes are defined as

$$C = \frac{N}{V} \tag{9.17}$$

where C is the concentration of solute (molecules/m^3), N is the number of particles in the solute, and V is the volume of the solute.

The solute flow rate through a membrane, J_s, due to diffusion is defined as

$$J_s = \omega RT (C_b - C_d)$$

$$J_n = \omega RT \left(\frac{N_b}{V_b} - \frac{N_d}{V_d} \right) \tag{9.18}$$

where R is the gas constant (J/(mol K)), T is the absolute temperature, 310 K, and ω is the solute permeability of the peritoneal membrane (1×10^{-5} moles/(N·s)). The designations of b and d stand for the body fluid and dialysate, respectively.

Through some algebra and knowing J_s, the surface area of the peritoneum and the initial concentrations, the concentration in the body determined as a function of time is

$$C(t) = C_o \left[\frac{V_d}{V_b + V_d} e^{-\alpha t} + \frac{V_b}{V_b + V_d} \right] \tag{9.19}$$

This relationship can be plotted to determine at what point the dialysate should be refreshed. Normally, 30 min cycles are done. After 30 min, the body fluid concentration is approximately $0.9524C_o$ using the values given above. The concentration after n cycles is simply computed as

$$C(t) = (0.9524)^n C_o \qquad (9.20)$$

and may be used to estimate the reduction in body fluid solute concentration following the dialysis. As mentioned previously, the rate of dialysis is monitored by weighing the dialysate and spent dialysate.

9.11 Kidney Function

9.11.1 Extracellular Water by Impedance

Human body composition is sometimes measured in order to assess nutritional deficiencies, loss of water during dialysis, after severe burns, or any condition where the fluid balance is in question. If the kidneys are not functioning properly, the body retains water. The 24 h urine test is the most common diagnostic tool for renal failure, it measures the amount of urine a person produces in one day. Doctors may use it with other tests to measure the excretion of certain substances, such as creatinine, into urine over a 24 h period.

One other method to detect excess retention is bioelectrical impedance analysis (BIA), which estimates total body water by passing about 800 µA at 50 kHz between electrodes on the wrists and ankles. A frequent equation used is

$$\text{TBW} = a\text{Ht}^2/R + b\text{Wt} + c \qquad (9.21)$$

where TBW is the total body water in m^3, Ht is the height of the person in m, R is the resistance in Ω, and Wt is the weight in N. a, b, and c are constants determined by calibrating against isotopic dilution techniques (see Section 10.4.1) or using other equations estimating lean body mass, from which body fat can be estimated.

The BIA method is controversial. Patterson (1989) points out that wrist-to-ankle measurement is influenced mostly by the impedance of the arm and leg and less than 5% of the total impedance is contributed by the trunk, which contains half the body mass. The National Institutes of Health (1994) concludes that there are many conditions in critical illness for which conventional BIA is a poor measure of TBW.

Another method that is used to estimate extracellular water is by deduction. There are three basic components of total body mass: water, lean body mass (muscle), and body fat. Section 10.4 gives many methods of estimating body fat. Other methods provide us with equations that estimate lean body mass. If one of each of these methods are used to obtain an approximation for lean body mass and body fat, these can be subtracted from the total body mass to obtain an estimate of the extracellular water.

9.12 Bones and Joints

Bones can last for centuries and in some cases for millions of years. They provide the anthropologist with a means of tracing both the cultural and the physical development of man. Because of the importance of bone to proper functioning of the body, a number of medical specialists such as the dentist, orthopedic surgeon, and radiologist are concerned with the health of bone. Bones are also of interest to medical physicists and engineers because they deal with engineering problems concerning static and dynamic loading forces that occur during standing, walking, running, lifting and so forth. Nature has solved these problems extremely well by varying the shapes of the various bones in the skeleton and the type of bone tissue of which they are made. Bone has six important functions in the body: support, locomotion, protection of various organs, storage of chemicals, nourishment, and sound transmission (e.g. the middle ear). Bone consists of two different types of materials plus water. They are collagen, the major organic fraction, which is about 40% of the weight of solid bone and 60% of its volume and bone mineral, the so called inorganic component of the bone, which is about 60% of the weight of the bone and 40% of its volume. Either of the components may be removed from the bone, and in each case the remainder, composed of only collagen or bone mineral, looks like the original bone. The collagen remainder is quite flexible, like a chunk of rubber, and can even be bent into a loop. When it is removed from the bone, the bone mineral remainder is very fragile and can be crushed with the fingers. Collagen is produced by the osteoblastic cells, and mineral is formed on the collagen to produce bone (Cameron, 1978).

9.12.1 Bone Mineral Density

The strength of the bone depends to a large extent on the mass of bone mineral present. In diseases like osteoporosis, the bone mineral mass is considerably reduced. Up to a few years ago osteoporosis was difficult to detect until a patient appeared with a broken hip or a crushed vertebra. By that time it was too late to use preventive therapy, thus, bone mineral is very important and commonly measured to detect bone diseases such as osteoporosis. The bone mineral content of specimens can be measured by ashing the specimen in a furnace or demineralizing it in a decalcifying solution (Ashman, 1989). The most commonly used technique for noninvasively measuring bone mineral content in the bone is dichromatic or dual photon absorptiometry (DPA).

In the early part of the 20th century, X rays were used to measure the amount of bone mineral present in the bone. Some major problems exist with using an ordinary X ray: the usual X-ray beam has many different bands of energy, and the absorption of X rays by calcium varies rapidly with energy in this range of energies; the relatively large beam contains much scattered radiation when it reaches the film; the film is a poor tool for making quantitative measurements since it is nonlinear with respect to both the amount and the energy of X rays. The net result of these problems is that a large change in the bone mineral mass (30% to 50%) must occur between the taking of two X rays of the same patient, before a radiologist can be sure that there has been a change. Figure

9.10 shows that in dual photon absorptiometry, three problems with the X-ray technique are largely eliminated by using an X-ray source filtered to yield two monoenergetic X-ray beams at about 30 keV and 70 keV, a narrow beam to minimize scatter, and a scintillation detector that detects all photons and permits them to be sorted by energy and counted individually. Tests are frequently made in the spine, hip, and forearm but can be done on the entire body.

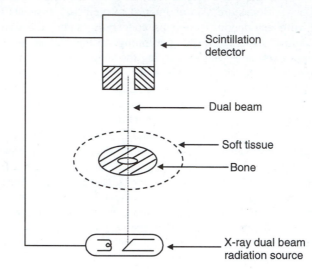

Figure 9.10 In a dual photon absorptiometer, an X-ray source is filtered to emit at two discrete energies.

Hologic (2003) improves clinical bone densitometry by integrating bone mineral density (BMD) with Instant Vertebral Assessment (IVA). Possible only with fan-beam technology, IVA generates a high-resolution image of the entire lateral spine in just 10 s, enabling physicians to visually assess vertebral status for a more accurate determination of fracture risk than just BMD alone. A different instrument measures the speed of sound (SOS, in m/s) and broadband ultrasonic attenuation (BUA, in dB/MHz) of an ultrasound beam passed through the heel, and combines these results to obtain the Quantitative Ultrasound Index (QUI). The output is also expressed as a T-score and as an estimate of the Bone Mineral Density (BMD, in g/cm^2) of the heel.

9.12.2 Stress and Strain

Tensile Loads

When forces are applied to any solid object, the object is deformed from its original dimensions. At the same time, internal forces are produced within the object. The relative deformations created at any point are referred to as strains. The internal force intensities (force/area) are referred to as stresses, at that point. When bone is subjected to forces, these stresses and strains are introduced throughout the structure and can vary in a very complex manner (Ashman, 1991).

Figure 9.11(a) shows a cylindrical bar of length L and a constant cross-sectional area A subject to a pure tensile force F. As the load is applied, the cylinder begins to stretch. This situation can be described by an equation that describes the stretching of a spring (Hooke's law)

$$F = kx \tag{9.22}$$

where F is the applied force, x is the change in length or elongation of the spring, and k is the spring constant or stiffness of the spring.

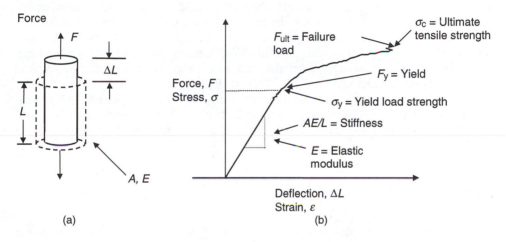

Figure 9.11 Tensile stress σ on a cylindrical bar causes tensile strain ε.

The analogous relation for stretching of the cylinder is

$$\Delta L = \frac{FL}{AE} \tag{9.23}$$

where ΔL is the elongation of the cylinder, L is the original unstretched length, A is the cross-sectional area, F is the force, and E is the elastic (Young's) modulus (18 GPa for compact bone).

Tensile, or uniaxial, strain, ε, can be calculated using the formula

$$\varepsilon = \frac{\Delta L}{L} \tag{9.24}$$

Similarly, tensile stress, σ, is calculated using the formula

$$\sigma = \frac{F}{A} \tag{9.25}$$

These tests can be performed on specimens with different lengths, cross-sectional areas, and under forces of varying magnitudes (Black, 1988).

Figure 9.11(b) shows that a cylinder of bone tested in tension yields a linear region (also known as the elastic region) where the atoms of the bone are displaced only slightly by reversible stretching of the interatomic bonds. This is followed by a nonlinear region where yielding and internal damage occurs, often involving irreversible rearrangement of the structure. After yielding, nonelastic deformation occurs until finally failure or fracture results. The load at which yielding occurs is referred to as the yield load, F_y. The load at which failure occurs is called the ultimate or failure load, F_{ult}. This curve describes the behavior of the structure since the curve differs for a different cross-sectional area or length. It also represents the mechanical behavior of the material as opposed to the behavior of the structure. In the stress–strain curve, the material yields at a stress level known as the yield strength and fractures at a stress level known as the fracture strength or the ultimate tensile strength. Toughness is the ability to absorb energy before failure and is calculated from the total area under the stress–strain curve, expressed as energy per unit volume in J/m^3 (Black, 1988).

The above quantities can be measured using any of the displacement type sensors (e.g. strain gage, LVDT (Linear Variable Differential Transformer), capacitive, and piezoelectric sensors). The most commonly used sensors are the strain gage and the LVDT.

Shear Loads

When forces are applied parallel to a surface or along the edge of an object, the object deforms in a way shown in Figure 9.12. The sides of the object perpendicular to the forces stretch and shear stresses and strains result.

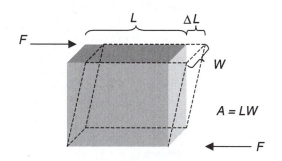

Figure 9.12 Shear stress τ, causes shear strain γ.

Shear strain, γ, can be calculated using the formula

$$\gamma = \frac{\Delta L}{L} \qquad (9.26)$$

Shear stress, τ, can be calculated using the formula

$$\tau = \frac{F}{A} \qquad (9.27)$$

where ΔL is the distance of shear deformation, L is the original length, A is the cross-sectional area, and F is the acting force.

The shear modulus, G, can be calculated using τ and γ using the relationship

$$G = \frac{\tau}{\gamma} \qquad (9.28)$$

See more information about measuring shear stress and strain in Chapter 6.

9.12.3 Strain Gage

The strain gage is a variable resistance sensor whose electric resistance is

$$R = \frac{\rho l}{A} \qquad (9.29)$$

where R = resistance in Ω, ρ = resistivity in $\Omega \cdot$ m, l = length of the wire in m, and A = cross-sectional area in m^2. An increase in length causes an increase in resistance. The sensitivity is expressed by the gage factor

$$G = \frac{\Delta R / R}{\Delta L / L} = (1 + 2\mu) + \frac{\Delta \rho / \rho}{\Delta L / L} \qquad (9.30)$$

where μ is Poisson's ratio, which can be expressed as

$$\mu = -\frac{\Delta D / D}{\Delta L / L} \qquad (9.31)$$

where D is the diameter of the cylindrical specimen. Poisson's ratio is the ratio between the lateral strain and axial strain. When a uniaxial tensile load stretches a structure, it increases the length and decreases the diameter.

Figure 9.13 shows four strain gage resistances that are connected to form a Wheatstone bridge. As long as the strain remains well below the elastic limit of the strain gage resistance, there is a wide range within which the increase in resistance is linearly proportional to the increase in length.

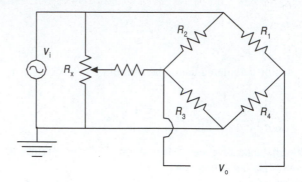

Figure 9.13 Four strain gage resistances R_1, R_2, R_3, and R_4 are connected as a Wheatstone bridge. v_i is the applied voltage with the bottom terminal grounded. v_o is the output voltage, which must remain ungrounded and feeds a differential amplifier. Potentiometer R_x balances the bridge. See Section 2.1.6 for more information on Wheatstone bridges and potentiometers.

The types of strain gages are dictated by their construction. Unbonded strain gages may be formed from fine wires that stretch when strained, but the resulting sensor is delicate. The wire's resistance changes because of changes in the diameter, length and resistivity. Bonded strain gages may be formed from wires with a plastic backing, which are glued onto a structural element, such as carefully dried bone. Integrated strain gages may be formed from impurities diffused into a silicon diaphragm, which form a rugged pressure sensor (Peura and Webster, 1998). A strain gage on a metal spring is useful for measuring force within a uniaxial tensile test machine.

9.12.4 LVDT

A transformer is a device used to transfer electric energy from one circuit to another. It usually consists of a pair of multiply wound, inductively coupled wire coils (inductors) that facilitate the transfer with a change in voltage, current or phase. A linear variable differential transformer (LVDT) is composed of a primary coil and two secondary coils connected in series opposition, as shown in Figure 9.14. The ac excitation is typically 5 V at 3 kHz. The coupling between these two coils is changed by the motion of the high permeability magnetic alloy between them. When the alloy is symmetrically placed, the two secondary voltages are equal and the output signal is zero. When the alloy moves up, a greater voltage is transformed to the top secondary coil and the output voltage is linearly proportional to the displacement. The LVDT is useful in determining the strain on tendons and ligaments (Woo and Young, 1991).

The most commonly used technique for measurement of stress and strain on bone specimens is the Uniaxial tension test using the LVDT. Figure 9.15 shows the set up for this test. It consists of one fixed and one moving head with attachments to grip the test specimen. A specimen is placed and firmly fixed in the equipment, a tensile force of known magnitude is applied through the moving head, and the corresponding elongation is measured. Then using Eqs. (9.24) and (9.25), the uniaxial stress and strain can be calculated.

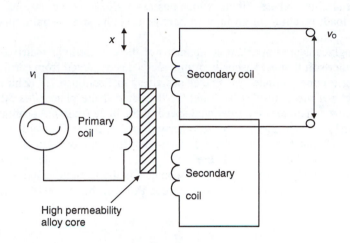

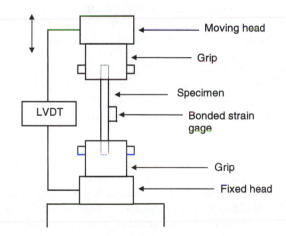

Figure 9.14 In a linear variable differential transformer, displacement of the high permeability magnetic alloy core changes the output voltage.

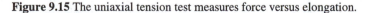

Figure 9.15 The uniaxial tension test measures force versus elongation.

9.12.5 Soft Tissue Strain

A direct method of measuring ligament strain is by mounting a strain gage load cell within cadaveric knees (Markolf et al., 1990). A noncontact method for measuring ligament strain is the video dimension analyzer (VDA) (Woo et al., 1990). Reference lines are drawn on the specimen using stain, and the test is videotaped. The VDA system tracks the reference lines, yielding strain versus time. Tendon and ligament tissues contain low-modulus elastin, which bears the majority of the load for strains up to about 0.07. At higher strains, collagen fibers, which possess a zigzag crimp, take an increasing portion of the load, resulting in an upward curvature of the stress–strain curve (Black, 1988).

Proteoglycans are important components of the extracellular matrix of articular cartilage and other soft tissues. Viscosity of proteoglycans extracted from cartilage can be measured using a cone-on-plate viscometer described in Section 6.7.1 (Zhu and Mow, 1990). The angle of the cone α is 0.04 and the diameter of the plate $2a$ is 50 mm. The shear rate $\gamma = \omega/\alpha$, where ω is the rotational speed of the plate (rad/s). Apparent viscosity $\eta_{app} = 3T/(2\omega a3\gamma)$, where T = torque. Another viscometer is the Ostwald capillary viscometer, which calculates the coefficient of viscosity η from the pressure gradient dp/dL, the volume rate of flow Q, and the tube radius R using the equation $\eta = [\pi R^4/(8Q)](dp/dL)$ (Fung, 1981). Another viscometer is the Couette coaxial viscometer, in which an outer rotating cylinder transmits torque through the test fluid to an inner coaxial cylinder.

9.12.6 Joint Friction

Diarthrodial (synovial) joints have a large motion between opposing bones. To diagnose disease and design artificial joints, we desire to measure friction between them. The resistance to motion between two bodies in contact is given by frictional force $F = \mu W$, where μ is the coefficient of friction and W is the applied load (Black, 1988). Surface friction comes either from adhesion of one surface to another due to roughness on the two surfaces, or from the viscosity of the sheared lubricant film between the two surfaces. Lubrication of bone joints is an important factor in determining coefficient of friction. Rheumatoid arthritis results in overproduction of synovial fluid in the joint and commonly causes swollen joints. The synovial fluid is the lubricating fluid that is used by the joints. The lubricating properties of the fluid depend on its viscosity; thin oil is less viscous and a better lubricant than thick oil. The viscosity of synovial fluid decreases under the large shear stresses found in the joint.

The coefficient of friction is measured in the laboratory using arthrotripsometers (pendulum devices). Here (in Figure 9.16), a normal hip joint from a fresh cadaver is mounted upside down with heavy weights pressing the head of femur into the socket. The weight on the joint is varied to study the effect of different loads. The whole unit acts like a pendulum with the joint serving as the pivot. From the rate of decrease of the amplitude with time, the coefficient of friction is calculated. It can be concluded that fat in the cartilage helps to reduce the coefficient of friction. When synovial fluid is removed, the coefficient of friction is increased considerably. Figure 9.16 shows an arrangement for measuring the coefficient of friction (Mow and Soslowsky, 1991).

We wish to measure wear in joints of artificial materials such as ultra-high-molecular-weight polyethylene (UHMWPE) (Dowson, 1990). For polymeric materials, the volume $V = kPX$ produced by wear during sliding against metallic or ceramic countersurfaces is proportional to the applied load P and the total sliding distance X. k is a wear factor indicative of the wear resistance of a material.

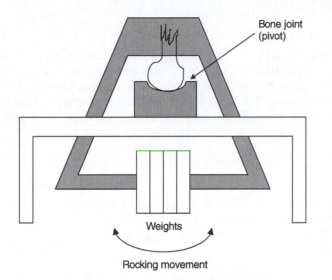

Figure 9.16 Decay of oscillation amplitude in the pendulum device permits calculation of the coefficient of friction of a joint.

9.12.7 Bone Position

Bone position is important for calculating the loading forces that act on it. The most complicated bones in the body, on which these forces are calculated, are in the spinal cord. Calculating forces for other bones, such as the femur, is relatively easy. These loading forces are static and dynamic in nature.

A goniometer is an electric potentiometer that can be attached to a joint to measure its angle of rotation. For working details of the goniometer refer to Chapter 10.

Human joint and gross body motions can be measured by simple protractor type goniometers, electrogoniometers, exoskeletal linkage devices, interrupted light or normal and high speed photography, television-computer, X-ray or cineradiographic techniques, sonic digitizers, photo-optical technique, and accelerometers (Engin, 1990).

9.12.8 Bone Strain-Related Potentials

Bending a slab of cortical or whole living or dead bone yields piezoelectric potentials of about 10 mV, which we can measure using electrodes (Black, 1988). If the strain is maintained, the potentials rapidly decay to a very low value, called the offset potential. Some workers hypothesize that these potentials have a role in directing growth, healing, and remodeling.

9.13 Skin

The primary function of the skin is to act as a buffer between the body and the outside world. It absorbs sunlight, softens blows, insulates from heat and cold, retains fluid within the body and keeps foreign organisms and extraneous chemicals out of it. It plays an important role in body's defense against infection. It also acts as the body's radiator, and as such, is the most significant element in the thermoregulation system.

9.13.1 Water Loss

Transepidermal Water Loss (TWL or TEWL) refers to the rate at which water migrates from the viable dermal tissues through the layers of the stratum corneum to the external environment. In the absence of profuse sweating, the TWL is predominantly controlled by the diffusion of water vapor in the stratum corneum caused by the difference in vapor concentration between the inside and outside surfaces.

TWL measurements have been used in studying the restoration of barrier function in wounded skin, as well as in evaluating occlusive properties of topical preparations. TWL can be measured using the direct or indirect measurement techniques. Flow hygrometry is a commonly used technique for direct measurement. The indirect measurement technique relies on establishing a boundary air layer over skin of known geometry. The two commonly used techniques of indirect measurement are the closed cup and the open cup methods (Marks, 1981).

Flow Hygrometry

In flow hygrometry, the flux of water vapor out of a fixed area of skin is determined by measuring the increase in water vapor concentration in the flowing gas stream. The increase in humidity is read at a sensor output once steady-state conditions have been reached. TWL can be calculated using the formula:

$$\text{TWL} = \frac{K \times V \times R}{A} \qquad (9.34)$$

where K is the instrument constant, V is the increase in the sensor output, R is the gas flow rate and A the skin area isolated by the measuring chamber. TWL is usually measured in $mg/(cm^2 \cdot h)$. Numerous possible sources of error in the flow hygrometry system include uncertainty in the gas flow rate, uncertainty in the actual area of skin exposed to flowing gas, absorption of water vapor in tubing connecting the skin chamber with the sensor, and leaks in the seal of the chamber to the skin site. These errors can be minimized by careful design of the system.

In this method, transfer of water vapor to the sensor is by convection, which normally requires gas tanks or pumps, valves, and tubing, in addition to the sensor and the skin chamber. In contrast, indirect measurement techniques, which are described later, require little auxiliary equipment. Figure 9.17 shows a flow hygrometer (Smallwood and Thomas, 1985).

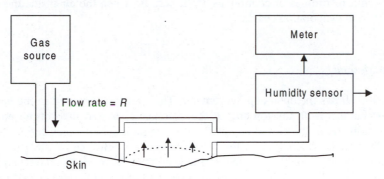

Figure 9.17 The flow hygrometer measures the increase in humidity of gas flowing over the skin.

Closed Cup Method

In the closed cup method, the humidity sensor is sealed into one end of a cylinder of known length. The cylinder is placed on the skin, trapping a volume of unstirred air between the skin and the detector as shown in Figure 9.18(a). Within a few seconds, the sensor voltage begins to rise steadily followed by a gradual decrease in rate of change. The application of diffusion principles to the water vapor in the volume of trapped air predicts this behavior and shows that the TWL is directly proportional to the slope of the transient linear portion of the detector output curve shown in Figure 9.18(b). The TWL is found from

$$\text{TWL} = K \times l \times \frac{dv}{dt} \tag{9.35}$$

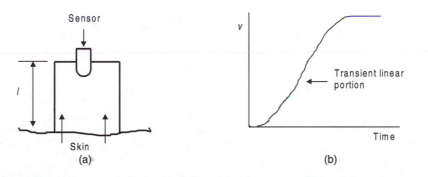

Figure 9.18 The closed cup hygrometer: (a) configuration of measuring cup, (b) typical sensor output curve.

where K is an instrument calibration constant, l is the distance between the detector and the skin surface, v is the detector voltage, and t is the time. The closed chamber method does not permit recordings of continuous TWL because when the air inside the chamber is saturated, skin evaporation ceases.

Open Cup Method

Figure 9.19(a) shows the open cup hygrometer. The sensor is mounted in the wall of the cylindrical chamber with the top end of the chamber open and the bottom end placed against the skin. As in the closed cup method, the cylindrical chamber defines a volume of undisturbed air in which the transfer of water vapor is controlled by diffusion in air. A few minutes after applying the chamber to the skin, the sensor reaches a new equilibrium in which the rate of flow of vapor out of the open end of the cylinder is equal to the rate of flow of vapor out of the skin, the TWL. In this condition the rate of flow can be calculated, according to the principles of diffusion, from the product of diffusion coefficient of water vapor in the air and the concentration gradient. For the configuration shown in Figure 9.19(a) the sensor is mounted at a distance l from the end of the cylinder where the humidity is determined by the room condition. Thus the increase in sensor response divided by the length determines the humidity gradient. The equation for TWL measurement with the open cup method is then

$$\text{TWL} = \frac{(V - V_0) \times K \times D}{l} \tag{9.36}$$

where V and V_0 are equilibrium and initial sensor voltages, respectively, K is an instrument constant, D is the diffusion coefficient of water in air, and l is the distance between the sensor and the end of the cylindrical chamber open to the room air. Air movement and humidity are the greatest drawbacks of this method. This method is currently used in commercially available devices.

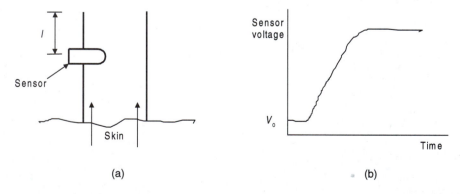

(a)   (b)

Figure 9.19 The open cup hygrometer: (a) configuration of measurement cylinder, (b) typical sensor output curve

In the flow hygrometry method, the investigator has the option of selecting the water content of the sweep gas within the limits of the humidity sensor. For indirect methods, unless measurements are made in a humidity-controlled room, the variable ambient condition can have a profound effect on the resulting measurements.

9.13.2 Color

Among the physical skin parameters, color is important in clinical dermatology. The experienced dermatologist uses color information in several ways. First, he frequently bases his first-view diagnosis or elaboration of differential diagnosis to a considerable amount on specific colors of the skin lesions themselves. Second, the color differences often allow the clinician to judge the distribution of lesions on the human body, even without close inspection and palpation. Finally, the intensity of color may provide useful information about the severity of a pathological process, and changes in color to normal tell the dermatologist if the treatment works.

The color under which we perceive the skin depends on a number of variables, including pigmentation, blood perfusion, and desquamation (shedding of skin). Because color perception is a subjective sensory and neurophysiological process, the evaluation of color is highly observer dependent. This has been a concern not only in dermatology, but also in industries such as dye production and printing in which highly consistent colors are necessary. In order to measure color objectively, color-measuring devices have been developed instead of having it judged by subjective observers.

These devices are useful in dermatology, skin pharmacology, toxicology, and cosmetology because they allow a quantitative measurement instead of a historically subjective grading of reaction.

Colors may be described by their hue (color position in the color wheel), lightness (called value), and saturation (called chroma). Any color is expressed as a combination of (H/V/C) (hue, value, chroma) (Berardesca, 1995).

Dermaspectrometer and Erythema Meter

The instruments used to measure the variations in skin color are the DermaSpectrometer and the Erythema meter. Their working principle can be explained as follows: Inflammatory skin erythema is the result of an increased presence of erythrocytes in skin structures. In inflammation, cutaneous vessels are dilated, and blood flow is increased. Because oxyhemoglobin has absorption in the spectral range from 520 to 580 nm, it absorbs green light, while red light is reflected to the absorber. Changes in skin redness affect the absorption of green light, but affect red light less. An erythema index of the skin can therefore be based on the ratio between the reflection of red and green light.

$$Erythema\ index = \log_{10} \frac{(Intensity\ of\ reflected\ red\ light)}{(Intensity\ of\ reflected\ green\ light)} \qquad (9.37)$$

Similarly, melanin pigmentation of the skin leads to an increased absorption both in the green and red parts of the spectrum. A melanin index may be defined as follows:

$$Melanin\ Index = \log_{10}\ (Intensity\ of\ reflected\ red\ light) \qquad (9.38)$$

The DermaSpectrometer emits green and red light (568 and 655 nm) from an LED source. The Erythema Meter emits green and red light (546 and 671 nm) from a tungsten lamp. Figure 9.20 shows that the light reflected from the skin is detected with a photosensor. A microprocessor calculates the erythema and melanin index, which is then displayed.

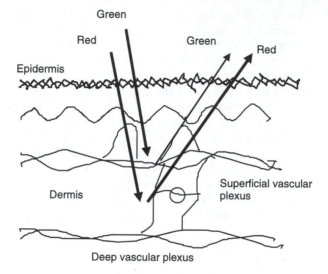

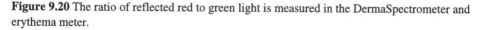

Figure 9.20 The ratio of reflected red to green light is measured in the DermaSpectrometer and erythema meter.

Chromameter and Micro Color

The Chromameter and Micro Color are another pair of instruments based on the same working principle as the DermaSpectrometer and Erythema meter. The perceived color of an object as the spectral distribution of reflected light depends on the light with which the object is illuminated. This refers both to the color of lighting and the angle under which the light hits the object. As the perceived color of the object may be expressed as the proportion of the remitted to the absorbed light, equal lighting over the whole visible spectrum is essential. Selective lighting results in limited color information; for example, a red object would appear black when perceived under a green lamp.

Lighting is achieved using a Xe flash lamp that emits an intensive white light covering the whole visible spectrum. In order to compensate for variation in illumination, part of the emitted Xe light is sent to a set of color sensors, whereas the rest illuminates the object (dual beam system, Figure 9:21(a)). The color sensors analyze the illuminating light, and through a microprocessor, the light reflected from the object is adjusted for variations in the illuminating light. Because the measuring unit is a hollow chamber, the light emitted from the flash lamp hits the surface of the object from all angles. In the Chromameter, only the light remitted at 0° to the axis of the instrument (90° to the object surface) is collected for color measurement (d/0 measurement principle, Figure 9.21(b)), whereas in the Micro Color, the light reflected at 8° to the axis (d/8 measurement) is collected. The reflected light is transferred to three photodiodes. Before each diode, a color filter ensures a specific spectral sensitivity with peaks at 450 nm (blue), 550 nm (green), and 610 nm (red). These photodiodes simulate the human eye with its three blue, green, and red sensitive cones in the central fovea. The light reaching the sensors is transformed into electric signals that are used by the microprocessor for the calculation of the color values.

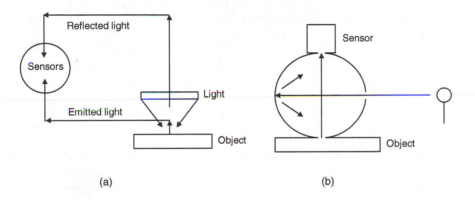

(a)   (b)

Figure 9.21 (a) The spectrum of object light is compared with the spectrum of the light source alone to allow for differences in lighting of the object (b) Illuminating light is diffuse whereas measured light is perpendicular to the object (d/0).

9.14 Problems

9.1   Sketch a spirometer output. Label and describe lung volumes that it can measure. Label and describe lung volumes that it cannot measure.

9.2   You attach a 10 kΩ rotary potentiometer with 5 V across it to a spirometer pulley so the potentiometer wiper voltage varies from 0 to 5 V. You notice that a 10 kΩ from the wiper to ground causes loading error. Draw an op amp circuit that prevents this loading error.

9.3   You wish to measure lung volume using a body plethysmograph. Sketch the equipment. Describe (1) sensor type and location, (2) system and method for calibration, (3) procedure for measurement, (4) results of measurement, (5) equation, and (6) any corrections required.

9.4   In a 3000 L body plethysmograph with closed shutter, the subject blows to achieve a lung pressure of 10000 Pa. The box pressure changes by 10 Pa. Calculate the lung volume, stating any approximations.

9.5   Give the equations and describe how to measure airway resistance.

9.6   Given resistivity of the thorax is 300 $\Omega \cdot$ cm. Assume electrode area and thoracic dimensions and estimate the impedance of the thorax for use in impedance plethysmography for measuring pulmonary volume changes.

9.7   Describe the importance of measuring pulmonary air-flow. List the different methods used for the measurement. Explain any one of them.

9.8   Sketch a graph of the results obtained when measuring diffusing capacity of the lung. Label axes and significant quantities.

9.9   The measured plasma concentration of creatinine is 1 mg/100 mL. The urine flow is 4 mL/min. The urinary concentration is 30 mg/100 mL. Calculate the glomerular filtration rate.

9.10   Describe a pyelogram. Explain its importance in diagnosing kidney disease. Explain how this test is performed clinically.

9.11   Describe safety devices required for hemodialysis.

9.12   Find the transmembrane pressure given KUF is 4 (mL/h)/mmHg. Total fluid to be removed is 2810 mL and hours on the dialysis is 4 h.

9.13   Find the negative pressure across the membrane given transmembrane pressure is 315 mmHg. The arterial pressure and venous pressure are 130 mmHg and 100 mmHg respectively.

9.14   Explain why we measure bone stress. List and describe different sensors that can be used for this measurement.

9.15   A bone specimen with a radius of 1.26 cm is tested in a uniaxial tension test. The original length of the specimen is 30 cm. After applying a force of 1000 N the length is changed to 31.5 cm. Calculate the stress and strain generated in the specimen. Calculate the elastic (Young's) modulus.

9.16   Describe arthrotripsometers, what they measure and their operating principle. Also explain how there is a exponential decay of the oscillations of the pendulum with respect to time.

9.17   Explain the importance of measuring water loss. Explain how water loss is measured using flow hygrometry.

9.18   Explain how skin color is important to a dermatologist. List different methods of skin color measurement and explain any one of them.

9.15 References

Ashman, R. B. 1989. Experimental techniques. In S. C. Cowin (ed.) *Bone Mechanics*. Boca Raton FL: CRC Press.

Berardesca, E., Elsner, P., Wilhem, K. -P. and Maibach, H. I. 1995. *Bioengineering of the Skin: Methods and Instrumentation*. Boca Raton FL: CRC Press.

Bhattacharya, A. and McGlothlin, J. D. 1996. *Occupational Ergonomics: Theory and Applications*. New York: Marcel Dekker.

Black, J. 1988. *Orthopaedic Biomaterials in Research and Practice*, New York: Churchill Livingstone.

Bronzino, J. D. (ed.) 2000. *The Biomedical Engineering Handbook*. 2nd ed. Boca Raton FL: CRC Press.

Cameron, J. R. and Skofronick, J. G. 1978. *Medical Physics*. New York: John Wiley & Sons.

Cohen, S. H. 1981. *Non-invasive Measurements of Bone Mass and Their Clinical Applications*. Boca Raton FL: CRC Press.

Dowson, D. 1990. Bio-tribology of natural and replacement synovial joints. In V. C. Mow, A. Ratcliffe and S. L.-Y. Woo (eds.) *Biomechanics of Diarthrodial Joints*. Vol. II. New York: Springer-Verlag.

Ducheyne, P. and Hastings, G. W. 1984. *Functional Behavior of Orthopedic Biomaterials Vol. 1: Fundamentals, Vol. 2: Applications*. Boca Raton FL: CRC Press.

Engin, A. E. 1990. Kinematics of human shoulder motion. In V. C. Mow, A. Ratcliffe and S. L.-Y. Woo (eds.) *Biomechanics of Diarthrodial Joints*. Vol. II. New York: Springer-Verlag.

Feinberg, B. N. 1986. *Applied Clinical Engineering*. Englewood Cliffs NJ: Prentice-Hall.

Fung, Y. C. 1993. *Biomechanics: Mechanical Properties of Living Tissues*. 2nd ed. New York: Springer.

Gregory, M. C. 1988. Kidney, artificial. In J. G. Webster (ed.) 1988. *Encyclopedia of Medical Devices and Instrumentation*. New York, John Wiley and Sons.

Hobbie, R. K. 1997. *Intermediate Physics for Medicine and Biology*. 3rd ed. New York: Springer-Verlag.

Hologic. 2003. Products. [Online] www.hologic.com/.

Kryger, M. H. 1994. Monitoring respiratory and cardiac function. In M. H. Kryger, T. Roth, and W. C. Dement (eds.) *Principles and Practices of Sleep Medicine*, 2nd ed. Philadelphia: W. B. Saunders.

Markolf, K. L., Gorek, J. F., Kabo, J. M., Shapiro, M. S. and Finerman, G. A. M. 1990. New insights into load bearing functions of the anteriaor cruiate ligament. In V. C. Mow, A. Ratcliffe and S. L.-Y. Woo (eds.) *Biomechanics of Diarthrodial Joints*. Vol. I. New York: Springer-Verlag.

Marks, R. and Payne, R. A. (eds.) 1981. *Bioengineering and the Skin*. Boston: MTP Press Ltd.

Vishay Measurements Group, Raleigh NC. 2003. Interactive guide to strain gage technology. [Online] www.vishay.com/company/brands/measurements-group/.

Mow, V. C. and Hayes, W. C. 1991. *Basic Orthopedic Materials*. New York: Raven Press.

Mow, V. C. and Soslowsky, L. J. 1991. Friction, lubrication, and wear of diathrodial joints. In V. C. Mow and W. C. Hayes (eds.) *Basic Orthopaedic Biomechanics*. New York: Raven Press.

National Institutes of Health. 1994. *Bioelectrical impedance analysis in body composition measurement, Technology Assessment Conference Statement*. December 12–14. 1–35.

Patterson, R. 1989. Body fluid determinations using multiple impedance measurements. *IEEE Eng. Med. Biol. Magazine*, **8** (1): 16–8.

Petrini, M. F. 1988. Pulmonary function testing. In J. G. Webster (ed.) *Encyclopedia of Medical Devices and Instrumentation*. New York: John Wiley and Sons.

Peura, R. A. and Webster, J. G. 1998. Basic sensors and principles. In J. G. Webster (ed.) *Medical Instrumentation: Application and Design*. 3rd ed. New York: John Wiley and Sons.

Primiano, F. P. 1998. Measurements of the respiratory system. In J. G. Webster (ed.) *Medical Instrumentation: Application and Design*. 3rd ed. New York: John Wiley and Sons.

Smallwood, R. H. and Thomas, S. E. 1985. An inexpensive portable monitor for measuring evaporative water loss. *Clin. Phys. Physiol. Meas.*, **6**: 147–54.

Webster, J. G. 1988. *Encyclopedia of Medical Devices & Instrumentation*. New York: John Wiley and Sons.

Webster, J. G. 1998. Measurement of flow and volume of blood. In J. G. Webster (ed.). *Medical Instrumentation: Application and Design*. 3rd ed. New York: John Wiley and Sons.

Woo, S. L. -Y., Weiss, J. A. and MacKenna, D. A. 1990. Biomechanics and morphology of the medical collateral and anterior cruciate ligaments. In V. C. Mow, A. Ratcliffe and S. L.-Y. Woo (eds.) *Biomechanics of Diarthrodial Joints*. Vol. I. New York: Springer-Verlag.

Woo, S. L.-Y. and Young, E. P. 1991. Structure and function of tendons and ligaments. In V. C. Mow and W. C. Hayes (eds.) *Basic Orthopaedic Biomechanics*. New York: Raven Press.

Zhu, W. B. and Mow, V. C. 1990. Viscometric properties of proteoglycan solutions at physiological concentrations. In V. C. Mow, A. Ratcliffe and S. L.-Y. Woo (eds.) *Biomechanics of Diarthrodial Joints*. Vol. I. New York: Springer-Verlag.

Chapter 10

Body Temperature, Heat, Fat, and Movement

Chao-Min Wu

This chapter deals with measurements of physical parameters from the total body, such as temperature, heat, fat, and movement. These four parameters are somewhat related because they all deal with energy. Heat is the thermal energy content of a body. It is defined as the vibratory motion of its component particles. Body heat is the result of a balance between heat production and heat loss, which occurs through conduction, convection, radiation and evaporation. Temperature is a measure of the tendency of a body to transfer heat from or to other bodies. Balancing heat production within the body against heat loss to the surroundings determines the body temperature. Body fat is energy related in that it is part of the heat storage process. We discuss how the human body regulates body temperature and how this regulation is related to heat production, heat loss, and heat storage in the form of body fat. We further discuss clinical temperature measurement techniques and common devices. Then we move our discussion to various methods of body heat and body fat measurements. We compare various techniques among these measurements as well. Finally, we cover the measurement of body movement. In each section, we explain why and how to measure these physical parameters.

10.1 Regulation of Body Temperature

Body temperature is regulated by a well-designed control system. The average normal body temperature remains almost constant, within ±0.6 °C for a healthy person. It is generally considered to be 37 °C when measured orally. When speaking of the body temperature, we usually mean the temperature in the interior—the core temperature. Deviation of the core temperature beyond the normal range is a signal of a variety of diseases and could have a great impact on a person's life if not treated appropriately. Temperature is an indicator of the health of a person. While the core temperature is maintained constant, the surface temperature, the temperature of the skin or subcutaneous tissues, varies with the temperature of the surroundings. This temperature is an important factor when discussing heat loss through sweating (evaporation) from the skin.

Regulation of body temperature uses feedback control, analogous to room temperature control (Milsum, 1966). In room temperature regulation, we preset the desired temperature on the thermostat as the reference temperature. The thermostat controls the furnace or air-conditioner (actuator) to increase or decrease the room temperature to the preset temperature. When the ambient air temperature drops, the furnace turns on to try and match the preset temperature. During this process, temperatures from sensors inside the thermostat and from the ambient air are fed back to the thermostat to speed up the response and allow fine control of room temperature.

Figure 10.1 shows a block diagram of body thermoregulation. The hypothalamus in the brain is the temperature-regulating center for the body. Functionally, it is divided into two regions called the heat loss control region and the heat maintenance region. The heat loss control is located in the anterior hypothalamus, which contains more warm-sensitive neurons than cold-sensitive ones. Temperature-sensitive neurons act like the temperature sensor in the thermostat. It is clear that the set-point temperature for a healthy person is 37 °C. All the temperature control mechanisms attempt to bring the body temperature back to this normal level. However, low skin temperature and the presence of substances released from toxic bacterial cell membranes, such as lipopolysaccharide toxins, cause the critical set-point temperature to rise. These substances are called pyrogens and cause fever during the period of illness (Guyton and Hall, 1996). This change of the set-point level acts as a unit step input to the body temperature control system, which is considered a first-order system. The time constant for our body to match this set-point change is about 3 h.

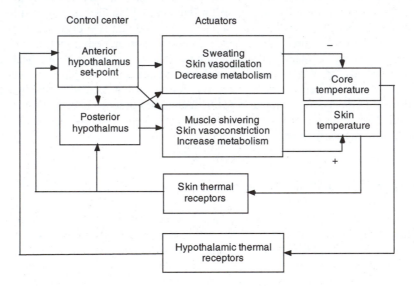

Figure 10.1 The human temperature regulation system can increase or decrease body temperature.

The heat maintenance region is located in the posterior hypothalamus where it combines sensory signals transmitted from the anterior hypothalamus and peripheral temperature sensors to control the temperature-increasing or temperature-decreasing mechanisms of the body. This part of the hypothalamus acts like a thermostat controller in room temperature regulation. When the body is too hot, the heat maintenance region triggers temperature-decreasing procedures to reduce body heat by sweating, skin vasodilation, and decreasing tissue metabolism. Sweating is the main mechanism by which the body causes heat loss via water evaporation. The effect of skin vasodilation is an increase in heat loss by increasing blood flow to the skin surface, and it is handled through the autonomic nervous system. Excessive tissue metabolism is inhibited to decrease heat production, which in turn decreases body temperature.

On the other hand, temperature-increasing procedures are instituted when the body temperature is too cold. They are: skin vasoconstriction, shivering, and increased thyroxine secretion. The reaction of skin vasoconstriction, in contrast to skin vasodilation, is caused by stimulation of the posterior hypothalamus. Shivering is a reaction of skeletal muscles to call for an increase in heat production of the body. It is excited by the cold signals from the skin receptors (Guyton and Hall, 1996). The primary motor center for shivering is located at the posterior hypothalamus (dorsomedial portion). The response time of skin vasomotor control and shivering is a matter of minutes. However, the increase of thyroxine secretion is a long-term process of heat production increase through the endocrine system. It usually takes several weeks to complete. Thyroxine is the hormone secreted by the thyroid gland to increase the rate of tissue metabolism.

In addition to the physiological control of body temperature, behavioral response to the change of environment plays an important role in body temperature control, especially in a very cold environment. It is usually an efficient way to maintain our body temperature. For example, we put on more clothes when we expect severe cold weather.

10.2 Clinical Temperature Measurement

There are two kinds of clinical temperature measurements: surface or skin temperature measurement and core temperature measurement. We measure temperature with a thermometer, which is a device that detects a change in temperature. This is done by detecting a change in the physical property of a sensor. This physical property can be a mechanical expansion of a temperature dependent liquid, such as mercury, or an electric resistance change of a thermally sensitive resistor (thermistor). The thermometer uses these materials as sensors to measure temperature. Traditionally, we measure temperature from an oral cavity, rectum, or under the arm by contact measurements. The measured temperature varies from site to site, ranging from the oral temperature, which is lower than the core temperature by about 0.5 °C to the rectal temperature that is lower than the core temperature by as much as 2 °C (Fraden, 1991). These temperature measurements need patient cooperation and are affected by the presence of food, drink, and physical activity.

The tympanic membrane that connects the outer ear (ear canal) and the middle ear (three small auditory bones) provides an ideal location for temperature measurement. Theoretically, temperature measurement taken near the tympanic membrane not only is close to the core temperature (proximate to the hypothalamus), but also is not affected by interfering factors. In practice, noncontact measurement of the tympanic membrane temperature is not superior to other measuring sites in reflecting the real core temperature. The measured temperature is still lower than the core temperature by about 0.5 to 1 °C. It all depends on how close the measurement is made to the tympanic membrane. However, it provides a convenient and comfortable site for reliable and accurate temperature measurement for a variety of patients.

10.2.1 Surface Temperature Measurement

Radiation from the skin surface is the main mechanism of heat loss from the body. It accounts for 60% of the body heat loss under normal conditions (Guyton and Hall, 1996). The relationship between the energy emitted from the skin surface and the skin surface temperature forms the basis of surface temperature measurement. The surface temperature is determined by the skin blood supply and the condition of subcutaneous tissues, and is a good indicator for bone fractures and inflammation. Abdominal placement of a thermal sensor (thermistor) for skin temperature measurement is important in neonatal monitoring against cold stress. Medical thermography is a technique to scan the skin surface and map its thermal distribution. Physicians have used this technique to diagnosis tumors and breast cancers. However, modern imaging techniques, such as MRI and mammography, are more popular diagnostic tools for these carcinomas.

Liquid Crystal Thermometer

Liquid crystal thermometers are made of chemical compounds with the property of reflecting light within a range of temperature (26 °C to 40 °C). For example, the crystals can be formulated to a change in color through red, yellow, green, and blue as the substance is heated from 38 °C through to 40 °C. They are constructed by impregnating Mylar strips with encapsulated liquid crystals with attached adhesive, and they are often used for surface temperature measurements, especially when monitoring the temperature trend in an operation room. Liquid crystal thermometers are easy to use, inexpensive, and disposable, but less precise than other thermometers (see the next section). In general, their resolution is around 0.2 °C to 0.5 °C.

10.2.2 Core Temperature Measurement

While surface temperature measurement is used by physicians as a diagnostic tool for specific diseases, core temperature measurement is a rather routine method used in hospitals, clinics, or even at home. It can detect fever caused by pyrogens that are released from virus, bacteria, and degeneration of old tissues. High and lasting fever above 41 °C can cause damage to the brain and other internal organs that are vital to human life. Therefore, core temperature measurement should be accurate and fast. Temperature measurement varies from site to site. While Bair and Davies (1995) reported that temperature measurement at the pulmonary artery reflects the best core temperature in critically ill patient monitoring, rectal or tympanic temperature measurement is also quite reliable and has less danger, less cost, and more accessibility. This section discusses basic principles, advantages, and disadvantages of different types of thermometers for core temperature measurements.

Mercury Thermometer

The mercury thermometer is a mechanical thermometer, based on the principle of volume expansion. The temperature-dependent mercury expands when the temperature increases, and the result of the change is read on a calibrated scale. The mercury thermometer is the most commonly used device to measure body temperature at home because of its low cost and ease of use. Its accuracy varies widely depending on how well the expansion of mercury is calibrated. The cool mercury pulls down the temperature of the tongue, after which it takes 3 to 5 min for new perfused blood to warm up the mercury. Measurement time is affected by how well the probe and tissue surface are coupled. Infection and cross–contamination are its main disadvantages, although constant decontamination of the thermometer can reduce the risk of infection. Other disadvantages are a potential damage to mucosal tissues when measured orally or rectally, and possible mercury intoxication due to the broken glass. Caution should be taken in handling and storing of the mercury thermometer. Accuracy of the mercury thermometer is usually manufacturer-dependent.

Electronic Thermometer

Thermometers that use temperature-responsive sensors are called electronic thermometers. Because their sensor mass is smaller and they use electronic circuits, their response time is faster than that of mercury thermometers. Sensors may vary their electric resistance or voltage. Thermistors are thermally sensitive resistors with either negative (NTC) or positive (PTC) temperature coefficients. The electric resistance of a NTC type thermistor decreases as the temperature increases. Figure 10.2 shows examples of resistance–temperature curves for NTC type thermistors. They are composed of dissimilar materials with temperature-dependent electrical resistance, such as oxides of nickel, manganese, or cobalt metal, to enhance their temperature sensitivity. Dissimilar materials are mixed homogeneously and produced in the form of beads, rods, disks, and other shapes.

The resistance–temperature curve shown in Figure 10.2 is based on Eq. (10.1). The resistance of the thermistor at temperature T, in degrees kelvin (K), is given as

$$R_T = R_0 \exp\left[\beta\left(\frac{1}{T} - \frac{1}{T_0} \right)\right]$$

$$(10.1)$$

where β is the characteristic temperature of the material, R_0 is the resistance at $T_0 = 298$ K. The β parameter is temperature dependent, however, it can be considered as temperature independent for medical applications because the range is only 37 $\pm$5 °C.

The temperature coefficient, or sensitivity, α, can be defined by differentiating Eq. (10.1), which yields

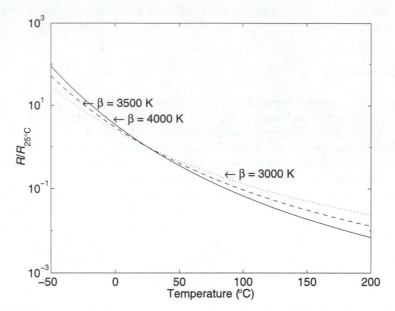

Figure 10.2 Examples of resistance–temperature curve for three NTC thermistors, β = 3000 K, 3500 K, and 4000 K.

$$\alpha = R_T \left(\frac{dR_T}{dT} \right) = -\frac{\beta}{T^2} \, . \tag{10.2}$$

where α is temperature dependent and exhibits the nonlinearity of thermistors. At 27 °C (300 K), given $\beta = 4000$ K, α is -4.4%/K.

In contrast to the homogeneous composite of dissimilar materials for the thermistor, a thermocouple is composed of dissimilar materials that are constructed as fused junctions of two materials (e.g., copper and constantan wires). Bare wires have a rapid response time, while wires contacting a metallic sheath have a slower response time. Wires enclosed in an insulator have the slowest response time. When the thermocouple is heated, a current flows from the measuring side (hot junction) to the reference side (cold junction). It develops an electric potential, which can be measured when we place a voltmeter between open ends, as shown in Figure 10.3. In 1823, Seebeck first reported this phenomenon, called the Seebeck effect. Figure 10.3 shows the measuring principle of a J type thermocouple. An approximate formula for the Seebeck voltage that is related to junction temperatures T_1 and T_2 is

$$V = C_1 (T_1 - T_2) + C_2 \left(T_1^2 - T_2^2 \right) \tag{10.3}$$

where C_1 and C_2 are constants that depend on the thermocouple pair with T in kelvins. If the temperature difference $T_1 - T_2$ is small, the second term of Eq. (10.3) can be dropped.

The sensitivity or thermoelectric power of the thermocouple is the derivative of Eq. (10.3) with respect to T_1 is

$$\alpha = \frac{dV}{dT_1} \qquad (10.4)$$

For example, the sensitivity for the copper–constantan thermocouple is 45 μV/°C at 20 °C (Cobbold, 1974). To prevent changes in $T_1 - T_2$ due to ambient temperature change, an electronic cold junction is used.

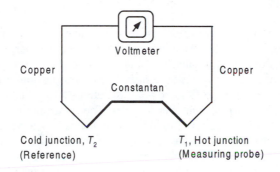

Figure 10.3 The J type thermocouple is formed from copper and constantan.

Electronic thermometers use thermistors or thermocouples as sensors to measure temperature in the mouth, rectum, or axilla (armpit). Thermistor type sensors are suitable for continuous temperature monitoring in patients during surgery and for infants' skin temperature monitoring on the abdomen. Microprocessor-based electronic thermometers use either the steady-state or predictive method to improve the response time for temperature measurement to less than 1 min. A thermometer with steady-state design monitors the temperature reading until it reaches a thermal steady state between the probe and measuring site. Once it is in the steady state, the display shows the temperature of the sensor. In contrast, a thermometer of the predictive type measures the temperature based on a manufacturer's algorithm that is developed from several experiments of the rate of rise in temperature to estimate the final temperature (Bair and Davies, 1995). Figure 10.4 shows a simplified block diagram for an electronic thermometer.

The major advantages of electronic thermometers are the small size of sensors, ease of use, and fast response time. Their disadvantages are similar to those of mercury thermometers, such as mucosal damage and cross–infection if not carefully handled. A disposable probe cover is used to prevent cross-infection. Sliding the probe from the floor of the mouth to the sublingual pocket warms up the probe and decreases oral temperature measurement error and response time.

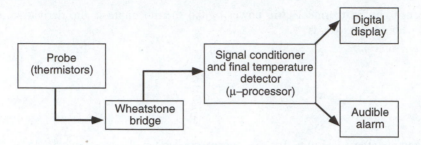

Figure 10.4 An electronic thermometer uses a thermistor sensor probe.

Infrared Thermometer

A basic infrared thermometer (IRT) design includes four parts: a waveguide (tube) to collect the energy emitted by the target, a sensor to convert the energy to an electric signal, an emissivity adjustment to match IRT calibration to the emitting characteristics of the object being measured, and a sensor temperature compensation circuit to ensure that temperature variations within the IRT are not transferred to the final output. Figure 10.5 shows a block diagram of an infrared thermometer. The tympanic membrane emits infrared energy. The collection of infrared radiation is enhanced by a waveguide, which improves reflection of the infrared radiation onto the sensor when the shutter opens. The pyroelectric sensor converts infrared radiation into electric signals, which are fed through the amplifier, multiplexer (MUX), and analog-to-digital converter (ADC) to the microprocessor for processing and display. The microprocessor handles emissivity adjustment and temperature compensation, and calculates the patient (target) temperature according to Eq. (10.5)

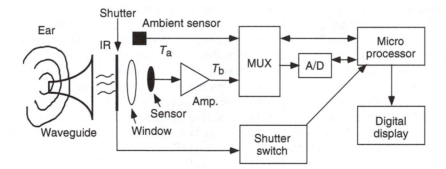

Figure 10.5 The infrared thermometer opens a shutter to expose the sensor to radiation from the ear.

$$T_b = \frac{\left[\chi(N_T - N_{T0}) + T_0^4\right]}{4} \tag{10.5}$$

where T_b is the patient temperature, χ a constant ($= 1/A\sigma\varepsilon_a$), T_0 is the reference temperature ($= 310$ K), N_{T0} is the net flux of the reference temperature, and N_T is the net flux of infrared radiation, which is shown as follows.

$$N_T = A\sigma\varepsilon_a \left(T_b^4 - T_a^4\right), \tag{10.6}$$

where A is the effective body (target) area, σ the Stefan–Boltzmann constant ($= 5.67 \times 10^{-8}$ W/m$^2\cdot$ K^4), ε_a the emissivity of surroundings (sensor), T_b the body temperature, and T_a the sensor temperature (T_a and T_b are both in kelvins).

The ambient sensor that monitors the sensor temperature (T_a), is a thermistor. The infrared sensor is a pyroelectric sensor followed by a current-to-voltage converter that is used for the pyroelectric sensor (Fraden, 1991). The pyroelectric sensor, like the piezoelectric sensor does not respond to steady inputs and in response to a step input it yields a high-pass first order output of the form $e^{-t/\tau}$. The whole measurement cycle, from shutter opening to display, is 1 s, and its accuracy and resolution are about ±0.1 °C. The main advantages of the infrared thermometer are that it only contacts the ear canal and not the eardrum (i.e. less infection), ease of use, and fast response. It is suitable for the temperature measurement of young or high-risk patients. However, inaccurate measurements occur when patients have curved ear canals or the cerumen (ear wax) obstructs the tympanic membrane.

10.3 Measurement of Body Heat: Calorimetry

Calorimetry is the measurement of heat given off by a subject. The term *direct calorimetry* is used for the direct measurement of heat loss from the subject by radiation, convection, conduction, and evaporation. Indirect calorimetry measures respiratory gas exchange to infer the amount of heat production. Heat production is calculated from oxygen (O_2) consumption and carbon dioxide (CO_2) production and converted to energy expenditure based on an empirical equation. Calorimetry provides the basis for the science of nutrition, and it is used to estimate nutritional requirements of humans and to evaluate different foods. It is also a powerful research tool for energy balance and thermal physiology studies. Finally, it is used clinically as a diagnostic tool for the investigation of metabolic disorders, energy requirements of major illness, and sports medicine. In this section, we will deal with the principles and applications of direct and indirect calorimetry.

10.3.1 Direct Calorimetry

Direct calorimetry has been used on animals and in humans for energy and thermal regulation studies since the beginning of the century. Many direct calorimeters have been built and they are usually classified under one of four principles of operation: gradient layer calorimeter, air–flow calorimeter, water flow calorimeter, and compensating heater calorimeter. The gradient layer calorimeter is usually used for studies in temperature regulation of human subjects while they are performing different work. The air–flow calorimeter is popular in studies of energy expenditure for humans and animals. The water flow calorimeter can be used for both applications without social isolation. The compensating heater calorimeter is commonly used in experiments for small animals. Webb (1985) provides detailed information on the historical development of these direct calorimeters and a list of human calorimeters.

Gradient Layer Calorimeter

Gradient layer calorimeters are isothermal calorimeters. They are chambers lined on all surfaces with a layer of insulating material and surrounded by a heat sink metal wall or a constant temperature water jacket. As sensible (i.e. nonevaporative) heat from the subject passes through the thin layer of insulating material to the heat sink (metal wall or water jacket), the temperature gradient across the layer of insulation is sensed by a network of thermocouples (e.g. copper–constantan). Benzinger and Kitzinger (1949) first introduced this type of calorimeter. These thermocouples are used to detect the temperatures of the inner and outer surfaces. The evaporative heat loss from the subject can be estimated from the increased moisture in the ventilating air that exits the chamber. The main objective of the gradient layer calorimeter design is to prevent any sensible heat loss from convecting out of the chamber by heating up the ventilating air. The design and construction of a gradient layer calorimeter is a complicated engineering task. It includes the control of the ventilation and cooling system, and instrumentation for the temperature measurement for a huge chamber. The performance of the calorimeter depends on initial calibration and routine calibration with known electric energy, and is controlled by a computer. McLean and Tobin (1987) provide a detailed description of how to design and construct a gradient layer calorimeter. They reported that the accuracy of total heat loss measurement is around ±1% or ±3 W within the operation range (10 to 40 °C). The response time of the measurement to a unit change of heat output (i.e. when a person is first placed inside) is 80% of the change is recorded within 3 min and 90% within 10 min, but the full response takes nearly an hour. Their calibration equations are listed as follows:

$$V_C = \frac{\left(\dot{Q}_N - \dot{Q}_C - \dot{Q}_L\right)}{k_c} \tag{10.7}$$

$$V_T = \frac{\dot{Q}_C}{\dot{V}k_t} \tag{10.8}$$

$$V_E = \frac{\left(\dot{Q}_E \frac{\dot{v}}{\dot{V}} + \dot{Q}_C \frac{\dot{v}}{\dot{V}} + \Delta p\right)}{k_e} \tag{10.9}$$

$$\dot{Q}_N = (V_C + V_T)k + \dot{Q}_L \tag{10.10}$$

$$\dot{Q}_E = (V_E - V_T)k - \Delta p \frac{\dot{V}}{\dot{v}} \tag{10.11}$$

where V_C, V_T, and V_E are measured voltages from the chamber, thermocouples, and evaporative heat, respectively; $\dot{Q}_N$ and $\dot{Q}_E$ are the rates of nonevaporative and evaporative heat production in the chamber; $\dot{Q}_C$ is the rate of nonevaporative heat convection out of the chamber; $\dot{Q}_L$ is the rate of heat leakage through any gaps in the heat-sensitive layer; $\dot{V}$ and $\dot{v}$ are chamber and evaporative heat measuring air flow rates; k_c, k_e, and k_t are constants (sensitivities); Δp is a term due to unequal sensitivities or unequal sampling rates; during the calibration, as performed with the chamber controlled at 25 °C and the evaporative heat measuring unit at 10 °C, the desired value $k = k_c = k_e \dot{V}/\dot{v} = k_t \dot{V} = 46.5$ W/mV (40 kcal/h·mV) is obtained by adjusting the output circuit of the calorimeter when the ventilation system is in operation, with no load in the chamber ($\dot{Q}_N = \dot{Q}_E = 0$).

The advantages of the gradient layer calorimeter are accuracy and fast response; however, they are expensive and not transportable. Figure 10.6 shows that they are complicated in design and construction.

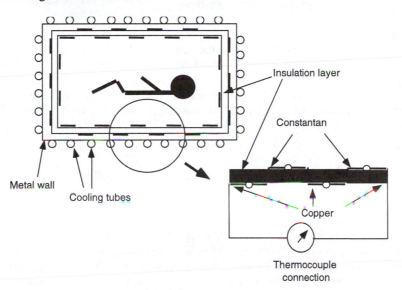

Figure 10.6 In a gradient layer calorimeter, thermocouples measure the difference in temperature across the wall. Ventilating system and measurements not shown.

Air-Flow Calorimeter

An air-flow calorimeter is a convection calorimeter where heat loss through the surface is prevented by insulation. Heat loss from the subject is carried away from the chamber by air-flow. The rate of heat removal by the air-flow is estimated from the mass flow rate and temperature rise of the air. An important objective for the design of an air-flow calorimeter is to prevent any water condensation on the surface of the chamber. Figure 10.7 shows the principle of an air-flow calorimeter. The seated subject is confined in a chamber with polyurethane insulation applied to the outsides. The heat loss from the subject is transferred to the ventilating air, whose temperature is measured at the inlet and the outlet of the chamber. The heat loss ($\dot{Q}_a$) from the subject is the product of air mass flow rate ($\dot{m}$), specific heat (c_a), and temperature change ($T_2 - T_1$) of the ventilating air:

$$\dot{Q}_a = \dot{m}c_a\left(T_2 - T_1\right) \tag{10.12}$$

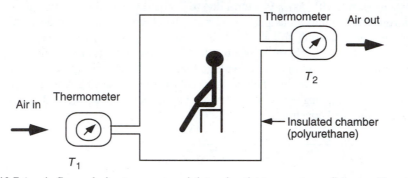

Figure 10.7 An air-flow calorimeter measures inlet and outlet temperatures, flows, and humidity.

The advantages of the air-flow calorimeter are simplicity of construction and fast response to heat change (7 to 20 min). However, they need very accurate humidity and temperature measurements.

Water-Flow Calorimeter

A water-flow calorimeter is a heat-sink calorimeter. Similar to a convection calorimeter, the water-flow calorimeter uses, instead of air, water as the heat exchanger. The removal rate of heat is estimated from the flow rate and temperature rise of the coolant (water). Figure 10.8 shows a suit calorimeter, which is a typical water flow calorimeter. Webb (1985) gives a detailed description of his suit calorimeter, which consists of a water-cooled undergarment, insulating outer garments, a controlled heater and cooling machinery for the water loop, and the related devices for control and data handling. The water-cooled undergarment includes a garment made of washable synthetic fibers, an elastic mesh material, and an array of polyvinyl chloride tubes in contact with the skin. Each 20% of the total of small tubes (a total length of 80 m with an inner diameter of 1.6 mm) covers the head and neck, torso, each leg, and both arms, respectively. The circulating water is pumped, at 1.5 kg/min into the inlet tube of the garment at the ankles, wrists, and neck and collected from the outlet tubes at the waist and neck. The subject's heat loss to the water ($\dot{Q}_w$) is estimated from the mass flow rate ($\dot{m}$), specific heat (c_w) and temperature difference between water outlet (T_2) and inlet (T_1). The formula is similar to Eq. (10.12). The advantages of the water flow calorimeter are its fast response (30 s) and it is fully transportable. In addition, the subject is not socially isolated and confined in the chamber or room. One main disadvantage reported by Webb (1985) is that the heat loss from the subject occurs by conduction, from skin to the cooled tubing. This method of heat loss is physiologically unusual.

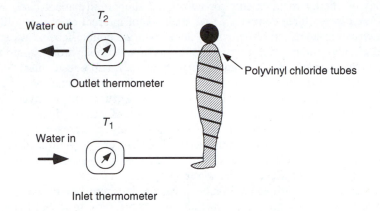

Figure 10.8 The water-flow calorimeter measures the inlet and outlet water temperature.

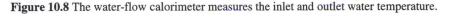

Compensating Heater Calorimeter

A compensating heater calorimeter supplies electric heater power to maintain a constant temperature inside an uninsulated chamber. The heater is needed to protect against heat loss to the environment. Figure 10.9 shows that when the subject's heat is added to the chamber, the heater needs less power to maintain the constant temperature of the chamber. The decreased power then can be used to calculate the subject's heat loss. The advantages of this type of calorimeter lie in its speed of response (on the order of 1 min) and lightweight construction. It is suitable for small caged animals and no longer used for human subjects.

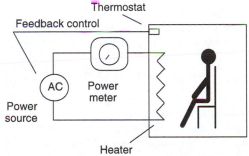

Figure 10.9 The compensating heater calorimeter requires less heater power when the subject supplies heat.

10.3.2 Indirect Calorimetry

Indirect calorimetry estimates heat production by quantitative measurements of the oxidative processes of substrates, especially the measurement of the rates of oxygen consumption ($\dot{V}O_2$) and carbon dioxide production ($\dot{V}CO_2$). Indirect calorimetry constitutes a noninvasive and nonintrusive technique that can be applied in basic research, clinical and field studies of energy requirements for nutritional support in patient care, and sports medicine. We discuss three commonly used methods of indirect calorimetry in this section: open-circuit systems, closed-circuit systems, and the double-labeled water method. The closed-circuit system and the open-circuit system are further grouped into the respiratory gas exchange method. Most ambulatory, laboratory, and bedside uses for the estimation of energy expenditure are based on measurements of respiratory gas exchange. Respiratory exchange methods depend mainly on the measurement of oxygen consumption ($\dot{V}O_2$), either alone or combined with measurement of carbon dioxide ($\dot{V}CO_2$) and methane production ($\dot{V}CH_4$), and sometimes urinary nitrogen excretion ($\dot{N}$). The apparatus that collects gas includes a mouthpiece with a nose clip, a ventilated hood or bed canopy, or a whole body chamber. The whole body chamber system sometimes is considered as a new category. Because the whole body chamber is also a closed-circuit system but conducted in a confined environment, we consider it as one of the closed-circuit systems. The differences among these methods are in response time, accuracy, duration of measurement, and the degree of confinement. In general, the whole body chamber system is used for physiological research over long periods of time (24 h or longer). The double-labeled water method allows a free living environment but requires expensive isotopes for the measurement. A detailed review of techniques for the measurement of human energy expenditure and commercially available product information are given by Murgatroyd et al. (1993) and Branson et al. (1995).

Open-Circuit System

The open-circuit system determines $\dot{V}O_2$ from the minute ventilation rate ($\dot{V}_E$), and the difference (ΔF) between inspired (FI) and expired (FE) gas concentrations. This calculation of $\dot{V}O_2$ includes a correction for change in barometric pressure, temperature, and relative humidity of the inspired air for adjustment of the volume to standard temperature, pressure, and dry air condition (STPD: 0 °C, 760 mmHg, dry). Energy expenditure (EE, kcal/day) is estimated by the standard (Weir, 1949) equation:

$$EE = \left[\left(\dot{V}O_2\right)(3.941)+\left(\dot{V}CO_2\right)(1.11)\right]\times 1440 \qquad (10.13)$$

where $\dot{V}O_2$ and $\dot{V}CO_2$ are expressed in L/min, and 1440 = the number of minutes in a day. It assumes that 12.3% of the total calories arise from protein metabolism. When urinary nitrogen excretion ($\dot{N}$) is measured, a third term ($-\dot{N}$ (2.17)) in g/min is added to the bracket of Eq. (10.13) to account for protein correction of energy expenditure.

Open-circuit calorimeters use either the mixing chamber or the dilution technique to measure inspired and expired concentrations of oxygen (FIO_2, FEO_2) and carbon dioxide ($FICO_2$, $FECO_2$), and inspired and expired minute ventilation ($\dot{V}_I$, $\dot{V}_E$). In an ideal situation, $\dot{V}O_2$ is determined by

$$\dot{V}O_2 = \dot{V}_I (FIO_2) - \dot{V}_E (FEO_2) \tag{10.14}$$

and $\dot{V}CO_2$ by a similar equation:

$$\dot{V}CO_2 = \dot{V}_E (FECO_2) - \dot{V}_I (FICO_2) \tag{10.15}$$

However, it is technically difficult to measure the difference between $\dot{V}_I$ and $\dot{V}_E$. The relationship between $\dot{V}_I$ and $\dot{V}_E$ is commonly calculated by equating the quantity of nitrogen in inlet and outlet air stream, i.e. the Haldane transformation. Then, $\dot{V}_I$ is obtained by the equation

$$\dot{V}_I = \frac{FEN_2}{FIN_2} \dot{V}_E \tag{10.16}$$

where FEN_2 = expired nitrogen concentration and FIN_2 = inspired nitrogen concentration. If only N_2, O_2, and CO_2 are analyzed, $\dot{V}_I$ can be obtained by

$$\dot{V}_I = \frac{1 - FEO_2 - FECO_2}{1 - FIO_2 - FICO_2} \dot{V}_E \tag{10.17}$$

where $FEN_2 = 1 - FEO_2 - FECO_2$ and $FIN_2 = 1 - FIO_2 - FICO_2$. Substituting Eq. (10.17) into Eq. (10.14), the formula for $\dot{V}O_2$ becomes

$$\dot{V}O_2 = \left[\frac{(FIO_2 - FEO_2 - FIO_2 FECO_2 - FICO_2 FEO_2)}{1 - FIO_2 - FICO_2} \right] \times \dot{V}_E$$

or

$$\dot{V}O_2 = \left[\frac{(1 - FEO_2 - FECO_2) \times FIO_2}{1 - FIO_2} - FEO_2 \right] \times \dot{V}_E \tag{10.18}$$

when $FIN_2 = 1 - FIO_2$, because $FICO_2 = 0.03\%$ in most situations. This equation is used by most commercially available systems. In general, $\dot{V}_I$ and $\dot{V}_E$ are of similar magnitude but not equal; the same is true for FIO_2 and FEO_2. The problem with using the Haldane transformation is that it becomes increasingly inaccurate as the FIO_2 increases above 0.6 (see next). Consequently, extreme precision in estimates of these four variables would be necessary if $\dot{V}O_2$ were to be calculated directly from Eq. (10.14). If we assume $\dot{V}_I$ is equal to $\dot{V}_E$, there will be significant errors in $\dot{V}O_2$ measurement, for example in a patient whose respiratory rate is 10 breaths per minute at a 1000 mL tidal volume ($\dot{V}_T$) and whose $\dot{V}O_2$ is 250 mL/min, the respiratory quotient (RQ, given by the ratio $\dot{V}CO_2/\dot{V}O_2$) of this patient is 0.7. This means that 25 mL per breath of O_2 produces 17.5 mL per breath of CO_2. In this case, the difference between $\dot{V}_I$ and $\dot{V}_E$ is 75 mL. If $\dot{V}O_2$ is measured at an O_2 concentration of 60% ($FIO_2 = 0.6$) assuming that $\dot{V}_I$ is equal to $\dot{V}_E$, an error of 18% would be obtained for $\dot{V}O_2$ measurement (205 mL/min). When RQ reaches 1.0, this error becomes small. Measurement of $\dot{V}CO_2$ is much simpler. When inspired air is room air, it contains only very small amount of CO_2 ($< 0.03\%$). $\dot{V}CO_2$ may be calculated accurately from a simpler form

$$\dot{V}CO_2 = \dot{V}_E \left(FECO_2\right) \qquad (10.19)$$

This simpler form is also applied to measurement of $\dot{V}CO_2$ for patients with mechanical ventilation where $FICO_2$ is 0.0.

Most commercially available open-circuit systems use mixing chambers as in Figure 10.10. Expired air from the patient passes through a facemask or mouthpiece and is directed to the mixing chamber. At the end of the mixing chamber, a small sample is drawn by a vacuum pump through O_2 and CO_2 sensors that measure FEO_2 and $FECO_2$. A pressure transducer and thermistor are necessary for pressure- and temperature-compensated gas measurements. A volume transducer is used to measure minute ventilation ($\dot{V}_E$). A microprocessor controls the instrument and provides minute-by-minute data on O_2 consumption, CO_2 production, RQ, and energy expenditure of the subject.

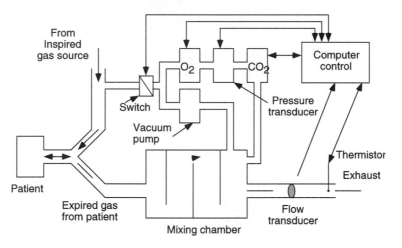

Figure 10.10 A microcomputer-based open-circuit system includes a mixing chamber, O_2 and CO_2 analyzers, and the various variables (pressure, flow, temperature) used to calculate $\dot{V}O_2$.

When measuring respiratory exchange gas, there are two broadly used methods: the air may be directly expired through a mask or mouthpiece, as mentioned above, or it may be drawn from a hood or canopy enclosing the subject's head. The latter is sometimes called a flow-by system. It needs a pump to draw constant air over a subject's head, and this air–flow is mixed with the patient's expired air. The process of $\dot{V}O_2$ and $\dot{V}CO_2$ measurements is the same as before. However, the patient's expired air is diluted by the constant air flow, the difference in O_2 and CO_2 concentrations are smaller than those measured by the former method. Consequently this demands that the analyzers be more precise and free of drift. The Deltatrac™ metabolic monitor (Datex–Engstrom Division, Instrumentarium Corp., PO Box 900, FIN-00031, Helsinki, Finland) and Vmax29n™ (SensorMedics Corp., Yorba Linda, CA) represent this type of system. It is suitable for use with a mechanical ventilator or a canopy for patients of all ages. The Deltatrac consists of an infrared CO_2 analyzer, a paramagnetic O_2 analyzer, a 4 L mixing chamber, and a CRT display (Weissman et al., 1990). When the Deltatrac is used with a mechanical ventilator (Figure 10.11), the expired gases enter the mixing chamber, from which gas is sampled and analyzed to determine FEO_2 and $FECO_2$. After the expired gases pass through the mixing chamber, the gases are diluted with room air driven by a flow generator (Q, 40 L/min). $\dot{V}CO_2$ is then calculated as

$$\dot{V}CO_2 = Q\left(F^*CO_2\right) \tag{10.20}$$

where F^*CO_2 is the concentration of CO_2 in the diluted gas sample. The respiratory quotient (RQ) is then calculated using the Haldane transformation.

$$RQ = \frac{1 - FIO_2}{\dfrac{FIO_2 - FEO_2}{FECO_2} - FIO_2} \tag{10.21}$$

where FIO_2 is the O_2 concentration from the inspired limb of the ventilator. $\dot{V}O_2$ is then determined by the formula

$$\dot{V}O_2 = \frac{\dot{V}CO_2}{RQ}. \tag{10.22}$$

A microprocessor controls the sampling of gases. CO_2 measurements are made by alternately sampling the diluted CO_2 (F^*CO_2) and the true CO_2 concentration ($FECO_2$). O_2 measurements occur by alternately sampling the difference between inspired and expired O_2 ($FIO_2 - FEO_2$) and the measurement of FIO_2. Weissman and colleagues (1990) reported that the accuracy of the Deltatrac for $\dot{V}O_2$ and $\dot{V}CO_2$ measurements is within ±7%. They also showed a different calculation procedure for canopy measurement of $\dot{V}O_2$ and $\dot{V}CO_2$ in spontaneously breathing patients.

In addition to the mixing chamber and the dilution techniques mentioned above, some mixing chamber systems use a breath-by-breath technique for $\dot{V}O_2$ and $\dot{V}CO_2$ measurements. The breath-by-breath technique was originally designed for exercise testing and has been adapted for use with a mechanical ventilator. It makes measurements for every breath of the patient's expiration and differs from the mixing chamber system in that the mixing chamber is absent. Gas samples are collected on a breath-by-breath basis and averaged over a preset time to obtain mixed expired values (Branson et al., 1995). Furthermore, bedside mixing chamber systems can be modified and used for ambulatory and field studies by dispensing with the CO_2 analyzer to simplify the procedure. However, this results in lower accuracy.

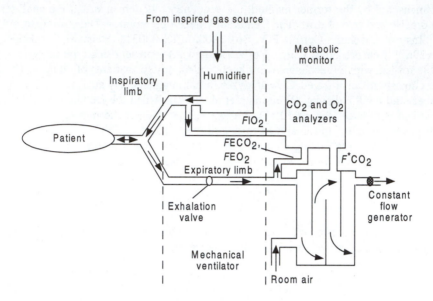

Figure 10.11 A dilution system used by Deltatrac uses mechanical ventilation. FIO_2 is the inspired oxygen concentration, $FECO_2$ is the true expired carbon dioxide concentration, FEO_2 is the expired oxygen concentration, and F^*CO_2 is the diluted carbon dioxide concentration.

Closed-Circuit System

Closed-circuit systems calculate $\dot{V}O_2$ by measuring over time the volumetric change from a source of oxygen. They use the volumetric loss and replenishing techniques to measure $\dot{V}O_2$. The former observes how long it takes to consume all the oxygen that is within a spirometer. The latter measures the amount of oxygen needed to keep the spirometer full. A generic volumetric loss closed-circuit system consists of a CO_2 absorber, a breathing circuit, a spirometer (see Figure 9.2) full of 100% O_2, and a CO_2 analyzer (Figure 10.12). Closed-circuit systems allow $\dot{V}O_2$ and $\dot{V}CO_2$ measurements during mechanical ventilation and during spontaneously breathing ventilation with a mouthpiece plus nose-clip or facemask. During measurements, the patient inspires O_2 from the spirometer, and the expired gases enter the mixing chamber. Samples of the expired gas are analyzed to obtain $FECO_2$ before passing through the soda lime scrubber. Movements of the spirometer are recorded as V_T and $\dot{V}_E$. The volume change in the spirometer as a function of time is used to calculate the rate of O_2 consumption. In a similar way as previously described, the rate of CO_2 production is determined from measurements of $\dot{V}_E$ and $FECO_2$. During mechanical ventilation, the bellows in Figure 10.12 is driven by the ventilator to deliver gas to the patient.

The whole body chamber is an early design of a volume-replenishing closed-circuit system. In the whole body chamber the subject is kept in a sealed chamber that is ventilated with a constant, measured supply of fresh air. The respiratory gas exchanges produced by the subject are measured in terms of their effects on the composition of chamber air. Samples of well-mixed air are drawn from the chamber for continuous analysis. The differences in oxygen and carbon dioxide concentrations in the in-going and out-going air are measured and used to calculate energy expenditure. McLean and Tobin (1987) stated that the Atwater and Benedict respiration chamber represents one of the earliest designs of this type of calorimeter. The fundamental design of a whole body chamber is similar to, and functions as, the mixing chamber in Figure 10.12. During the experiment, air is drawn from the chamber by an air pump and passes through an external circuit. This external circuit includes one CO_2 absorber, two moisture absorbers, a spirometer, and an oxygen supply. Moisture in the air is absorbed by passage through a container of sulfuric acid, and this is followed by absorption of CO_2 in soda lime; the moisture given off from the soda lime is then absorbed by a second container of sulfuric acid. The pressure drop in the system, caused by the uptake of oxygen by the subject and the absorption of any expired CO_2, is detected by the spirometer. As pressure in the chamber decreases, the volume of spirometer decreases. This then releases a valve on the oxygen inlet line. As the oxygen supply replenishes the chamber with oxygen, the spirometer bell rises and the valve closes again. A meter is used to measure the delivery of oxygen. The amount of CO_2 production is obtained from a weight change of the soda lime container plus the moisture collected in the second sulfuric acid container.

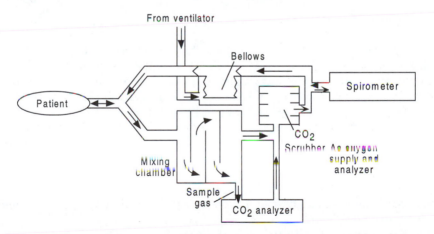

Figure 10.12 A closed-circuit system uses the volumetric loss principle. The spirometer is used as an oxygen supply. The spirometer volume change, as a function of time, is used to calculate the rate of oxygen consumption.

The whole body chamber offers fast response, accuracy, precision, and allows comparative measurements during sleep, at rest, before and after meals, and during controlled periods of exercise. Ravussin and Rising (1992) reported that the response time for their respiration chamber (a total volume of 19,500 L) measured during CO_2 dilution was 3 min for 93% of the total response and 4 min for 99% of the total response. In their review of techniques for the measurement of human energy expenditure, Murgatroyd et al. (1993) reported that an accuracy of $\pm 1\%$ in gas exchange measurement by the whole body chamber is achievable. They further reported that an accuracy of 1.5 to 2% of energy expenditure was obtained if the precision of the gas exchange rate is better than 4 mL/min over a period of 30 min in a chamber with a volume of 10,000 L. Although the whole body indirect calorimeters provide fast response, accuracy and precision of measurement, satisfactory results are only obtained with careful chamber design and construction, and by the choice of the best possible instrumentation. Nevertheless, expensive cost and artificial environment (the subject is not under free-living conditions) are still its disadvantages. Current design of the closed-circuit volumetric replenishment indirect calorimeter employs an ultrasonic transducer to monitor the volume level of a bellows that is used as an oxygen reservoir. As oxygen is depleted, a microcomputer-controlled calibrated volumetric pulse of oxygen is delivered to maintain the bellows at the preset reference level. The $\dot{V}O_2$ is calculated by multiplication of the pulse volume by the number of pulses per minute. In addition, the ultrasonic transducer is also used to monitor V_T. Measurement of $\dot{V}O_2$ made by this technique has been found to be quite accurate. A good correlation ($r > 0.99$) between measured and actual $\dot{V}O_2$ has been reported by several authors (Branson et al., 1988; Keppler et al., 1989). Advantages of the closed-circuit indirect calorimeters are that they are not affected by fluctuations of FIO_2 and by an $FIO_2 > 0.6$. Details of technical considerations have been discussed in Branson et al. (1995).

Doubly-Labeled Water Method

The doubly-labeled water (DLW) method was first introduced by Lifson, Gordon, and McClintock in the 1950s (Lifson et al., 1955). They used the principles of indirect calorimetry to measure total energy expenditure from the disappearance rates of two stable isotopes: deuterium (^2H) and oxygen 18 (^{18}O). Since the early 1980s (Schoeller and van Santen, 1982), the use of doubly labeled water has provided a technique where the total energy expenditure can be measured in a free-living individual over 1 to 2 weeks. It also has considerable appeal for field studies to estimate the energy cost of activities and thermogenesis by subtracting the basal metabolic rate from the total energy expenditure. The principle of the method is based on six assumptions listed as follows:

1. The body is in steady-state; i.e. the volume of total body water pool remains constant over time.
2. The turnover rates of water and carbon dioxide are constants over time.
3. The volume of labeled water distribution is only equal to total body water.
4. The isotopes are lost only as water and carbon dioxide.
5. Losses of labeled water and carbon dioxide in urine and exhaled carbon dioxide have the same level of enrichment as in body water.
6. The background levels of the isotopes remain constant.

Figure 10.13 shows the principle of the method. After a subject drinks a dose of DLW, the deuterium mixes with the body water pool and the ^{18}O mixes with both water and bicarbonate pools. The production rate of CO_2 is calculated as the difference between ^{18}O and ^{2}H disappearance rates. The formula for this calculation is

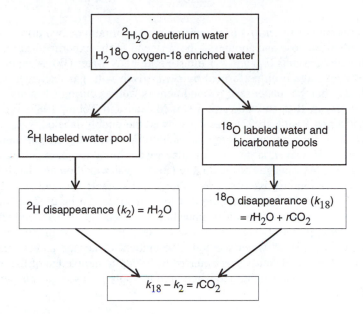

Figure 10.13 Principle of the doubly-labeled water method. r is the production rate, k represents rate constants determined from the experiment.

$$rCO_2 = \dot{V}CO_2 = \left(\frac{W}{2}\right) \times (k_{18} - k_2) \times 22.4 \qquad (10.23)$$

where r (in L/h) is the production rate, W is the size of total body water in mol, and k_{18} and k_2 are fractional disappearance rates for ^{18}O and ^{2}H, respectively. They are computed from measurements of the isotopic concentrations c_{18} and c_2 using $k = (1/c) \times (\Delta c / \Delta t)$. The factor 2 in the denominator of Eq. (10.23) is the molar proportion of oxygen in carbon dioxide in relation to water. In practice, Lifson et al. (1955) modified Eq. (10.23) by including the fractional factor: f_1 (0.93) for $^{2}H_2O$ gas/$^{2}H_2O$ liquid, f_2 (0.99) for H_2 gas/$H_2{}^{18}O$ liquid, and f_3 (1.04) for $C^{18}O_2$ gas/ $H_2{}^{18}O$ liquid to take a natural fraction of the isotopes between different body fluids into account. Eq. (10.23) becomes

$$rCO_2 = \dot{V}CO_2 = \left[\frac{W}{2f_3}(k_{18} - k_2) - \left(\frac{f_2 - f_1}{2f_3}\right)\frac{\dot{m}}{18}\right] \times 22.4$$

or

$$rCO_2 = \dot{V}CO_2 = \left[\frac{W}{2.08}(k_{18} - k_2) - \frac{\dot{m}}{624}\right] \times 22.4 \qquad (10.24)$$

where $\dot{m}$ is the rate of water loss as vapor in g/h. A minimum of two urine samples (one a few hours after the dose and another at the very end of the experiment) are collected to compute the rate constants (k_{18} and k_2) and the total body water (W). A number of cross-validation studies have been performed by comparison with gas exchange analysis and the whole body chamber under resting condition, as well as during sustained heavy exercise (see reviews in Ritz and Coward, 1995; Mclean and Tobin, 1987; Prentice et al., 1991; Ravussin and Rising, 1992). Results showed the doubly-labeled water method is a suitable technique for assessing energy expenditure in free-living circumstances with an accuracy of ±5%. However, large underestimates were found in obese subjects (Ravussin and Rising, 1992). Advantages of the doubly labeled water method are that it is noninvasive (oral dosing), nonintrusive (urine sampling), and suitable for noncompliant subjects and field studies. In addition, deuterium distribution space can be used to estimate total body water for an estimate of body composition (see Section 10.4.1). Nevertheless, the high cost for the expense of ^{18}O-enriched water and technical complexity are disadvantages of the doubly-labeled water method. The amount of dosage and its resulting cost depend on how precise the mass spectrometer is. In-depth knowledge of the assumptions of the doubly labeled water technique is required to apply it to various physiological and pathological situations.

10.4 Measurement of Body Fat

Body fat measurement is performed during body composition studies. It is used in a wide variety of fields, including human biology, medicine, sports science, and nutrition. Fat measurements may be used to assess obesity in children and adults, and to investigate the effects of malnutrition. Body fat mass is defined as pure fat—triglyceride fat in the body. Adipose tissue consists of approximately 83% fat, 2% protein, and 15% water (Jensen, 1992). Fat free mass (FFM) is defined as lean body mass plus nonfat components of adipose tissue. The assumption used by most body fat assessment methods is that the body is made of two compartments, fat and fat free weight. The chemical composition of the fat-free body is assumed to be relatively constant. The fat free body has a density of 1.1 g/cm^3 at 37 °C (Behnke et al., 1942) with water content of 72 to 74% (typical value 73.2%, Pace and Rathburn, 1945). Body fat, or stored triglyceride, is also assumed to be constant with a density of 0.9 g/cm^3 at 37 °C (Mendez et al., 1960). All body fat methods use either this two–component model or a further division of the fat and fat-free weight into four compartments: water, protein, bone mineral, and fat (Brozek et al., 1963). The various body fat measurement techniques may be further divided into two categories. Some techniques are *direct* measures of body fat such as deuterium oxide dilution, densitometry, and dual photon absorptiometry (DPA)/dual energy X-ray absorptiometry (DEXA), whereas others are *indirect* and require cross–validation with other methods. Examples of indirect methods that will be discussed here are anthropometry and bioelectrical impedance analysis (BIA). Although MRI and CT are able to directly measure regional fat distribution, especially interabdominal fat content, high cost and long scanning time still limit these two techniques to research studies.

10.4.1 Direct Measurement of Body Fat

Direct measurement techniques of body fat such as densitometry (underwater weighing) and deuterium oxide dilution are commonly used as reference methods for measuring percent body fat. They are based on the two-compartment model and a constant hydration factor of FFM, respectively. DPA/DEXA is another direct measuring technique that has yet to be tested as a reference method. Underwater weighing provides accurate measurement, but it requires a well-trained technician, complicated equipment, and laborious procedures. Consequently, it is not practical for field studies. In addition, underwater weighing is not suitable for groups such as the handicapped, diseased people, and young children. In contrast, the deuterium oxide dilution method is applicable for a variety of groups. However, it also requires sophisticated equipment to measure the final concentration of the isotope.

Deuterium Oxide Dilution

The use of the deuterium oxide dilution technique to measure body water to predict FFM uses the fact that fat is mainly anhydrous and water is a relatively constant component (73.2%) of FFM. It is based on the assumption that the isotope has the same distribution volume as water, is exchanged by the body in a manner similar to water, and is nontoxic in the amount used. During the experiment, subjects are asked to drink or be injected with a known amount of the labeled water (deuterium oxide water). Then the amount of the labeled water in saliva, urine, or plasma is determined after an equilibrium period (3 to 4 h). The calculation of total body water (TBW) volume is based on the simplified relationship

$$C_1 V_1 = C_2 V_2 \qquad (10.25)$$

where $C_1 V_1$ is the amount of the isotope given, and C_2 is the final concentration of the isotope in saliva, urine, or plasma, and V_2 is the volume of TBW. Available techniques for the assay of deuterium in aqueous solution are gas chromatography (see Section 3.15.2), mass spectrometry (see Section 3.12), and filter-fixed infrared absorptiometry. The quantity of the isotope given depends on the analytical system used and the objective of the research. Lukaski (1987) reported that a 10 g dose of deuterium oxide (99.7% purity) mixed with ~300 mL of distilled-deionized water is used routinely in his laboratory for the filter-fixed infrared absorptiometry. Some investigators have given an oral dose of deuterium oxide of 1 g/kg body weight (Mendez et al., 1970) for gas chromatography and an oral dose of 2 g D_2O has been used for mass spectrometry (Halliday and Miller, 1977).

For adult white individuals (Wellen et al., 1994), the calculation of fat free mass from TBW volume is done by dividing V_2 by 1.04 and 0.732 to account for the estimated 4% nonaqueous H^+ exchange and water fraction of the FFM respectively.

$$\text{FFMTBW(kg)} = \frac{V_2}{(1.04 \times 0.732)} \qquad (10.26)$$

$$\%\text{BFTBW} = \left[\frac{(\text{weight} - \text{FFMTBW})}{\text{weight}} \right] \times 100 \qquad (10.27)$$

where FFMTBW is FFM derived from TBW (V_2), weight is the subject's body weight, and %BFTBW is the derived percent body fat from FFMTBW. For populations of low (children) or high (obese) TBW, water fraction of the FFM (0.732) in Eq. (10.26) should be adjusted to maintain an accurate estimation of %BFTBW. This change of TBW is attributed to the fact that 15% of adipose tissue is water. In general, the accuracy associated with measuring body fat by the TBW method is about 3% to 4% (Lukaski, 1987).

Densitometry (Underwater Weighing)

The most accurate method for estimating body fat is the direct measurement of body density from underwater weighing. This method is often used as the reference method for evaluating new indirect methods such as bioelectrical impedance analysis. Densitometry assumes that the human body has two distinct compartments: fat mass and fat free mass (Siri, 1961; Brozek et al., 1963). The measurement of body density is accomplished by measuring body volume (according to Archimedes' principle) and body weight (by the difference between the subject's weight in air and weight in water). Archimedes' principle states that the volume of a submerged object is equal to the amount of water the object displaces. In a human subject, the apparent volume of the subject is determined by correcting for the volume in the lung and the density of the water using the water temperature at the time of the measurement.

The weighing device consists of a plastic chair suspended from a scale into a huge tank containing water maintained at an average temperature of 35 °C. The tare weight for the chair and water temperature are measured before the experiment. During the measurement, the subjects sit on the chair with minimal clothes and are submerged under water. While remaining under water, the subjects exhale completely and hold their breath. Underwater weight is recorded after bubbles stop coming from the subject's mouth and oscillation of the scale due to the waves has subsided. Underwater weight is recorded 10 times for each subject and the mean of the three highest weights is used. Residual lung volumes are measured on land to account for its contribution to the subject's buoyancy. The body density (BD) is expressed in g/cm^3 and is calculated from Eq. (10.28) (Guo et al., 1987):

$$BD = \frac{W_a}{\dfrac{W_a - (W_w - T)}{D_w} - RV} \tag{10.28}$$

where W_a = weight in air in grams, W_w = weight in water in grams, D_w = density of water in g/cm^3 at actual water temperature, RV = residual lung volume in cm^3, and T = tare weight in g. Density is then used to estimate the body fat with standard equations (Siri, 1961; Brozek et al., 1963). The equation used by Siri to calculate the percent body fat (%BF) is

$$\%BF = \frac{4.95}{BD} - 4.50 \tag{10.29}$$

and Brozek's equation is

$$\%BF = \frac{4.57}{BD} - 4.412 \tag{10.30}$$

   Lukaski (1987) reported that Eqs. (10.29) and (10.30) give results within 1% body fat if the body density is within 1.10 to 1.03 and that Siri's equation yields higher values than the Brozek's equation for subjects with more than 30% body fat. Due to the errors associated with variations of water and bone mineral contents, an error of 1 to 3% in prediction of percent body fat with this method is expected (Lohman, 1990). In general, the error is greater in populations with lower bone mineral content (children, women, and old people) and populations with high bone mineral content (athletes). Noninvasive and accurate measurement of body density and good estimation of percent body fat make the underwater weighing a standard method for evaluation of other new and portable techniques such as bioelectrical impedance analysis.

Dual Photon Absorptiometry (DPA)/Dual Energy X-ray Absorptiometry (DEXA)

Another technique for direct measurement of body fat is dual photon absorptiometry or dual energy X-ray absorptiometry. DPA/DEXA has been used in clinical settings to diagnose osteoporosis and also to evaluate types of therapy for osteoporosis by directly measuring bone mineral contents of the patients. Photon aborptiometry was first introduced by Cameron and his colleagues (1962, 1963) to measure bone mineral content (BMC) in the limbs with a single isotope (Iodine-125 at 27.3 keV or Americium-241 at 59.6 keV). This technique is called single photon absorptiometry (SPA) and is based on the principle that bone mineral content is directly proportional to the amount of absorbed photon energy by the target bone (see Figure 9.9). The necessity of constant thickness of absorber and uncontrollable variations in transmission due to the presence of fat tissue limit the measurement of BMC to bones only in the extremities. Mazess and his colleagues (Mazess et al., 1970; Judy, 1970) developed dual photon absorptiometry to measure contents of bone mineral and soft tissue in the appendicular parts of the skeleton (e.g. femur) with the isotope Gadolinium-153 (^{153}Gd, 44 and 100 keV). This method has been applied to measure BMC in the axial skeleton (e.g. vertebrae) with the isotopes Americium-241 (^{241}Am, 59,6 keV) and Cesium-137 (^{137}Cs, 662keV) by Roos and Skoldborn (1974). The underlying principle of DPA/DEXA for the measurement of body fat is that the ratio of absorbance of the two different energy level photons ($R = \mu_L / \mu_H$) is linearly related to the percent of fat in the soft tissue of the body. The mass attenuation coefficients for bone mineral and soft tissues are constant and are not affected by the amount of bone and soft tissues along the path of the photon beam. They depend on the energy level of the beam. The basic equations for DPA are as follows (Sorenson et al., 1989):

$$\ln\left(\frac{N_{0L}}{N_L}\right) = \mu_{sL} M_s + \mu_{bL} M_b, \tag{10.31}$$

$$\ln\left(\frac{N_{0H}}{N_H}\right) = \mu_{sH} M_s + \mu_{bH} M_b, \tag{10.32}$$

where N_{0L} and N_L are the number of incident and transmitted low energy photons, N_{0H} and N_H are the number of incident and transmitted high energy photons, μ_{sL} and μ_{sH} are the mass attenuation coefficients of soft tissue at the low and high energies, μ_{bL} and μ_{bH} are the same quantities for bone mineral, and M_s and M_b are the area densities in g/cm^2 of soft tissue and bone mineral in the beam path. By solving Eqs. (10.31) and (10.32) simultaneously, M_b can be derived as:

$$M_b = \frac{\left[R \ln\left(\frac{N_{0H}}{N_H}\right) - \ln\left(\frac{N_{0L}}{N_L}\right) \right]}{(R\mu_{bH} - \mu_{bL})} \tag{10.33}$$

where $R = \mu_{sL} / \mu_{sH}$. Based on the same basic equations, DEXA uses a constant voltage X-ray generator and k-edge filters (made of cerium) to separate the X-ray beam into two energy levels, or a switching-pulse system that rapidly alternates the constant voltage X-ray generator to produce two beams with high and low energy simultaneously. Current examples of the former type of densitometer include Norland XR-36™ (Norland Medical System, Inc., Fort Atkinson, WI) and Lunar DPX™ series (Lunar Radiation Corp., Madison, WI). An example of the latter type of densitonometer is the Hologic QDR-1000™ series (Hologic, Inc., Waltham, MA). In recent years, DEXA has become the main stream in the densitometer industry. DEXA scans faster and has better image quality over DPA.

Most of the current commercially available densitometers usually include a scanning and examination table, X-ray system (scanner and scintillation detector), a computer system with additional software to control and calculate BMC and body composition, and positioning or calibration accessories. During the measurement, the patient needs to lie supine on the table and a series of transverse scans are made from head to toe at 1 cm intervals (Mazess et al., 1990). Measurements take 10 to 20 min for a total body measurement or 1 to 5 min for a regional scan. Typical transverse scan image area is around 60 cm × 200 cm. The value of R (weighted for mass of tissue in each pixel) is determined for the number of pixels containing the soft tissue alone. The %BF can be derived from the R value through application of a regression equation of %BF and the R value resulting from the calibration phantom provided by the manufacturer. Total body bone mineral content is calculated from the bone-containing pixels. These calculations are usually processed by the software provided by the manufacturer. Technology of automatic exposure control is also applied to DEXA to assure that each pixel receives the proper exposure regardless of tissue thickness. In general, the precision error for the percent body fat is within approximately 2% (Lukaski, 1987; Mazess et al., 1990; Jensen, 1992; Prentice, 1995). The advantages of DPA/DEXA include portability, capability of measuring total body or regional composition of bone mineral and body fat, high reproducibility (3% to 5%), and less dependence on the technician's skills and experience. Narrow width table and size limitation prohibit using the DPA/DEXA technique to assess very obese subjects. Radiation exposure of DPA/DEXA is between 2 and 10 mrem.

10.4.2 Indirect Measurement of Body Fat

Indirect measurement techniques of body fat such as anthropometry and BIA require empirically derived regression equations to estimate body fat or total body water (TBW). These techniques are suitable for population studies where individual distinction is less critical. They often need to be validated by reference methods such as densitometry and the isotope dilution method. Anthropometry is a rapid and inexpensive way to evaluate nutritional status for a population in a field study, but it requires a skilled technician and anthropometrist to achieve accurate measurement. Single or multiple frequency BIA also provides a rapid and simple method to allow the investigator to predict TBW and extracellular water (ECW). In general, both techniques are less likely to be useful for prediction of change in body composition.

Anthropometry (Skinfold Thickness)

Anthropometry is defined as measurement of body size. It includes measurement of body circumferences, skinfolds, weight, stature, etc. Measurement of skinfolds is one of the most commonly used anthropometric techniques. It is based on two assumptions: the subcutaneous adipose tissue represents a fairly constant proportion of the total body fat and the selected measurement sites represent the average thickness of subcutaneous adipose tissue (Lukaski, 1987). Skinfold thickness is the thickness of a double layer of skin and subcutaneous adipose tissue that is pulled away from the muscle at selected sites. Skinfold thickness can be measured using special devices such as the Lange caliper (Lange and Brozek, 1961, see Figure 10.14(a)) and the Holtain (the Harpenden) caliper (Tanner and Whitehouse, 1955). The main uses for skinfold thickness are to estimate body fat (percent body fat, or %BF) and the anatomical distribution of fat tissue for population surveys. Common sites for measurement of skinfold thickness are the chest, midaxilla, abdomen, supraillium, triceps, subscapular, and thigh.

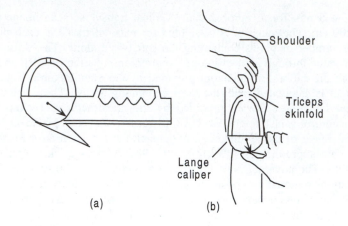

Figure 10.14 (a) Lange skinfold caliper used for assessing thickness of subcutaneous fat. (b) Illustration of an example of skinfold measurement, triceps skinfold taken on the midline posterior surface of the arm over the triceps muscle.

During the measurement of skinfold thickness, a double layer of skin and subcutaneous fat tissue is grasped between the thumb and forefinger and pulled forward from the site by 1 cm to allow the jaw of the caliper, maintaining a constant jaw pressure of 10 g/mm^2, to impinge on the skin. A distance of at least 1 cm should be maintained between the fingers and the caliper arms to prevent any interference from the pressure of the fingers. The calipers should be applied perpendicular to the body surface and perpendicular to the longitudinal axis of the skinfold (see Figure 10.14(b)). The release of the pressure of the caliper arms requires the measurement to be made within 4 s to standardize differences among subjects in compressibility of skinfolds. The sequence of measurements is repeated three times to reduce intrameasurer bias. For a more complete review of specific anthropometric measurement techniques, see the *Anthropometric Standardization Reference Manual* (Lohman et al., 1988).

Many regression equations are available for the prediction of body density and thus body fat from skinfold thickness measurements. These equations are generally valid for adult Caucasian populations, but overestimate body fat in elderly individuals. For example, Jackson and Pollack (1978) and Jackson et al. (1980) published equations for estimating body density in adult men, Eq. (10.34), and women, Eq. (10.35), aged 20 to 50 years.

$$BD = 1.1093800 - 0.0008267(X_1) + 0.0000016(X_1)^2 - 0.0002574(X_3) \quad (10.34)$$

$$BD = 1.0994921 - 0.0009929(X_2) + 0.0000023(X_2)^2 - 0.0001392(X_3) \quad (10.35)$$

where BD is body density, X_1 is the sum of chest, abdomen, and thigh skinfolds (mm), X_2 is the sum of triceps, thigh, and suprailium skinfolds in mm, and X_3 is the age. Correlation coefficients (r) for Eqs. (10.34) and (10.35) are 0.905 and 0.842, respectively. These body densities can subsequently be used for percent body fat estimation with Eq. (10.27).

The advantages of body fat prediction from measurement of skinfold thickness are that it is simple, inexpensive, and noninvasive. However, the accuracy and reliability of skinfold thickness measurement depends on the skill of the measurer, interaction between the measurer and the subject, and the characteristic of the subject. The potential error associated with the use of skinfolds, assuming the correct equation is used, the measurer is well trained, and correct skinfold sites chosen, is 3 to 4% (Lohman, 1990). In addition, body fat of edematous (excess fluid in the tissues) patients is often overestimated and very obese subjects cannot fit in the standard calipers.

Bioelectric Impedance Analysis

Bioelectrical impedance analysis (BIA) for body fat measurement is used for measurement of fat free mass of the body. It is also based on the two-compartment model (in parallel) as mentioned earlier. Fat free mass is comprised mainly of electrolyte-containing water (73.2%) in the body; it represents the main path of conduction for the applied current. In contrast, fat mass behaves as an insulator. Therefore, the impedance of the body is controlled by fat free mass. The method of BIA is based on the principle that the impedance of an isotropic conductor is related to the length and cross-sectional area of the conductor for a constant signal frequency. Prediction equations are often generated using BIA values to estimate fat free mass. With a constant signal frequency and a relatively constant conductor configuration, the impedance to an ac current can be related to the volume of the conductor

$$Z = \frac{\rho L^2}{V} \qquad (10.36)$$

where Z is the impedance in ohms (Ω), ρ is the resistivity in $\Omega \cdot$ cm, L is the conductor length in cm, and V is the volume in cm^3. A general procedure to measure the whole body impedance is to let the subject, 2 h after eating and 30 min after voiding, lie supine on a table with arms and legs sufficiently apart. Current injector electrodes are placed in the middle of the dorsal surfaces of the right metacarpal–phalangeal (hand) and metatarsal–phalangeal (foot) joints, respectively. Detector electrodes are positioned in the midline of the dorsal surfaces of the right wrist and ankle. Measurements of body resistance and reactance are made on ipsilateral and contralateral sides of the body (as shown in Figure 10.15). The lowest value of resistance for each subject is used to predict fat free mass (FFM). Hoffer et al. (1969) first used four surface electrodes (tetrapolar) for the study of TBW. The tetrapolar technique is used to minimize contact impedance and skin–electrode interaction. They introduced 100 µA of ac current at 100 kHz in 20 healthy subjects and 34 patients whose TBW were measured with the isotope dilution method (see Section 10.4.1) and showed that (stature)^2/Z was the best predictor of TBW. The correlation coefficients (r) of (stature)^2/Z against TBW for 20 healthy subjects and 34 patients were 0.92 and 0.93, respectively. Lukaski et al. (1985) demonstrated in his total body impedance study in 37 men that the resistance, R, is a better predictor than the impedance (square root of $R^2 + X_c^2$) when the reactance, X_c, is small. They used a current source of 800 µA at 50 kHz (RJL Systems, Detroit, MI). Consequently, the best predictor of TBW ($r = 0.95$) and FFM ($r = 0.98$) was (stature)^2/R. Kushner and Schoeller (1986) cross-validated the prediction of TBW with BIA by generating a prediction equation from a sample of 40 nonobese adults and applying it to 18 obese patients. They found that (stature)^2/R was a good predictor of TBW ($r = 0.96$). Segal et al. (1985) also confirmed the significant role of (stature)^2/Z as a predictor of FFM ($r = 0.912$) with 75 male and female subjects ranging from 4.9 to 54.9% body fat, but reported overestimation of FFM for obese subjects. Because fat free mass includes virtually all water in the body, the formula that is used to predict TBW via BIA is often used to estimate FFM. A general model that is often used to predict FFM or TBW in kg is

$$\text{FFM or TBW} = C_1 \left(\frac{\text{Stature}^2}{R} \right) + C_2 \left(\text{Weight} \right) + C_3 \qquad (10.37)$$

where C_1 and C_2 are regression coefficients of the independent variables Stature^2/R in cm^2/Ω and Weight in kg, respectively. C_3 is the intercept of this regression equation. These constants vary with different sample groups. The equation

$$\%\text{BF} = 100 - \left(\frac{\text{FFM}}{\text{Weight}} \right) \qquad (10.38)$$

has been used to estimate percent body fat. Segal et al. (1988) and Gray et al. (1989) have suggested that prediction of FFM can be enhanced by the use of gender and fatness specific equations.

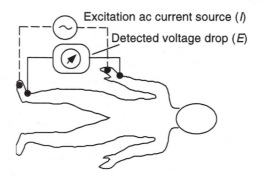

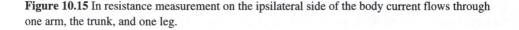

Figure 10.15 In resistance measurement on the ipsilateral side of the body current flows through one arm, the trunk, and one leg.

Current commercially available BIA instruments (RJL Systems, Detroit, MI; Dieto-System, Milan, Italy; etc.) are often equipped with multiple frequency function and specialized software. They provide empirically derived regression equations and are capable of estimating TBW, FFM, %BF, and extracellular water (ECW). Theoretically TBW is modeled as a parallel R–C circuit with the cell membrane as a capacitor to separate intracellular and extracellular fluids. At low frequency (~1 to 5 kHz), the current only reaches the extracellular fluid while at high frequency (~50 kHz and higher) it penetrates both fluids. Therefore, at low frequency the impedance can be used to estimate ECW. Deurenburg (1995) reported in his study of 103 healthy males and females, aged 19 to 51 years, ECW was best predicted by resistance measured at 1 kHz ($r = 0.93$) and TBW was best estimated by resistance measured at 100 kHz ($r = 0.95$). Regression coefficients (C_1, C_2, and C_3) estimated from his study for TBW and ECW were 0.368, 0.133, 2.1 and 0.177, 0.073, 2.0, respectively. Segal et al. (1991) also confirmed the significant role of resistance measured at 5 kHz and 100 kHz for prediction of ECW ($r = 0.93$) and TBW ($r = 0.95$) in 36 healthy males. Regression coefficients (C_1, C_2, and C_3) estimated from their study for TBW and ECW were 0.455, 0.14, 3.43 and 0.284, 0.112, –6.115, respectively. They cross-validated on two randomly selected subsets (18 each) and showed that the prediction equations were reproducible and valid.

BIA offers the advantage of rapid, safe, noninvasive measurements of percent body fat. In addition, multifrequency BIA can differentiate extracellular water (ECW) and total body water (TBW). Problems with the BIA method include no generally accepted regression equation as a reference for all sample groups and BIA instruments, and that accurate prediction of a change of ECW is quite difficult.

10.5 Measurement of Body Movement

Body movement is studied to learn how to analyze the movement of the body and to understand its underlying principles. It is useful for improving performance of motor skills (such as skating) and providing efficient methods of using the body in daily life skills (e.g., walking and running). Analysis of body movement requires observation and measurement. Many methods are available, ranging from simple visual observation to complicated computerized systems. Typical research or clinical gait analysis laboratory setups include goniometers, accelerometers, electromyography (EMG, see Chapter 7), force plates, and kinematic (motion) analysis systems. A kinematic analysis system varies from simple and subjective visual observation to more expensive and complicated video or optoelectronic systems that can provide objective, quantitative data of three-dimensional motion of selected points on the body. With additional force plate outputs, the kinematic analysis system can be used as a kinetic analysis system to provide force and moment information for human limb movement. In this section, we discuss working principles and applications for goniometers, accelerometers, and video and optoelectronic kinematic analysis systems. For additional information, refer to texts by Allard et al. (1995), Whittle (1996), Winter (1990), and Medved (2001).

10.5.1 Goniometers and Accelerometers

Goniometers

A goniometer is a device that measures the relative angle of a joint that connects two body segments. The goniometer (see Figure 10.16(a)) consists of two attachment arms, which are fixed to two limb segments of the subject on either side of a joint to measure the joint angle without interfering with natural movement. Traditionally, a goniometer uses a resistance potentiometer to convert changes in rotation to an electric output proportional to the positions of two attachment arms. The potentiometer is aligned with the joint axis to establish the zero position and is calibrated to measure the joint angle in degrees. Multiaxial measurements of the joint are possible with two or three potentiometers mounted orthogonally in different planes. The output of the goniometer provides continuous analog data of joint motion at relatively low cost. However, to ensure accurate measurements, fixation to the body with cuffs around the soft tissues must be performed carefully to prevent off-axis problems (Chao, 1980). The measurement error caused by the translation of the joint center during movement can be improved by replacing the rigid arms of the traditional goniometer with a flexible parallelogram linkage designed to accommodate the motion change outside the measurement plane (Thomas and Long, 1964). A triaxial parallelogram electrogoniometer (see Figure 10.16(b)) has been used to record motions in more than one plane for the lower limb (Isacson et al., 1986). Strain gages have also been used as an alternative to the potentiometer. They deform when they are stretched or compressed and the electric output is proportional to the change in joint rotation. The working principle of strain gage and its related circuitry is discussed in Section 8.3.1.

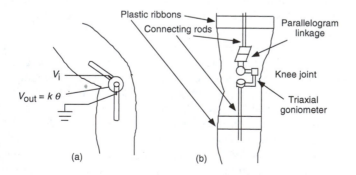

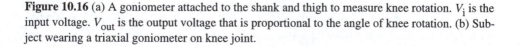

Figure 10.16 (a) A goniometer attached to the shank and thigh to measure knee rotation. V_i is the input voltage. V_{out} is the output voltage that is proportional to the angle of knee rotation. (b) Subject wearing a triaxial goniometer on knee joint.

Accelerometers

Accelerometers are instruments that measure acceleration. Basically, they all contain a small mass and make use of Newton's law ($F = ma$), measuring the force that is required to accelerate the known mass, as shown in Figure 10.17. There are two classes of accelerometers used in the analysis of body movement: strain gage (usually piezoresistive) and piezoelectric accelerometers. A strain gage accelerometer consists of strain gages (wires) bonded to a cantilevered mass and a base to which the cantilever beam is attached. When the base is accelerated, the cantilever beam deforms due to the inertia of the mass, thus changing the strain in the wires, which changes their resistances. We measure the change of resistance with a Wheatstone bridge circuit (see Section 9.11.2), which requires a differential amplifier. The resulting electric output is proportional to the acceleration of the mass. A piezoresistive accelerometer works under the same principle as a strain gage accelerometer. Instead of using strain sensitive wire, it uses piezoresistive strain elements as sensors. Piezoelectric accelerometers measure the force directly. The piezoelectric sensor converts the force produced by the acceleration to an electric charge or voltage (see Eq. (8.7)) and Section 8.5.2 for its charge amplifier and the working principle). For the analysis of body movement, the accelerometer is attached to a body segment at a specific point to measure acceleration in one direction. Two- or three-dimensional measurements are possible with several accelerometers grouped together orthogonally. The use of accelerometers for the analysis of body movement provides an alternative way to measure the velocity and displacement of the limb segment when the initial values of the limb's velocity and displacement are known (Morris, 1973). Some practical limitations prevent widespread use of the accelerometer in body motion analysis: necessity to exclude the effect of the field of gravity from true kinematic acceleration, difficulty in extracting the rotational acceleration, and low-frequency noise caused by the baseline drift in the measurement output.

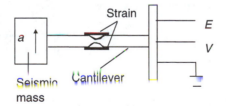

Figure 10.17 Vertical acceleration of the accelerometer frame bends the cantilever beam because the seismic mass remains at rest. Voltage output (V) is proportional to the acceleration (a). E is the supply voltage.

10.5.2 Video and Optoelectronic Systems

Gait can be measured on a single film using sequential strobe lights and a camera or by a video recorder. Figure 10.18 shows that a video or optoelectronic kinematic analysis system consists of active or passive markers placed at selected skin surface locations on bony landmarks, sensors that track markers' positions, and a computer system to control the timing for active marker display and signal processing of markers' tracings. The sensors, either optoelectronic circuitry or a video system, collect kinematic data from active (infrared LED) or passive markers and low-pass filter the data before an analog-to-digital converter (ADC), which feeds to a computer for later computation of the position and velocity of the rigid body in three-dimensional (3-D) space. A computerized analysis system often yields graphic displays that can provide information on joint positions, joint angles, segment angles, velocity, and acceleration as a function of time.

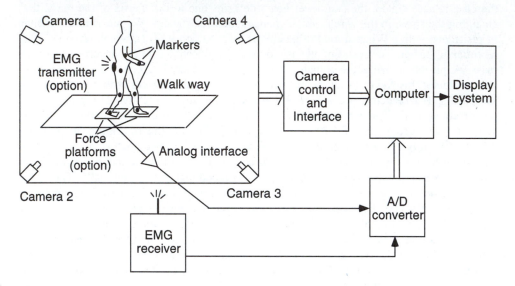

Figure 10.18 An example of a gait analysis setup includes a four-camera kinematic system, two force platforms, and an electromyogram (EMG) telemetry system.

There are many commercially available video or optoelectronic kinematic analysis systems (Walton, 1990), and they are different in the way that they combine video or optoelectronic systems with passive or active markers. For examples, *Vicon*™ (Oxford Metrics Ltd., Botley, Oxford, UK) and ELITE *Plus*™ (Bioengineering Technology & Systems, Milan, Italy) are video systems with passive (reflective) markers, and Selspot II™ (Selcom Inc., Southfield, MI), OPTOTRACK™ (Northern Digital Inc., Waterloo, Ontario, Canada), and CODA™ (Charnwood Dynamics Ltd., Leics, UK) are optoelectronic systems with active (infrared emitting diode) markers. All current kinematic analysis systems are precalibrated and have full 3-D motion tracking and display features in real time. In order to achieve real-time tracking of markers in 3-D, problems of real-time 2-D tracking with marker identity maintained and 2-D to 3-D transformations in real time have to be solved. The speed and convenience of tracking differs from one system to another, and also between software options from the same manufacturer. They all depend on how many video cameras and markers they use and how often they sample the data. Passive markers that allow automatic processes are recognized by contrast (*Vicon*), shape (ELITE *Plus*), and color (Videomex, Columbus Instruments, Columbus, OH) features during data analysis. They are suitable for sports studies or whenever large and fast movements are to be recorded because the subject is not hampered by attached wires and battery packs carried on belts as active markers. Active markers are infrared light-emitting diodes (IRLED, 800 nm) that allow simple labeling and high frequency sampling. The IRLEDs are often arranged in sequence and pulsed at different times (only one at a time) by a control unit. Most of the video systems use either stroboscopic illumination or CCD cameras (see Section 6.2.3 for more information) or both to avoid the smearing when the markers are moving. They are often synchronized by the camera control and interface unit to collect data with analog devices such as force plates, and EMG for movement kinetics.

Optoelectronic systems are similar, in principle, to video systems. They use photodetectors to focus the pulsed infrared signals from body markers on associated body landmarks. The determination of 3-D spatial coordinates from several 2-D sets of information is often accomplished by direct linear transformation (DLT) method (Abdel–Aziz and Karara, 1971). For m markers and n cameras, the DLT method determines the 3-D spatial coordinates of a marker from the 2-D coordinates of the marker, i ($i = 1, \dots , m$) on camera j, ($j = 1, \dots , n$) by

$$x_{ij} = \frac{A_j x_i + B_j y_i + C_j z_i + D_j}{I_j x_i + J_j y_i + K_j z_i + 1} \tag{10.39}$$

$$y_{ij} = \frac{E_j x_i + F_j y_i + G_j z_i + H_j}{I_j x_i + J_j y_i + K_j z_i + 1} \tag{10.40}$$

where for a given marker i:
$x_{ij} = x$ coordinate of marker i on camera j,
$y_{ij} = y$ coordinate of marker i on camera j,
$x_i = x$ coordinate of marker i in 3-D space,
$y_i = y$ coordinate of marker i in 3-D space,
$z_i = z$ coordinate of marker i in 3-D space,
A_j to K_j = coefficients in the transformation of marker i.

Figure 10.18 shows an example of a gait analysis laboratory, including a four–camera kinematic system, two force platforms, and an electromyogram (EMG) telemetry system. For more information, refer to texts in Allard et al. (1995) and Walton (1990).

10.6 Problems

10.1   Calculate the value of the thermistor temperature coefficient α for $T = 298$ K and $\beta = 3000$ K.

10.2   Design a circuit that converts the resistance change of a thermistor to a voltage, and then amplifies it to measure temperature.

10.3   Explain the Seebeck effect of a thermocouple and why and how to use it for temperature measurement.

10.4   Design a circuit that uses an electronic cold junction and thermocouple, then amplifies it to measure temperature.

10.5   Explain the operation of a pyroelectric infrared thermometer, sketch its block diagram, and sketch the response after the shutter opens.

10.6   Compare direct and indirect calorimetry.

10.7   Explain how to use the gradient layer calorimeter for heat measurement.

10.8   Figure 10.6 shows a simple gradient layer calorimeter. Sketch an expanded gradient layer calorimeter that includes ventilation. Show location of all sensors and explain how they work. Give a symbol for the output of each sensor. Explain by means of equations, block diagrams and/or words how to obtain the quantities given in equations 10.7 through 10.11 from your sensor outputs.

10.9   The oxygen consumption is 0.25 L/min. The carbon dioxide production is 0.06 L/min. Calculate the energy expenditure. Give units.

10.10   Define %BF and explain its purpose.

10.11   Explain how to measure %BF using densitometry.

10.12   Describe how to use the bioelectrical impedance method for body fat measurement.

10.13   Draw the circuit for a resistance potentiometer goniometer with an amplifier with a gain of 2.

10.14   Draw the circuit for a 4-arm piezoresistive strain gage accelerometer and amplifier.

10.15   Draw the circuit for a piezoelectric accelerometer and amplifier to pass frequencies higher than 1 Hz.

10.16   You want to analyze human locomotion; list the devices you need and explain why.

10.7 References

Allard, P., Stokes, I. A., and Blanchi, J.-P. 1995. *Three-Dimensional Analysis of Human Movement.* Champaign IL: Human Kinetics.

Abdel-Aziz, Y. I. and Karara, H. M. 1971. Direct linear transformation from comparator coordinates into object space coordinates in close-range photogrammetry. *Proc. Close-range Photogrammetry.* Am. Soc. of Photogrammetry. 1–18.

Bair, D. N. and Davies, A. 1995. Temperature probes and principles of thermometry. In R. L. Levine and R. E. Fromm, Jr. (eds.) *Critical Care Monitoring: From Pre-Hospital to the ICU.* St. Louis: Mosby.

Behnke, A. R., Feen, B. G., and Welham, W. C. 1942. Specific gravity of healthy men. *JAMA,* **118**: 495–98.

Benzinger, T. H. and Kitzinger, C. 1949. Direct calorimetry by means of the gradient principle. *Rev. Sci. Instrum.,* **20**: 849–60.

Branson, R. D., Lacy, J., and Berry, S. 1995. Indirect calorimetry and nutritional monitoring. In R. L. Levine and R. E. Fromm, Jr. (eds.) *Critical Care Monitoring: From Pre-Hospital to the ICU.* St. Louis: Mosby.

Branson, R. D., Hurst, J. M., Davis, K. Jr., and Pulsfort, R. 1988. A laboratory evaluation of the Biergy VVR calorimeter. *Resp. Care*, **33**: 341–7.

Brozek, J., Grande, I., Anderson, J. T., and Keys, A. 1963. Densitometric analysis of body composition: revision of some quantitative assumption. *Ann. NY Acad. Sci.*, **40**: 110–3.

Cameron, J. R., Grant, R., and MacGregor, R. 1962. An improved technique for the measurement of bone mineral content *in vivo*. *Radiology*, **78**: 117.

Cameron, J. R. and Sorenson, J. 1963. Measurement of bone mineral *in vivo*: an improved method. *Science*, **142**: 230–2.

Chao, E. Y. 1980. Justification of triaxial goniometer for the measurement of joint rotation. *J. Biomechanics*, **13**: 989–1006.

Cobbold, R. S. 1974. *Transducers for Biomedical Measurements: Principles and Applications*. New York: John Wiley & Sons.

Deurenburg, P. 1995. Multi-frequency impedance as a measure of body water compartments. In P. S. W. Davies and T. J. Cole (eds.) *Body Composition Techniques in Health and Disease*. Cambridge: Cambridge University Press.

Fraden, J. 1991. Noncontact temperature measurements in medicine. In D. L. Wise (ed.) *Bioinstrumentation and Biosensors*. New York: Marcel Dekker.

Gray, D. S., Bray, G. A., Gemayel, N., and Kaplan, K. 1989. Effect of obesity on bioelectrical impedance. *Am. J. Clin. Nutr.*, **50**: 255–60.

Guyton, A. C. and Hall, J. E. 1996 *Textbook of Medical Physiology*. 9th ed. Philadelphia: W. B. Saunders.

Halliday, D. and Miller, A. G. 1977. Precise measurement of total body water using trace quantities of deuterium oxide. *Biomed. Mass Spectrom*, **4**: 82–7.

Hoffer, E. C., Meador, C. K., and Simpson, D. C. 1969. Correlation of whole body impedance with total body water. *J. Appl. Physiol.*, **27**: 531–4.

Isacson, J., Cransberg, L., and Knuttsson, Z. 1986. Three-dimensional electrogoniometric gait recording. *J. Biomech.*, **19**: 627–35.

Jackson, A. S. and Pollock, M. L. 1978. Generalized equations for predicting body density of men. *Br. J. Nutr.*, **40**: 497–504.

Jackson, A. S., Pollock, M. L., and Ward, A. 1980. Generalized equations for predicting body density of women. *Med. Sci. Sports Exerc.*, **12**: 175–82.

Jensen, M. D. 1992. Research techniques for body composition assessment. *J. Am. Dietetic Assoc.*, **92**: 454–60.

Judy, P. E. 1970. *A Dicromatics Attenuation Technique for the in vivo Determination of Bone Mineral Content*. Ph.D. Thesis, University of Wisconsin, Madison, WI.

Keppler, T., Dechert R., Arnoldi, D. K., Filius R., and Bartlett, R. H. 1989. Evaluations of Waters MRM-6000 and Biergy VVR closed-circuit indirect calorimeters. *Resp. Care*, **34**: 28–35.

Kushner, R. F. and Schoeller, D. A. 1986. Estimation of total body water by bioelectrical impedance analysis. *Am. J. Clin. Nutr.*, **44**: 417–24.

Lange, K. O. and Brozek, J. 1961. A new model of skinfold caliper. *Am. J. Phys. Anthropol.*, **19**: 98–9.

Lifson, N., Gordon, G. B., and McClintock, R. 1955. Measurement of total carbon dioxide production by means of $D_2{}^{18}O$. *J. Appl. Physiol.*, **7**: 704–10.

Lohman, T. G. 1990. Body composition assessment in sports medicine. *Sports Med. Digest.*, **12**: 1–2.

Lohman, T. G., Roche, A. F., and Martorell, R. (eds.) 1988. *Anthropometric Standardization Reference Manual*. Champaign, IL: Human Kinetics Books.

Lukaski, H. C., Johnson, P. E., Bolonchuk, W. W., and Lykken, G. I. 1985. Assessment of fat-free mass using bioelectrical impedance measurements of the human body. *Am. J. Clin. Nutr.*, **41**: 410–17.

Lukaski, H. C. 1987. Methods for the assessment of human body composition: traditional and new. *Am. J. Clin. Nutr.*, **46**: 537–56.

Levine, R. L. and Fromm, R. E., Jr. (eds.) 1995. *Critical Care Monitoring: From Pre-Hospital to the ICU*. St. Louis: Mosby.

Mazess, R. B., Ort, M., Judy, P., and Mather, W. 1970. Absorptiometric bone mineral determination using ^{153}Gd. *Proc. CONF-700515, Bone Measurement Conf.*, Chicago, IL, 308.

Mazess, R. B., Barden, H. S., Bisek, J. P., and Hanson, J. 1990. Dual-energy X-ray absorptiometry for total-body and regional bone-mineral and soft-tissue composition. *Am. J. Clin. Nutr.*, **51**: 1106–12.

McLean, J. A. and Tobin, G. 1987. *Animal and Human Calorimetry*. Cambridge, UK: Cambridge University Press.

Medved V. 2001. *Measurement of Human Locomotion*. Boca Raton, FL: CRC Press.

Mendez, J., Keys, A., Anderson, J. T., Grande, F. 1960. Density of fat and bone mineral of mammalian body. *Metabolism*, **9**: 472–7.

Mendez, J., Procop, E., Picon-Reategui, F., Akers, R., and Buskirk, E. R. 1970. Total body water by D_2O dilution using saliva samples and gas chromatography. *J. Appl. Physiol.*, **28**: 354–7.

Milsum, J. H. 1966. *Biological Control System Analysis*. New York: McGraw-Hill.

Morris, J. R. W. 1973. Accelerometry—a technique for the measurement of human body movements. *J. Biomechanics*, **6**: 729–36.

Murgatroyd, P. R., Shetty, P. S., and Prentice, A. M. 1993. Techniques for the measurement of human energy expenditure: a practical guide. *Int. J. Obesity*, **17**: 549–68.

Pace, N. and Rathburn, E. N. 1945. Studies on body composition. III The body water and chemically combined nitrogen content in relation to fat content. *J. Biol. Chem.*, **158**: 685–91.

Prentice, A. M. 1995. Application of dual-energy X-ray absorptiometry and related techniques to the assessment of bone and body composition. In P. S. W. Davies and T. J. Cole (eds.) *Body Composition Techniques in Health and Disease*. Cambridge UK: Cambridge University Press.

Prentice, A. M., Diaz, E. O., Murgatroyd, P. R., Goldberg, G. R., Sonko, B. J., Black, A. E., and Coward, W. A. 1991. Doubly labeled water measurements and calorimetry in practice. In R. G. Whitehead and A. Prentice (eds.) *New Techniques in Nutritional Research*. New York: Academic Press.

Ravussin, E. and Rising, R. 1992. Daily energy expenditure in humans: measurements in a respiratory chamber and by doubly labeled water. In J. M. Kinney and H. N. Tucker (eds.) *Energy Metabolism: Tissue Determinants and Cellular Corollaries*. New York: Raven Press.

Ritz, P. and Coward, W. A. 1995. Doubly labeled water measurement of total energy expenditure. *Diabetes & Metabolism*, **21**: 241–51.

Roos, B. O. and Skoldborn, H. 1974. Dual photon absorptiometry in lumbar vertebrae. I. theory and method. *Acta Radiologica: Therapeutics*, **13**: 266–80.

Schoeller, D. A. and van Santen, E. 1982. Measurement of energy expenditure in humans by doubly-labeled water method. *J. Appl. Physiol.*, **53**: 955–9.

Segal, K. R., Gultin, B., Presta, E., Wang, J., and van Itallie, T. B. 1985. Estimation of human body composition by electrical impedance methods: a comparative study. *J. Appl. Physiol.*, **58**: 1565–71.

Segal, K. R., Van Loan, M., Fitzgerald, P. I., Hodgdon, J. A., Van Itallie, T. B. 1988. Lean body mass estimated by bioelectrical impedance analysis: a four-site cross-validation study. *Am. J. Clin. Nutr.*, **47**: 7–14.

Segal, K. R., Burastero, S., Chun, A., Coronel, P., Pierson, R. N., and Wang, J. 1991. Estimation of extracellular and total body water by multiple-frequency bioelectrical-impedance measurement. *Am. J. Clin. Nutr.*, **54**: 26–9.

Siri, W. E. 1961. Body composition from fluid spaces and spaces, analysis, of methods. In J. Brozek and A. Henschel (eds.) *Measuring Body Composition*. Washington DC: National Academy Sciences, 223–44.

Tanner, J. M. and Whitehouse, R. H. 1955. The Harpenden skinfold caliper. *Am. J. Phys. Anthropol.*, **13**: 743–6.

Thomas, D. H. and Long, C. 1964. An electrogoniometer for the finger—a kinesiologic tracking device. *Am. J. Med. Elect.*, **3**: 96–100.

Walton, J. S. (ed.) 1990. *Mini-Symposium on Image-Based Motion Measurement*. **SPIE-1356**, Bellingham, WA: The Society of Photo-Optical Instrumentation Engineers.

Webb, P. 1985. *Human Calorimeters*. New York: Praeger.

Webster, J. G. (ed.) 1998. *Medical Instrumentation: Application and Design*. 3rd ed. New York: John Wiley & Sons.

Weir, J. B. de V. 1949. New methods for calculating metabolic rate with special reference to protein metabolisms. *J. Physiol.*, **109**: 1–9.

Weissman, C., Sardar, M. S., and Kemper, M. 1990. In vitro evaluation of a compact metabolic measurement instrument. *JPEN*, **14**: 216–21.

Wellens, R., Chumlea, W. C., Guo, S., Roche, A. F., Reo, N. V., and Siervogel, R. M. 1994. Body composition in white adults by dual-energy X-ray absorptiometry, densitometry, and total body water. *Am. J. Clin. Nutr.*, **54**: 547–55.

Whittle, M. W. 1996. *Gait Analysis: An Introduction*. 2nd ed. Oxford: Butterworth-Heinemann.

Winter, D. A. 1990. *Biomechanics and Motor Control of Human Movement*. 2nd ed. New York: John Wiley & Sons.

Index